Remote Sensing in Mangroves III

Remote Sensing in Mangroves III

Guest Editor

Chandra Giri

MDPI

Basel • Beijing • Wuhan • Barcelona • Belgrade • Novi Sad • Cluj • Manchester

Guest Editor
Chandra Giri
US Environmental Protection Agency
Durham, NC
USA

Editorial Office
MDPI AG
Grosspeteranlage 5
4052 Basel, Switzerland

This is a reprint of the Special Issue, published open access by the journal *Remote Sensing* (ISSN 2072-4292), freely accessible at: https://www.mdpi.com/journal/remotesensing/special_issues/99A1R0483K.

For citation purposes, cite each article independently as indicated on the article page online and as indicated below:

Lastname, A.A.; Lastname, B.B. Article Title. *Journal Name* **Year**, *Volume Number*, Page Range.

ISBN 978-3-7258-5219-2 (Hbk)
ISBN 978-3-7258-5220-8 (PDF)
https://doi.org/10.3390/books978-3-7258-5220-8

Contents

About the Editor

Chandra Giri

Chandra Giri currently serves as a Senior Science Advisor at the U.S. Environmental Protection Agency (EPA) and is an Adjunct Professor at Duke University. Dr. Giri is an internationally recognized scientist and author. He is best known for creating the first high-resolution global map of mangrove forests, a landmark contribution that has shaped global mangrove conservation policies. He has written over 100 peer-reviewed scientific papers, dozens of articles in leading Nepalese newspapers, and six books in both English and Nepali. Dr. Giri has also composed and released many Nepali songs, bridging science, art, and culture through creative expression.

remote sensing

MDPI

Review

Accelerated Adoption of Google Earth Engine for Mangrove Monitoring: A Global Review

K. M. Ashraful Islam [1,2,*], **Paulo Murillo-Sandoval** [3], **Eric Bullock** [4] and **Robert Kennedy** [1]

[1] College of Earth, Ocean, and Atmospheric Sciences, Oregon State University, Corvallis, OR 97331, USA
[2] Department of Urban and Regional Planning, Chittagong University of Engineering and Technology, Raozan, Chattogram 4349, Bangladesh
[3] Departamento de Infraestructura y Geomática, Facultad de Ciencias del Hábitat, Diseño e Infraestructura, Universidad del Tolima, Ibagué 730001, Colombia
[4] US Forest Service, Rocky Mountain Research Station, Riverdale, UT 84401, USA
* Correspondence: islamkm@oregonstate.edu or ashraful.urp@cuet.ac.bd

Abstract: Mangrove forests support coastal resilience, biodiversity, and significant carbon sequestration, yet they face escalating threats from climate change, urban expansion, and land-use change. Traditional remote sensing workflows often struggle with large data volumes, complex preprocessing, and limited computational resources. Google Earth Engine (GEE) addresses these challenges through scalable, cloud-based computation, extensive, preprocessed imagery catalogs, built-in algorithms for rapid feature engineering, and collaborative script sharing that improves reproducibility. To evaluate how the potential of GEE has been harnessed for mangrove research, we systematically reviewed peer-reviewed articles published between 2017 and 2022. We examined the spectrum of GEE-based tasks, the extent to which studies incorporated mangrove-specific preprocessing, and the challenges encountered. Our analysis reveals a noteworthy yearly increase in GEE-driven mangrove studies but also identifies geographic imbalances, with several high-mangrove-density countries remaining underrepresented. Although most studies leveraged streamlined preprocessing and basic classification workflows, relatively few employed advanced automated methods. Persistent barriers include limited coding expertise, platform quotas, and sparse high-resolution data in certain regions. We outline a generalized workflow that includes automated tidal filtering, dynamic image composite generation, and advanced classification pipelines to address these gaps. By synthesizing achievements and ongoing limitations, this review offers guidance for future GEE-based mangrove studies and conservation efforts and aims to improve methodological rigor and maximize the potential of GEE.

Keywords: Google Earth Engine (GEE); mangrove; remote sensing; review

1. Introduction

The world would not be the same without mangroves. They are vital to protecting coastlines from natural disasters [1] and their loss would significantly impact coastal resilience, biodiversity, and ecological balance [1]. Although only about 3% of the tropical forests in the world are mangroves, they house unique and diverse animal species [2], contribute to maintaining coastal ecosystems and food chains [3], help combat natural disasters such as tropical cyclones and coastal erosion [4], and provide timber [5]. The evaluation of ecosystem services of mangroves has been estimated to be $194,000 per hectare per year [6].

However, mangrove degradation is a global concern [7], as around 4.3% of the world's mangroves disappeared between 1996 and 2016 [8]. The increase in population, climate change, aquaculture, and the growth of industrialization and urbanization threaten the global existence of mangroves [3]. The extirpation of substantial mangrove areas can result in the release of toxic substances into the environment, potentially damaging both upstream and downstream regions [9]. In addition, their loss removes breeding and feeding sites for a variety of marine life, including oysters, shrimp, and crabs [9]. Because mangroves play an important role in the sequestration of carbon, nitrogen, and heavy metals, their decline could contribute to increased global warming if not addressed [4]. Hence, it is crucial to closely monitor their condition and depletion rate.

Due to their significance, mangroves have been the subject of substantial research. Scientific studies on managing, restoring, conserving, mapping, and monitoring mangroves have been carried out by organizations such as the United Nations Educational, Scientific and Cultural Organization (UNESCO), United Nations Development Programme (UNDP), United Nations Environment Programme (UNEP), International Society for Mangrove Ecosystems (ISME), International Union for Conservation of Nature (IUCN), World Wildlife Fund (WWF), and Food and Agriculture Organization (FAO) [10–14]. Recently, scholarly attention has turned to mangroves for their role in the carbon cycle [15–21]. Due to their high carbon sequestration capacities, the loss of mangroves can have a substantial impact on global carbon budgets [3,22]. To participate in United Nations Framework Convention on Climate Change (UNFCCC) mitigation initiatives (e.g., Reduced Emissions from Deforestation and Degradation (REDD+), Clean Development Mechanism (CDM)), it is imperative to quantify the baseline carbon stock of ecosystems such as mangroves, which are known to store significant proportions of carbon per unit area [4,23–26].

Remote sensing (RS) can play a key role in mangrove science. RS techniques can reduce data acquisition-related difficulties and financial limitations associated with traditional field research [27–30]. RS can help detect early signs of mangrove deterioration and also plays a key role in mangrove management [31]. The use of RS has facilitated a shift in emphasis from project-scale to regional- and global-scale mangrove investigations [32]. Moreover, the environmental and economic advantages associated with the carbon sequestration and storage capabilities of mangroves are widely acknowledged by carbon markets [21], leading to a greater focus on the use of RS technologies to quantify carbon stocks in mangroves [33].

Compared with other forest ecosystems, mangrove RS poses more challenges. The presence of cloud cover is especially problematic in coastal zones, where mangroves are the most common [28,34]. As mangroves are found mainly in fragmented patches along coastal belts, high-resolution multitemporal images are needed to effectively monitor them [35]. The detection of submerged mangroves is another significant challenge of this task [36]. To do this, a thorough collection of available satellite images is required, as well as an evaluation of the image metadata to establish the specific timing of high and low tides [37,38].

In recent years, mangrove management and conservation initiatives have evolved from localized field surveys to programs that require long-term monitoring and rapid response to emerging threats [39,40]. Coastal agencies require multidecadal datasets to assess gradual processes in a mangrove ecosystem, as well as near-real-time analyses to respond quickly to extreme events (e.g., cyclone-driven defoliation, unplanned land use conversions) [39–41]. Generating and maintaining massive multisensor time series along an extensive mangrove coastline spanning thousands of kilometers requires images in the petabyte range. Thus, processing big data archives on local workstations causes delays and inconsistencies, preventing timely decision-making [42]. Modern mangrove conservation requires a solution that automates data ingestion, preprocessing, and analysis to ensure broad spatial and temporal coverage and quick turnaround for management [27,38].

However, mangrove studies have historically been limited by data availability, the shortage of computational resources, and the lack of automation in image processing [27]. In traditional remote-sensing approaches, researchers often needed to download large volumes of raw satellite data locally, manually apply cloud masks, stitch multiscene mosaics, and perform classification on desktop machines. These steps were time-consuming, error-prone, and difficult to scale beyond regional projects [27]. There was an increasing demand for more frequent and accurate monitoring of mangroves and the use of high-resolution multitemporal images to improve the accuracy of change detection and reduce the manual workload associated with data download and processing [43]. Furthermore, global mangrove studies required improved and fast computational platforms [28]. Despite advances in local computing power, the workflows for global studies could not efficiently handle petabyte-scale archives [27] (readers are referred to Section 3 for additional information).

The debut of Google Earth Engine (GEE) changed the way forest research is conducted using remote sensing by shifting all heavy data management and computation to a cloud-based platform [44]. Instead of manually downloading and preprocessing imagery, users can access a vast preprocessed catalog of Landsat, Sentinel, MODIS, and other sensors [44]. It also opened up advanced analytical options [45,46]. GEE provides an extensive array of tools for forest research, including image overlay, change vector analysis, indices computation, classification, transformation, and principal component analysis [46]. Furthermore, GEE allows users to obtain statistics of forest losses and gains over time using all available images [46]. In essence, GEE is invaluable for identifying spatial and temporal changes in forest characteristics and cover [46]. Therefore, it is no surprise that through its wide range of preprocessed satellite images, robust cloud computing capabilities, and interactive result dissemination flexibilities, GEE has become an invaluable tool for supporting and strengthening large-scale research initiatives in mangrove forest ecosystems [47]. The use of GEE has the potential to influence mangrove conservation and management practices in countries such as China [35], Vietnam [48], Bangladesh [49], Indonesia [50], and Myanmar [38], where data is often scarce and resources are limited to conduct field investigations.

However, there is still a knowledge gap in the current trends in mangrove research using GEE, specific use cases of GEE, and an understanding of how the many tools accessible on GEE can be leveraged collectively efficiently for mangrove monitoring. Furthermore, there are limited documents that highlight the constraints and considerations when using GEE in mangrove research, as well as studies that highlight specific case studies in which GEE has made major contributions to the area. Although several papers [29,30,33] have reviewed articles that focused on mangrove RS, reviews of how GEE was used for mangrove studies are scarce. Jahromi et al. [46] while reviewing the use of GEE for forest sciences, listed five articles that used GEE for mangrove mapping/classification. Gilani et al. [51] reviewed seven scientific articles that explicitly used GEE for mangrove mapping and reported on the study sites, modeling techniques, and the accuracy of the classification routines. However, since its inception in 2017, there has been a notable increase in the use of GEE for the study of mangroves. As described in the second section, numerous studies have employed GEE for applications related to mangrove mapping and analysis. Critical reviews of existing research can contribute significantly to the advancement of scientific knowledge in a given field [52]. Given the growing body of work on GEE applications in mangrove research, a dedicated review has become both timely and necessary. Furthermore, current knowledge regarding opportunities, limitations, and prospects of utilizing GEE for mangrove studies is still at a rudimentary level.

GEE supports a broad array of mangrove research tasks, and organizing these applications into clear categories is essential for grasping their underlying principles. By surveying existing efforts, we can distill the prevailing practices in GEE-driven mangrove analysis. An in-depth, worldwide review will enrich our insight into the principal challenges, methodologies, and considerations associated with using GEE for remote-sensing–based mangrove research. Finally, identifying emerging themes in recent GEE-based mangrove studies will help anticipate future directions in the field.

The purpose of this present article is to offer an extensive review of scientific efforts carried out by mangrove researchers utilizing GEE from 2017 to 2022. This article has four objectives: (1) to document how GEE has improved the remote sensing of mangroves and to investigate the many applications of GEE; (2) to find out research domains that leveraged GEE for mangrove studies; (3) to document the challenges encountered and identify the common practices for mangrove analysis using GEE; and (4) to provide insight into future research directions in this area.

2. Methods

For our literature review, we relied on the Web of Science. The keywords used on the platform were *Google Earth Engine* (or *GEE*) and *Mangrove*. On the Web of Science Core Collection, we used this query: (((TS = (*Google Earth Engine*)) OR TS = (*GEE*))) AND TS = (*Mangrove*). Our inclusion criteria were as follows:

- Peer-reviewed journal articles published in English between January 2017 and December 2022.
- Explicit use of GEE for any stage of data collection or analysis.
- Focus on mangrove ecosystems (e.g., extent mapping, species classification, change detection, carbon accounting).

We did not apply any filter based on journal impact factor. Instead, we relied on Web of Science's own editorial and impact-based selection process. It is important to acknowledge that there exist several more databases, such as Google Scholar and Scopus, which serve as platforms for hosting academic literature [53]. However, for the purpose of our current analysis, we limited our selection of journals to those included in the Web of Science database. This is because the Web of Science team employs rigorous evaluation criteria to ensure that quality works are hosted in its repository [54]. They have, for example, 24 quality criteria aimed to choose for editorial integrity, as well as 4 impact criteria that seek out the most prominent journals in their respective disciplines [55]. However, future systematic reviews could incorporate complementary databases such as Scopus or Google Scholar to capture any additional studies that may not appear within Web of Science's indexing portfolio.

Based on these search terms (searched on 28 March 2023), we found 75 articles that were published from 2017 to 2022. We carefully read all 75 papers and selected only those that used GEE, at least for data collection, and specifically studied mangroves. A total of 54 scientific journal articles were selected for final evaluation. If we had any questions about a specific method, analysis result, or use of GEE in an article, we contacted the paper's corresponding authors for clarification. We organized the papers according to theme (mangrove RS subfield), motivation (driving factors) and specific applications of GEE in mangrove studies.

3. Pre-GEE Era: Problems Encountered

During the pre-GEE era, researchers investigating mangroves faced numerous obstacles. According to Giri et al. [28], the issue of cloud cover in satellite imagery posed a significant obstacle for researchers. Moreover, in a separate study, Giri et al. [43] used Landsat images and had to select only cloud-free images. The raw images further required them to convert digital numbers into top-atmospheric reflectance. They also employed the maximum likelihood classification algorithm but observed poor change detection accuracy. As a future direction, they pointed to the need for frequent and credible monitoring of mangroves by utilizing high multitemporal images to improve accuracy and change detection capabilities. Ibharim et al. [56] conducted their investigation on mangroves using only three multispectral satellite data sources. Consequently, this restriction hindered the ability to capture temporal variations and limited the scope to conduct a broad analysis of mangrove spatiotemporal change dynamics. Another study conducted by Jean-Baptiste and Jensen [57] focused on the utilization of a single (clearest non-cloudy) scene. Younes Cárdenas et al. [27] emphasized this limitation, observing that the use of a single image yielded limited results and did not demonstrate the persistence of relationships between spectral indices of mangroves and biophysical variables. Consequently, they expressed concern regarding the limited use of all available satellite images. According to them, most studies analyzed fewer than ten images, which increased uncertainty, ignored the impact of extreme weather events on mangrove loss, and may have obscured sections of mangrove forests due to tidal height or water. Therefore, adequate observations are required to detect subtle or precipitous changes in mangroves, as this could make a substantial difference in preventing catastrophic diebacks and allowing for prompt responses [27].

The influence of water under the mangrove canopy, specifically the height of the tide, posed an additional significant challenge [27,58]. Younes Cárdenas et al. [27] highlighted the importance of examining how tidal height affects the spectral signature of mangroves. However, they observed that few mangrove studies incorporated tidal influences, which likely have affected the results, especially in regions with significant tidal fluctuations such as mangrove forests in Malaysia, Southern China, and Vietnam. To accurately map submerged mangroves, it is important to collect enough data, possibly leveraging all available satellite data, including images taken during high and low tides.

In addition, automation was limited during the pre-GEE era of mangrove research. Younes Cárdenas et al. [27] observed that a limited number of articles reported automation of tasks, such as calculating spectral indices for large image volumes. This limitation emphasized the need for tools and/or techniques that could facilitate and automate data analysis processes to manage the immense volume of data that is currently available. Furthermore, the need for remote sensing expertise and specialized software for image processing presented obstacles for researchers working with optical and SAR imagery [29,30]. Commercial software packages (e.g., ERDAS Imagine, ENVI, ArcGIS) often carry hefty license fees, which limited accessibility.

In addition to data limitations, the pre-GEE era was beset by computational inadequacies and a lack of local machines with sufficient image-processing capacity [27]. Considering this, Younes Cárdenas et al. [27], Giri [59], and Granell et al. [60] highlighted the significance of incorporating open-source cloud computing resources in future mangrove research. This concept originated not only from the need for adequate data storage and processing capabilities but also from the search for cost-effective solutions for large-scale, long-term mangrove research.

In light of these obstacles, Wang et al. [61] suggested that researchers should leverage powerful cloud computing platforms such as GEE. GEE maintains a vast quantity of preprocessed imagery from multiple sources (including a full archive of Landsat, Sentinel, and MODIS) [44], which can facilitate the development of large-scale mangrove studies. The availability of petabytes of data, as well as a variety of processing tools and algorithms, can enable researchers to surmount previous limitations and conduct in-depth analyses of mangrove ecosystems.

4. GEE Testbeds: Study Sites and Scales

Mangroves are most abundant in Asia, followed by West Africa, Australia, and Central and South America (Figure 1). Approximately 75% of mangrove area exists in the top 15 mangrove–rich countries [28]. GEE-based studies mostly focused on the top mangrove-rich countries (Figure 1). However, there were a few countries with no GEE-based mangrove studies, including Papua New Guinea (8th mangrove-rich country, 3.5% of global total), Cuba (10th, 3.1%), Mozambique (13th, 2.3%), and Madagascar (14th, 2.0%). The distribution of study areas based on the affiliated country of the author is given in Appendix A, Table A1. We observed that the studies cover mangrove ecosystems on almost every inhabited continent, with a strong focus on China. Apart from China, first-author affiliations included countries from South and Southeast Asia, Iran, and various European and African nations. It should be noted that 30 studies (out of 54) solely examined the country of affiliation of the first author.

Some papers stated whether their research focused on mangrove regions at the regional or project level. However, where it was unclear, the corresponding authors were contacted. In the context of our present study, a project-scale study is defined as one that examines a smaller portion of a larger mangrove forest present on a national or regional scale. Out of 54 studies, about 65% (35 studies) focused on country- or regional-scale, while around 33% (18 studies) were local- or project-based. There was only one article that focused on both regional and project-scale study areas. Again, although not in the top 15 mangrove-rich countries list, the majority of studies that used GEE were concentrated in China (Figure 2). There were eight country-wide, three regional-scale, and two project-scale studies that considered the Chinese territory. Six studies considered the Bangladesh part of the Sundarbans mangrove forest, and three studies considered the Indian part. Three regional and two project-scale studies were conducted in Vietnam. It is to be noted that despite being the most mangrove-rich country, Indonesia had only one regional-scale and three local-scale studies. Myanmar was examined in three regional and one project-scale studies. Australia was the subject of two regional and one project-scale studies. In other parts of the world, two or fewer studies were conducted. Most articles were published in *Remote Sensing*, an open-access journal by MDPI (Multidisciplinary Digital Publishing Institute; Figure 3).

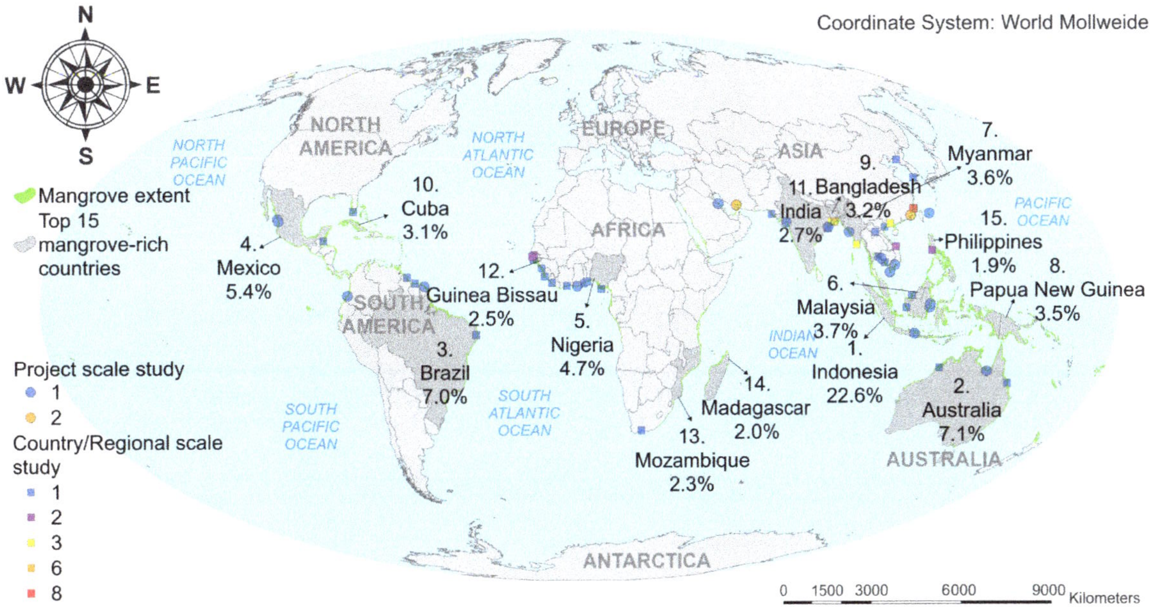

Figure 1. Study areas where GEE was used for mangrove RS. The 2020 Global Mangrove Watch (GMW) Version 3.0 data was used to show mangrove extent [62].

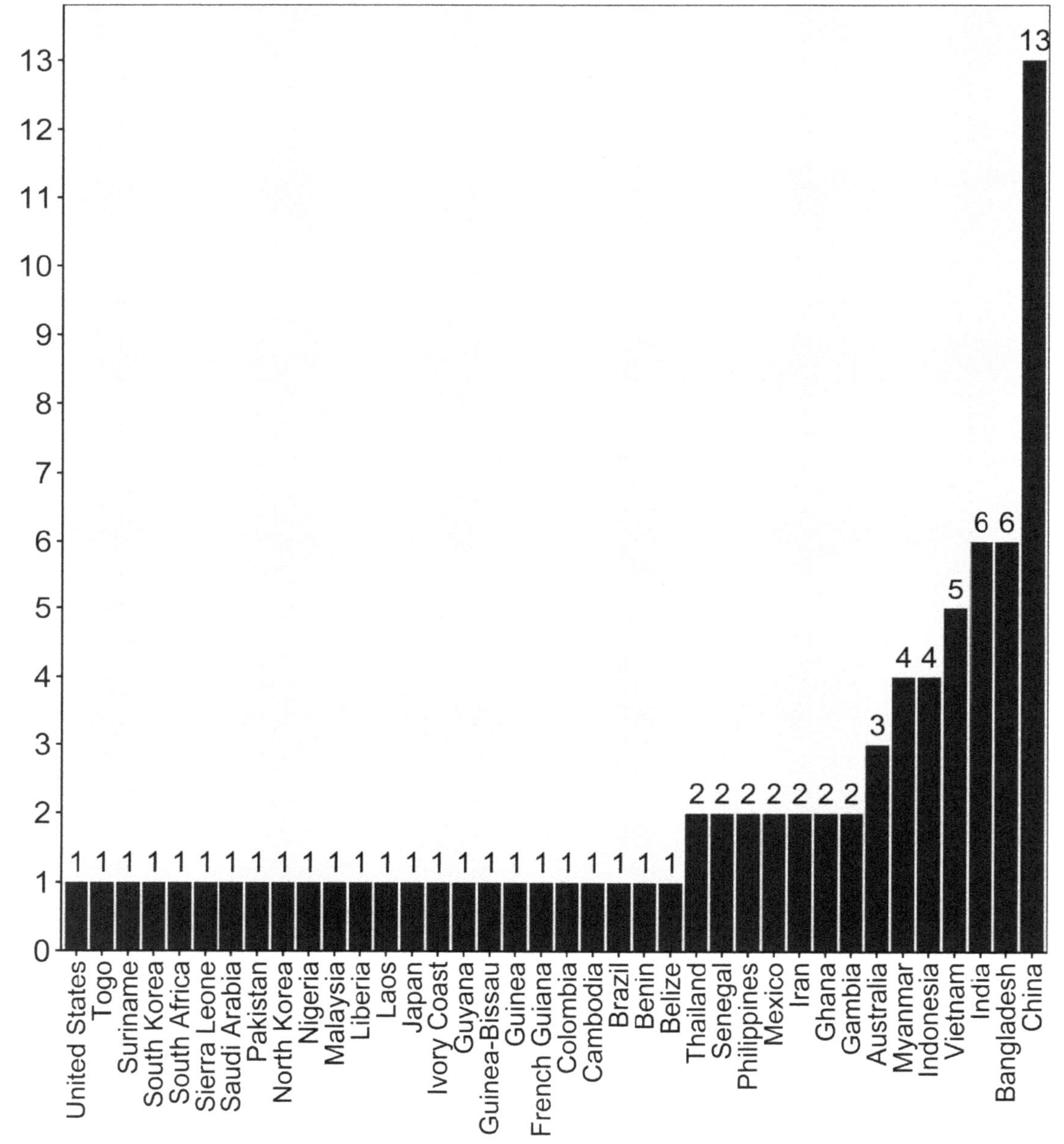

Figure 2. The number of articles that examined mangroves in various countries. Note that some papers considered more than one study site.

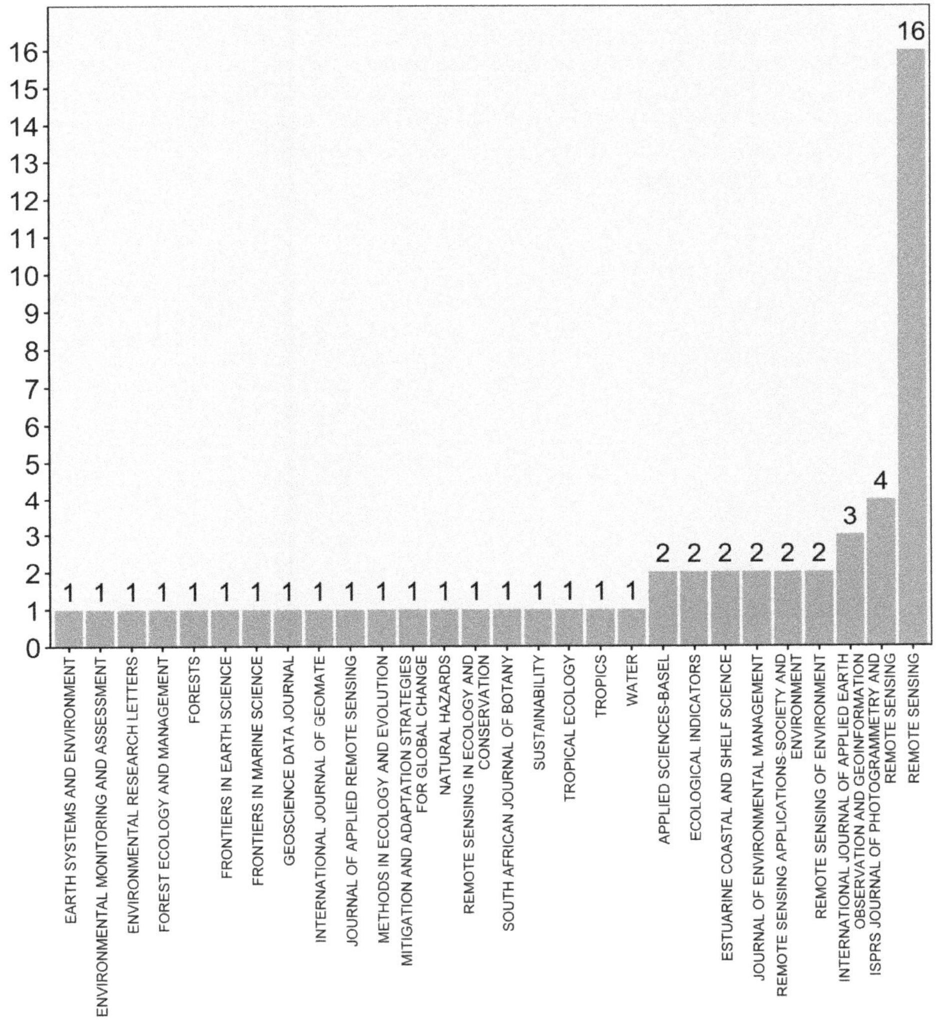

Figure 3. The names of the journals that published the 54 articles.

5. Use of GEE: Broad Categories

Mangrove studies have utilized GEE for: (1) data extraction, filtering and/or preprocessing; (2) training models or running algorithms; (3) assessing the accuracy of the models or estimations; (4) generating interactive web tools; or (5) adopting a combination of these usages (Figure 4). The driving factors that propelled these studies can also be divided into four types. Researchers have tried to either accurately map the extent of mangroves, understand the change dynamics, to assess the damage caused by calamities, or constrain leaf biochemical properties.

These 54 studies can be further grouped based on the subfield of mangrove RS and the type of study conducted. Wang et al. [61] divided the mangrove RS subfield into three major groups: (1) distribution mapping; (2) extracting biophysical properties; and (3) modeling ecosystem processes. Distribution mapping can further be separated into two interconnected components: (a) mapping the entire area, and (b) mapping individual species. Works that studied biophysical properties estimated either the height and biomass of mangroves, or the leaf area index (LAI) [61]. Furthermore, ecosystem process modeling can take two routes: understanding the carbon flux and "process modeling" of evapotranspiration [61]. Wang et al. [61] also identified two more mangrove RS topics: health condition analysis and effect of climatic condition on mangroves.

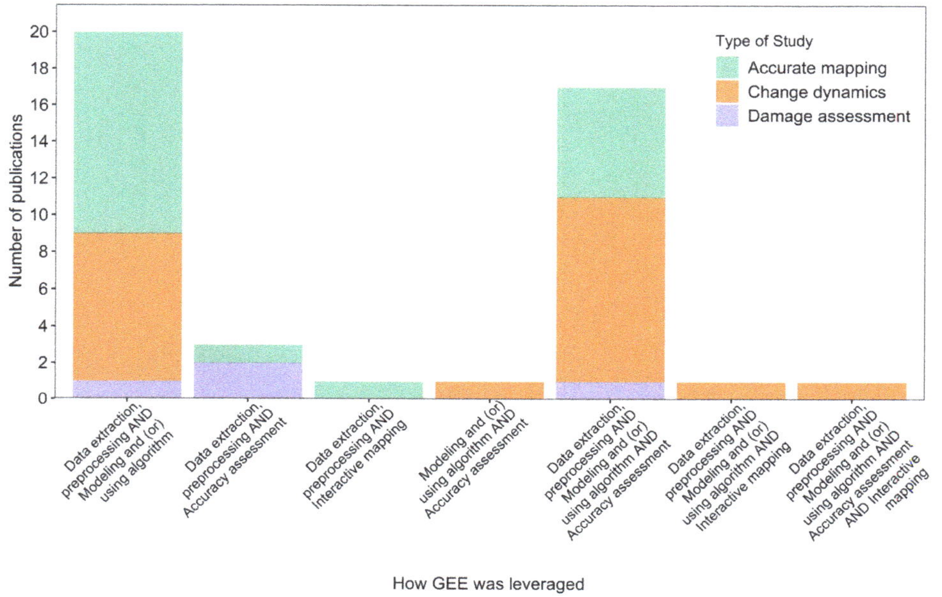

Figure 4. RS-based mangrove studies that had more than one use of GEE.

GEE-based mangrove studies have primarily focused on data extraction, preprocessing, modeling, and accuracy evaluation (Table 1). These areas have garnered the most attention and recognition within the field, as evidenced by the number of publications and citations. However, despite its capabilities, the utilization of GEE for interactive mapping of mangrove data remains limited. As previously mentioned, GEE was utilized for more than one purpose in several mangrove research studies. The most used combinations were data extraction/preprocessing and modeling/algorithms, as well as data extraction/preprocessing and accuracy assessment (Figure 4). While accurate mapping dominates the literature, there is also notable interest in change dynamics analysis, and fewer studies on damage assessment. It is noteworthy that while certain subfields have received little attention, they nonetheless provide essential insights on how GEE has been used all over the world for studying mangroves. In particular, GEE has proven highly effective for detecting subtle phenological shifts, quantifying biomass changes, and informing locally tailored conservation strategies across diverse mangrove regions [63]. Also, it is worth mentioning that even though functions to perform a particular task (like accuracy assessment) are available on GEE, some researchers utilized third-party GIS software or open-source applications instead. Figure 4 presents specific tasks that researchers carried out using GEE tools and functions.

Table 1. Classification of mangrove RS articles by subfield and application of GEE. The number in parentheses represents the number of citations.

Mangrove RS Subfield	How GEE Was Utilized	Accurate Mapping	Change Dynamics	Damage Assessment	Leaf Biochemical Property Inversion
Extent mapping	Data extraction, preprocessing	3 (35)	2 (7)	0 (0)	0 (0)
	Data extraction, preprocessing AND Modeling and (or) using algorithm	10 (107)	6 (24)	1 (1)	0 (0)
	Data extraction, preprocessing AND Accuracy assessment	1 (206)	0 (0)	2 (24)	0 (0)
	Data extraction, preprocessing AND Interactive mapping	1 (48)	0 (0)	0 (0)	0 (0)
	Modeling and (or) using algorithm AND Accuracy assessment	0 (0)	1 (0)	0 (0)	0 (0)
	Data extraction, preprocessing AND Modeling and (or) using algorithm AND Accuracy assessment	5 (69)	8 (160)	1 (1)	0 (0)
	Data extraction, preprocessing AND Modeling and (or) using algorithm AND Interactive mapping	0 (0)	1 (15)	0 (0)	0 (0)
	Data extraction, preprocessing AND Modeling and (or) using algorithm AND Accuracy assessment AND Interactive mapping	0 (0)	1 (16)	0 (0)	0 (0)
Extent mapping and height estimation	Data extraction, preprocessing	0 (0)	0 (0)	1 (0)	0 (0)
	Data extraction, preprocessing AND Modeling and (or) using algorithm	0 (0)	1 (7)	0 (0)	0 (0)
Species mapping	Data extraction, preprocessing	1 (21)	1 (9)	0 (0)	0 (0)
	Data extraction, preprocessing AND Modeling and (or) using algorithm	1 (0)	0 (0)	0 (0)	0 (0)
	Data extraction, preprocessing AND Modeling and (or) using algorithm AND Accuracy assessment	1 (34)	0 (0)	0 (0)	0 (0)
Species mapping and height estimation	Data extraction, preprocessing AND Modeling and (or) using algorithm	0 (0)	1 (6)	0 (0)	0 (0)
Biomass estimation	Data extraction, preprocessing AND Modeling and (or) using algorithm AND Accuracy assessment	0 (0)	1 (0)	0 (0)	0 (0)
Health condition	Data extraction, preprocessing	0 (0)	1 (0)	0 (0)	0 (0)
Effect of climate	Data extraction, preprocessing AND Modeling and (or) using algorithm AND Accuracy assessment	0 (0)	1 (18)	0 (0)	0 (0)
Ecosystem process	Data extraction, preprocessing	0 (0)	0 (0)	0 (0)	1 (5)
Total		23 (520)	25 (262)	5 (26)	1 (5)

5.1. GEE Used for Data Extraction and Preprocessing

Because mangroves grow near the low-lying coastal belt, cloud filtering and topographic masking are often essential steps in any workflow. Computing and aggregating data on a per-pixel level is also important for deriving mangrove phenology. GEE is being used for these purposes. It is also becoming more popular for large-scale mapping and processing due to its speed and ease of use.

Almost all 54 papers have at least leveraged the basic utility of GEE, that is, data extraction and preprocessing. A wide range of GEE-based preprocessing techniques were used to improve the usability of satellite imagery (Figure 5). Clearly, the most popular was generating image composites (47 out of 54 articles). As indicated by their occurrence in 42 and 40 papers, respectively, calculating indices and cloud and shadow masking were the next most used preprocessing steps. The significance of topographic masking in mangrove research was addressed in 19 articles. Speckle filtering, a method intended to mitigate noise in radar images, has been adopted in four papers. Other preprocessing approaches include extracting texture information via Grey Level Co-occurrence Matrix (GLCM), scan-line gap filling, additional atmospheric correction, and subtracting dark objects.

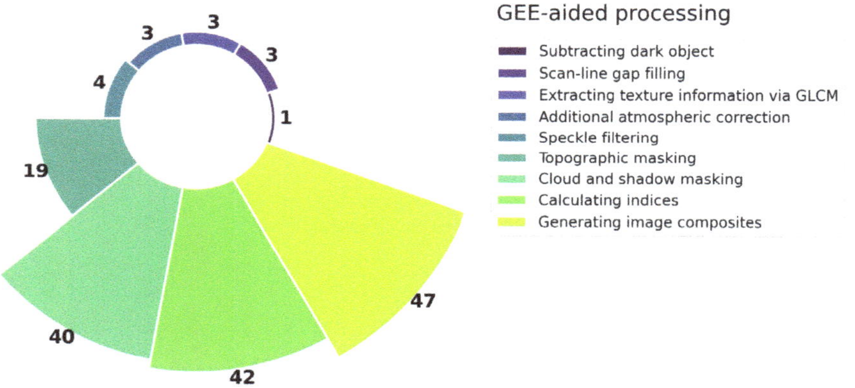

Figure 5. Techniques for preprocessing within GEE framework.

Often, a single article has applied multiple preprocessing techniques, indicating GEE's applicability in various steps. For example, Chen et al. [47] pioneered the application of GEE for mangrove remote sensing by precisely measuring mangrove extent along China's coastline region in 2015. They used cloud and shadow masking techniques and calculated indices like NDVI, EVI, LSWI, and mNDWI to estimate greenness, canopy coverage, and tidal inundation. In a separate study, Diniz et al. [34] used cloud and shadow masking techniques, composite images, and the Modular Mangrove Recognition Index (MMRI) to describe mangrove habitats. Please see Section 6 for a more detailed review of various preprocessing approaches used by different scholars within the GEE framework.

The number of publications and citations of various mangrove RS studies employing GEE for data extraction and preprocessing has increased over time (Figure 6). Accurate mapping, change dynamics, and damage assessment are all receiving a growing amount of scholarly interest. GEE is becoming more and more popular in mangrove studies because it can quickly run preprocessing steps for data from multiple years in a matter of seconds. This level of efficiency is reached at the pixel level, which makes it easy for researchers to work with large temporal records. The easy-to-use interface of GEE makes implementation simple, and the visualization tools make it easy to examine the results.

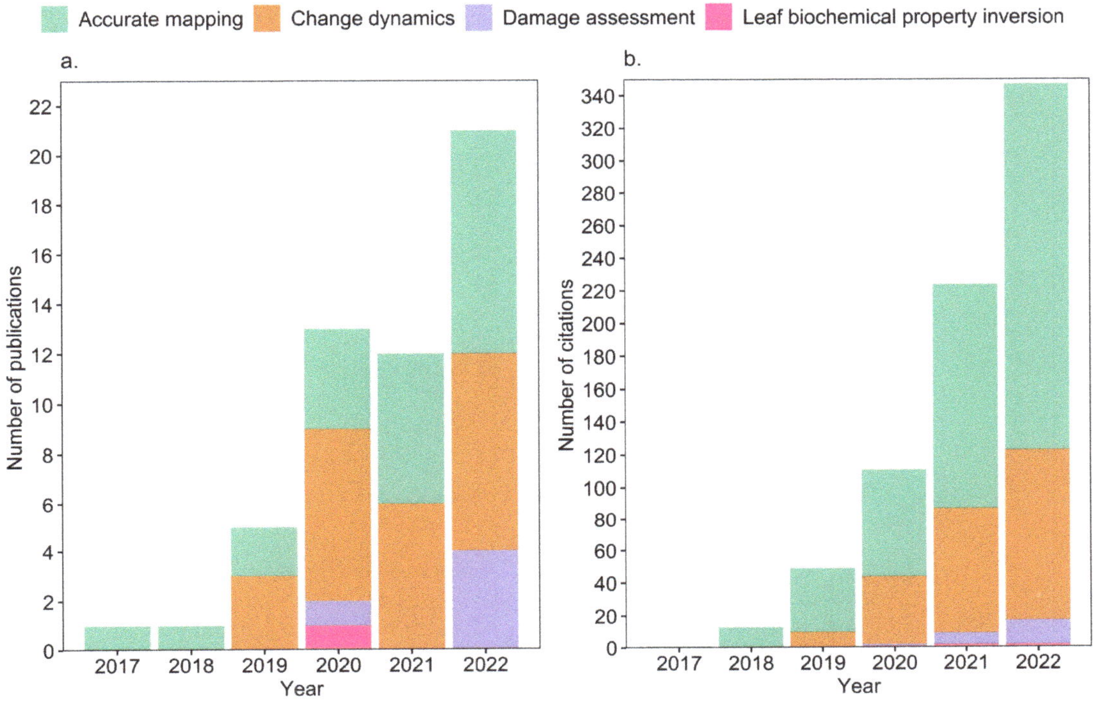

Figure 6. RS-based mangrove studies that utilized GEE for data extraction and preprocessing; (**a**) number of publications per year from 2017 to 2022, and (**b**) number of citations per year from 2017 to 2022.

5.2. Leveraging GEE-Based Algorithms

A key area where the cloud-based processing framework of GEE has been leveraged is through the application of sophisticated image processing algorithms (Figure 6). Learning-based algorithms for mangrove modeling (or classification) include Classification and Regression Trees [64], Gradient Tree Boosting [65], K-means clustering [66], Random Forests [67], Support Vector Machine [68], and Support Vector Data Description [69]. Other algorithms have been structured in GEE by authors from scratch (Harmonic Analysis of Time Series Algorithm [70]; Otsu Thresholding Algorithm [71]) or have built-in GEE (LandTrendr [72]; Linear Regression [73]). Some authors also used user-defined threshold/decision trees [74].

Random Forest (RF) [67] has proven effective in classifying mangroves and hence is the most widely used algorithm (Figure 7). Murray et al. [75], for example, used RF in GEE to incorporate predictor variables such as topography, climate, spectral bands, and spectral indices into mangrove classification in the Gulf of Carpentaria, Australia. In a study on temporal mangrove change in Brazil, Diniz et al. [34] used both RF and K-means clustering techniques. The work of Bhargava and Friess [76], as well as Li et al. [77], demonstrates an efficient use of GEE in training classifiers for estimating mangrove loss. The initiative undertaken by Nedd et al. [78] to extract biomass from the mangroves of Guyana by using RD shows the effectiveness of GEE through the seamless integration of Landsat, ALOS PALSAR, and Sentinel-1 SAR imagery, which was effective in distinguishing between mangroves and non-mangroves.

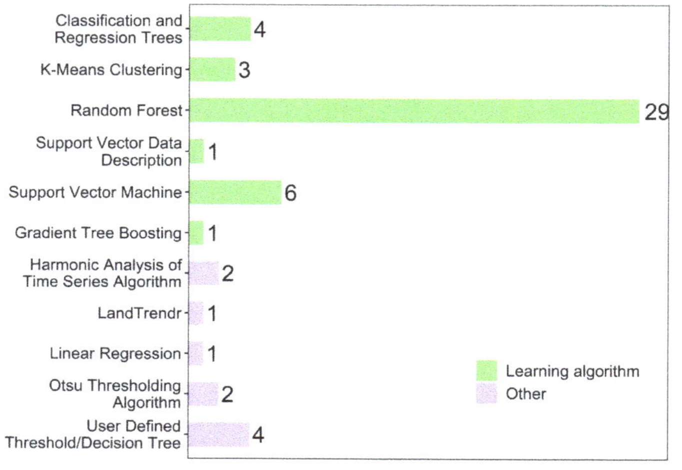

Figure 7. Number of studies that utilized various GEE-based algorithms for mangrove RS. Note that some studies have utilized more than one algorithm or modeling technique.

GEE makes it easier for researchers to optimize and visualize results by fine-tuning algorithm hyperparameters [75]. In addition, the GEE-based stratified random sampling method makes it easier to sample training data prior to classification [50]. After classification, different GEE-based post-processing techniques can be applied, such as temporal, spatial, and frequency filters and gap filling [34]. Based on the rising volume of publications and references to such pieces, it is evident that the scientific community is becoming more interested in classifications and algorithms available on GEE (Figure 8).

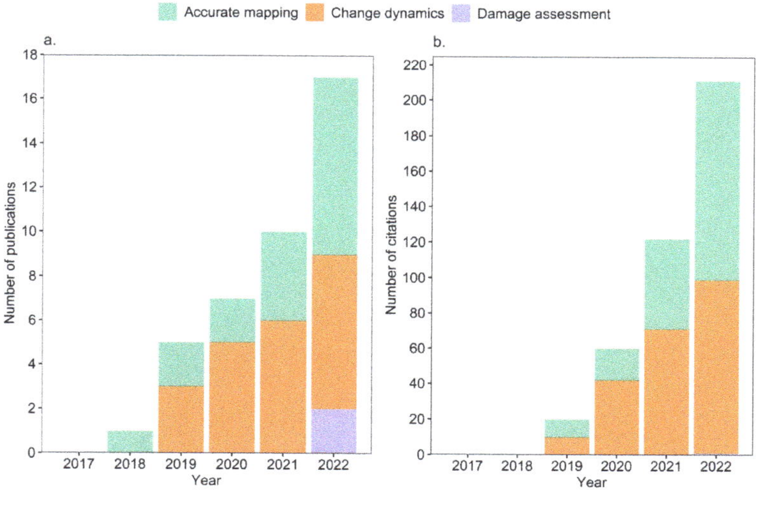

Figure 8. RS-based mangrove studies that utilized GEE-based modeling and (or) algorithm: (**a**) number of publications per year from 2017 to 2022 and (**b**) number of citations per year from 2017 to 2022.

5.3. GEE Used for Accuracy Assessment

Accurately reporting results is critical for researchers and policymakers to confidently rely on the data when making decisions about the mangrove ecosystem. This, in turn, depends on reference data to evaluate the maps. GEE facilitates the acquisition of accuracy reports based on the reference data for author-generated mangrove maps. Accuracy assessments performed on GEE tend to center on extent mapping, specifically on accurately mapping mangrove for a single year, or on analyzing the trend in change over time. For instance, to assess accuracy, Chen et al. [47] compared their 2015 mangrove map with data from field surveys, in situ data from the China Mangrove Conservation Network, very high spatial resolution images from Google Earth captured during 2015, and a different set of in situ data from previous studies. The authors used a sampling grid to assess map accuracy. The random function in GEE was used to produce a total of 4100 random points, each with a buffer zone with a radius of 15 m. Similarly, Diniz et al. [34] used GEE to randomly assign sample points to test the resulting mangrove maps, generating a confusion matrix and reporting different accuracy parameters. Multiple reference data sources were utilized by Shrestha et al. [79] as training data to calculate the accuracy of mangrove mapping in their study. They also used GEE-based stratified random sample points for assessing the accuracy. Thus, GEE can be leveraged to create confusion matrices and readily obtain a report of overall accuracy, producer accuracy, user accuracy, and kappa coefficient [47,80].

GEE has been widely used by researchers for accuracy assessment (in different sub-fields) (Figure 9). Except for 2021, the number of publications using GEE for accuracy reporting has increased every year. In addition, citations of papers reporting accuracy are on the rise.

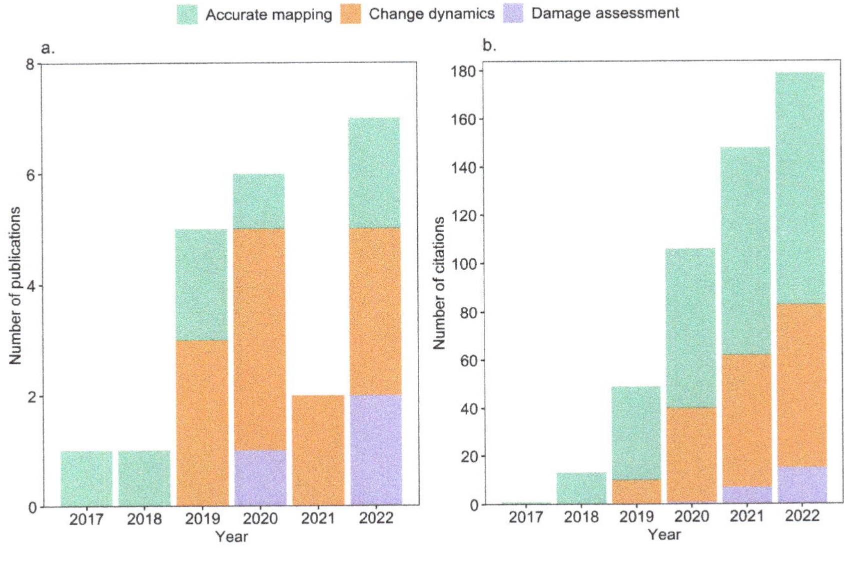

Figure 9. RS-based mangrove studies that utilized GEE for accuracy assessment: (**a**) number of publications per year from 2017 to 2022 and (**b**) number of citations per year from 2017 to 2022.

5.4. GEE Used for Interactive Mapping/Web Tool

Rapid processing and result displaying are key for any interactive tool targeted at people with less experience with coding or geospatial data [81]. Through GEE's interfaces, users may quickly and inexpensively extract map- and GIS-based products from satellite and/or other gridded data hosted within GEE [82]. However, works that introduced an interactive web-based mangrove mapping method are rather limited (Figure 10). Baloloy et al. [83] developed a GEE-based graphical user interface and an interactive mapping system that does not require coding expertise to generate mangrove maps. Users do not need to download large volumes of satellite images to use this web interface. Users are only required to define their region of interest, the start and end dates of images they are interested in, and the minimum and maximum MVI (mangrove vegetation index) thresholds. The workflow is completely automated and produces RGB and false color composites, as well as MVI vegetation and mangrove rasters. Following the application run, users can view all available images within the specified date range. Users can easily utilize the interactivity, zoom in and out, and select the MVI map with the least amount of cloud, cloud shadow, and smog coverage. This GEE-based web application has the potential to empower stakeholders to promptly and globally map mangroves. Yancho et al. [38] developed a GEE mangrove mapping methodology that allows stakeholders to perform temporal analysis and generate their own change maps by tailoring parameters and/or options to their specific needs. Users can define the region of interest, set input variables, calculate workflow thresholds, generate imagery composites, and compute chosen indices. Several masking features can be utilized to refine the classification scope and create maps of high- and low-tide events. Modern and historical image composites can be generated by users. Users can specify classification variables and hyperparameters, investigate correlation and spectral separability, and perform classification to generate classified maps and perform accuracy assessments. In cases where initial map classes were utilized to capture variability but where confidence in class borders may be inadequate, users are given the option to combine map classes after classification. The spectral separability of land cover classes may be investigated by users in reference to many possible spectral indices. Finally, the tool permits users to calculate dynamics and perform optional quality assurance using selected classification schemes. However, users must change/modify a few lines of code to tune the tool, and Yancho et al. [38] are developing a graphical user interface to make it more user-friendly for stakeholders and rid themselves of the need for coding knowledge. An increasing trend of studies using GEE for interactive mapping in RS-based mangrove research is evident (Figure 10). Although GEE-based interactive mapping has grown in popularity over the years, it has been the focus of fewer papers and consequently fewer citations than other use cases of GEE.

In short, GEE has become a cornerstone in mangrove remote sensing, primarily for data extraction, preprocessing, modeling, and accuracy assessment. Its strengths include efficient data handling, algorithm implementation, and validation at multiple spatial and temporal scales. Interactive mapping remains underexplored, despite its increasing relevance, as the majority of studies concentrate on mapping extent, tracking change, and assessing damage. This gap suggests future research on stakeholder engagement and decision support using accessible, user-driven mapping tools on GEE.

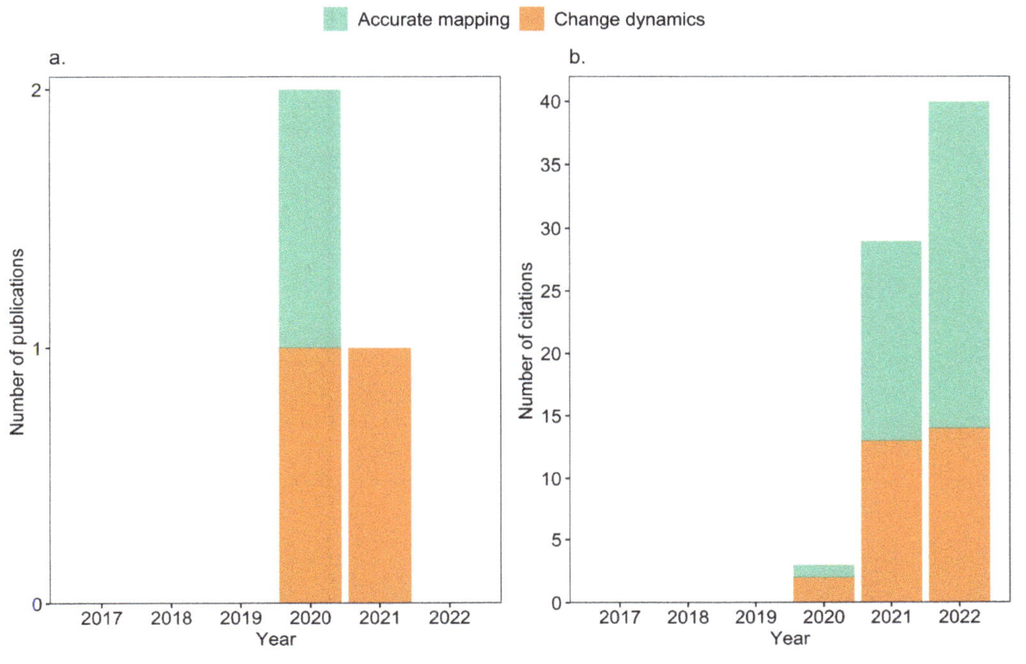

Figure 10. RS-based mangrove studies that utilized GEE for interactive mapping: (**a**) number of publications per year from 2017 to 2022 and (**b**) number of citations per year from 2017 to 2022.

6. Approaches for Studying Mangroves Using GEE

In this section, we describe a broad framework for mangrove study using GEE based on the synthesis of the papers we reviewed. First, we discuss how GEE functions can be used to filter, preprocess, and classify data, and we then describe the mangrove-specific GEE tools researchers have utilized.

Image processing for mangroves begins with the same basic steps as image processing for any other application domain. Firstly, GEE functions can filter images based on factors such as area of interest, time periods, cloud cover, cirrus, image bands, or other metadata properties [36,51,74,75,84–92]. There are GEE-based gap-filling algorithms that can fix data gaps in Landsat 7 scenes [93] and reduce noise in radar imagery like in Sentinel-1 and ALOS PALSAR [94]. Numerous classic and new indices and band ratios can be easily calculated with arithmetic operations [34,50,74,75,80,83,95–99]. By utilizing all available cloud-free images within a particular time-frame (e.g., for a specific year), temporally consistent mean/median image composites can be created [37,50,75,84,88,89,92,95,97,100–102].

GEE is used by researchers worldwide to classify mangroves and other important land cover types [100,103]. It has tools to make use of training and testing sample points for classification algorithms and offers drawing tools to collect random points from a base map [47,103]. GEE can also be used to combine radar and optical imagery to improve classification [76,86,88,92,104]. Post-processing of classified images can be performed in GEE [36,91]. Furthermore, accuracy assessment can be performed using a confusion matrix [75,80]. Finally, GEE allows for easy sharing of results and map products, allowing stakeholders, end users, and policymakers to analyze them [38,49,83].

That being stated, there are certain specific applications of GEE that are focused on mangroves, and are explained in more detail below:

6.1. Analyzing Indices and Spectral Signature

Using GEE, species diversity and structural components of mangroves can be understood from differences in spectral signature or indices values in different points in time, as demonstrated by several past works [74,83]. Furthermore, these differences can also be used to obtain an optimal time window for distinguishing mangroves and other invasive plants like *S. alterniflora* [96,99].

6.2. Generating Mangrove-Specific Image Composites

A mangrove-specific yearly highest-index-value composite can be created using GEE functions. For example, NDVI can be used to create a year-long greenest mangrove pixel composite [80,105]. Furthermore, standard deviation and percentile composites can be created to detect anomalies in mangroves [106]. Additionally, monthly composites (e.g., 3-, 4-, 6-month composites) can be created to understand seasonal patterns and variation of mangroves [50,51,86]. These composites can be further used to analyze mangrove disturbances or damages over time [76].

6.3. Developing and Testing Decision-Making Thresholds

Many mangrove studies have demonstrated that it is easy to implement user-defined rule-based logics in GEE [36,90]. As an example, Solanki et al. [98] categorized mangrove health based on NDVI thresholds. An NDVI value of more than 0.4 was considered to be healthy vegetation in their study. Similarly, user-defined rule-based decisions can be used to detect non-mangrove plants, mudflats, and permanent water [38,84]. Mangroves are generally found in low-elevation coastal zones, and different digital elevation data hosted on GEE can be utilized to create a topographic mask [38,78,102,107]. This can be performed using GEE by defining an elevation threshold beyond which mangroves are rarely found [84,107].

6.4. Creating Buffer Zones for Separating Mangroves from Water

GEE has functions for generating buffers by a user-defined distance from a geometry (i.e., polygon, polyline, or point), which is useful for separating coastal mangroves from other plants [38,90,95]. To create a distinction between mangrove and sea, buffer zones can also be created and tested to obtain the optimum intersecting line [47]. Moreover, it has been reported that ponds and other inland water bodies used for shrimp and fish farming can be found in close proximity to mangroves [48]. As Zhang et al. [91] demonstrated, GEE-based algorithms make it possible to successfully remove these inland water pixels and distinguish them from the sea water. This is achieved by examining tide-level and phenological patterns. This method uses RF classifiers to identify unvegetated wetlands, mangroves, and salt marshes, which effectively distinguishes between inland and seawater pixels.

6.5. Considering Tide and Submerged Mangroves

Coastal mangroves get inundated periodically and at varying frequencies [47,90]. User-defined rule-based logic in GEE can help to produce inundation frequency maps. For example, Chen et al. [47] and Dong et al. [96] created frequency maps by applying user-defined thresholds to image indices to understand inundation rates. Furthermore, GEE can be utilized to generate high tide and low tide composites for detecting submerged mangroves by employing a per-pixel ordering function and/or specifying a threshold pixel value for aggregation [38,91,99]. By leveraging GEE, quantile composites can also be generated to detect submerged mangroves to discern tidal inundation states [99,108]. More details on the application of this approach are described in the work of Xia et al. [99]. They estimated the tidal datum from a pixel of time series images by assigning values to the tenth and ninetieth quantiles, respectively, to detect low and high tide scenes. Also, intertidal zones can be identified with this approach [108]. Additionally, Otsu's binary thresholding method can be used to differentiate submerged mangroves from regular ones [97,99].

6.6. Time Series Analysis

For a particular mangrove patch, the entire area of interest or a single-pixel-based time series can be generated and analyzed using GEE [36,86,99,108,109]. This helps to understand seasonal variation in mangrove canopy greenness [80,83]. Phenological features can be readily extracted from the time series [47,74,80,108,110]. Harmonic analysis of time series can be applied by leveraging time series produced in GEE to better understand the dynamics of phenological changes and seasonality of mangroves [37,96]. Moreover, temporal segmentation algorithms like LandTrendr, which effectively uses time series data, are useful for investigating patterns in mangrove degradation, growth, and settlement [85].

6.7. Spectral Unmixing

Spectral unmixing separates mixed pixels and, for example, can aid in determining the portion of water pixels within mangrove patches near shore [90]. GEE has a spectral unmixing function that can be used for this purpose (https://developers.google.com/earth-engine/apidocs/ee-image-unmix (accessed on 2 January 2023)).

6.8. Mangrove Classification

GEE can be used to improve mangrove species mapping with few reference points, which is especially helpful for inaccessible forests. GEE-provided classification algorithms can be used to generate a preliminary species distribution map, which can then be used to improve the decision-making surface of the algorithm by iteratively refining the initial map through the inclusion of negative samples [92].

6.9. Comparing Maps and Results

GEE helps to readily map and monitor mangrove changes over time for various geographical locations. Comparing regions of interest helps to understand prevailing unique processes for different mangrove settings [49,83]. Moreover, different mangrove map products (raster/vector) can be imported directly to GEE and the new mangrove maps can be compared with other products to understand the spatial differences at pixel level [34,38,50,78,108]. This comparison can be both qualitative and quantitative as GEE helps to visualize data products as layers and perform arithmetic calculation to detect differences [34,49,83,90].

6.10. Assessing Drivers of Change

GEE has been used to understand the influence of various land cover changes on mangroves over a long period of time [37,95,109]. For example, Tinh et al. [48] showed how expansion of aquaculture ponds and croplands is negatively impacting mangrove forests. GEE also hosts meteorological datasets. The differences in phenological features of mangroves and/or other plants within the mangrove ecosystem and their relationship with climatic variables such as precipitation and temperature can be analyzed with the help of GEE [37,79,83,87].

In summary, GEE provides an exhaustive array of tools for mangrove remote sensing, including image filtering, preprocessing, classification, time series analysis, and change detection. Researchers can employ GEE to amalgamate optical and radar imagery, evaluate spectral indices, establish decision thresholds, and differentiate mangroves from adjacent terrestrial and aquatic features, including submerged or tidal zones. It additionally facilitates effective map sharing and stakeholder involvement. GEE enables qualitative and quantitative analyses across various regions and across the globe, allowing mangrove monitoring in data-scarce or inaccessible areas. This section aims to assist new users—be they researchers or enthusiasts—by outlining the key functions and practical procedures for utilizing GEE in mangrove studies.

7. Current Shortcomings of GEE for Mangrove Studies

In this section, we explore existing shortcomings when using GEE for mangrove studies. These shortcomings fall into three categories: *user-related shortcomings*, which stem from gaps in expertise or familiarity that prevent users from fully taking advantage of GEE functionalities; *platform-specific constraints*, referring to missing or underdeveloped tools and processing capabilities needed for mangrove-focused analyses; and *data limitation*, arising from insufficient high-resolution or regionally relevant datasets within the GEE catalog.

7.1. User-Related Shortcomings

GEE has been used in numerous studies for mangrove mapping, but many have used selective years of satellite imagery [49,86,88] even though GEE hosts petabytes of a complete collection of datasets. Furthermore, despite GEE hosting 10 m resolution Sentinel images [111], 26 studies have relied on coarse- or mid-resolution imagery, such as Landsat, which may limit success in accurately delineating mangrove patches [85,89,94,104]. Moreover, many studies lack sufficient ground validation, introducing uncertainties and limiting the reliability of the results [34,36,51,85,104,112,113]. Tidal effects in open mangrove canopies can also introduce uncertainties in mangrove classification. Although Yancho et al. [38] demonstrated that GEE can be leveraged to understand the tidal influence while delineating mangrove extent, most GEE-based studies have not adequately considered this influence [74,84,88,92,94,100]. GEE can be used to reduce the salt–pepper effect detected in SAR images [114]. The salt–pepper effect can introduce noise and reduce the precision of mangrove classification outcomes [99]. However, studies have not adequately addressed this issue [96,105,108]. Additionally, GEE offers interactive mapping and graphical interfaces, but as discussed before, only three studies have used it to empower stakeholders.

7.2. Platform-Specific Constraints

Working with large regions of interest in GEE can pose a challenge in mapping mangroves. Processing very large areas can lead to memory errors in GEE during global studies [38,75]. As a result, very few studies have attempted to conduct a global analysis [115]. This limitation stems from various constraints within the GEE cloud computing environment, including limitations on requests, processing time, memory space, and data storage [116]. This limitation hinders the ability to compare global patterns of mangrove erosion and progradation [49,94]. Furthermore, distinguishing water from non-water pixels is an important step for detecting submerged mangroves, and it can exceed GEE's processing limits at global scales, risking memory errors during training [97,117]. While some studies use manual, visual delineation [91], this is labor intensive and not scalable. Implementing automated GEE algorithms requires additional tiling or composite strategies to stay within platform quotas, and this highlights a platform-specific constraint.

GEE has limited supervised and unsupervised classification algorithms to offer for classifying land cover, including mangroves [118]. The training set size is limited to 100 MB, which can only accommodate datasets with a limited number of bands. Inference phase predictions require fewer than 400 bands of imagery, and the model must be less than 100 MB in size [118]. Additionally, the JavaScript API platform does not support advanced models—such as artificial neural networks and deep learning—thereby limiting researchers to traditional classification methods in mangrove studies.

7.3. Data Limitations

The GEE platform utilized for mangrove species mapping studies, but in limited numbers (Table 1). The absence of quality and multitemporal hyperspectral and Lidar data on GEE is likely the reason. It is worth noting that the hyperspectral dataset from Earth Observing-1 (EO-1) is accessible on GEE [119], but it does not have global coverage. Additionally, Bessinger et al. [100] and Cissell et al. [84] note the importance of elevation and height data while mapping mangroves, especially the need to consider relief and shadows. However, there are few high-resolution elevation datasets on GEE that have a global and multitemporal footprint.

Some studies have overlooked socio-economic factors, such as anthropogenic disturbances, that can influence mangrove mapping [79,80]. This indicates the need to include socio-economic gridded products on GEE to infer anthropogenic influences.

8. Lessons Learned

In this section, we identify and discuss the main findings on how GEE is improving mangrove mapping. We discuss how GEE has transformed mangrove mapping by facilitating regional and global monitoring and by reducing atmospheric outliers and cloud interference. We also discuss how GEE simplifies classification in varied geographic regions, overcomes tidal fluctuation issues, and empowers non-experts with user-friendly interfaces.

GEE has completely changed the way mangrove mapping is performed by offering methodical solutions to problems, including atmospheric outliers, cloud and shadow interference, and changing weather-related issues. Researchers have attempted to develop mangrove maps since the 1980s, albeit in an unsystematic manner, whereas systematized mangrove mapping methodologies were developed with the introduction of GEE [34,83,86]. Before the application of GEE, unfiltered clouds and shadow residues were classified as non-mangrove, resulting in incorrect results, which GEE-based cloud and shadow filtering addressed [34]. Using GEE, it is easy to generate median composites that mitigate the impact of atmospheric outliers on long-term monitoring of mangrove status [92]. In addition, GEE makes it easy and fast to perform band ratio operations, which can further eradicate the noise caused by varying illumination, topography, and atmospheric conditions [92].

With the high degree of automation provided by GEE, mangroves can be mapped often and precisely at different scales [94]. Furthermore, mangroves can now be monitored at both regional and global scales thanks to ongoing updates by GEE with the latest images and new satellite data [120]. Moreover, cross-boundary comparison is easy with GEE. For instance, Baloloy et al. [83] leveraged GEE to compute a new mangrove index and compared the results for different mangrove ecosystems in the Philippines, Japan, Thailand, Vietnam, Indonesia, Cambodia, South America, Africa, and Australia. Rather than classifying the entire study region, GEE can help segment it into tiles and then run multiple classification models, ensuring that GEE's calculation time and memory are not exhausted [91,121].

GEE helps researchers overcome the problem of tidal fluctuation by allowing them to utilize all available observations [90]. Low and high tide images are important for distinguishing submerged mangroves and can be performed using GEE [38,91]. The spectral characteristics of high-tide conditions allow for the differentiation between tidal flats and submerged mangroves [120]. In addition, images taken at low tide can help determine the extent of the mudflat. Leveraging GEE, the mean of the calculated indices can help to infer the tide level for every available satellite image to obtain the highest and lowest tide scenes [91]. Moreover, GEE aids in detecting scenes with similar tide conditions, which is essential to accurately detect mangrove disturbances within the intertidal zones [120].

GEE-enabled GUIs and tools can empower nonexpert users and encourage policy makers to provide constructive feedback using local knowledge related to mangroves to improve maps that suit their own specific requirements [38]. Tools and GUIs developed using GEE can be made publicly available online, which may improve the reliability of procedures while opening up new possibilities for personalization and modification [38].

Beyond our original study period, we sought to determine whether our work remained current by extending our literature search through December 2023. This resulted in 25 additional GEE-based mangrove studies. To ensure scientific rigor and focus on the most impactful work, we used inclusion criteria. Each candidate paper needed at least six total citations and a citation rate of at least two per year, and hence, selected sixteen new studies for review (summarized in Appendix A, Table A2). The key findings from these 2023 papers reveal a few distinct trends. GEE is increasingly using multisource, time series remote sensing, combining optical and SAR inputs to improve classification accuracy. Researchers are integrating temporal segmentation algorithms and phenological composites into GEE to detect long-term changes in mangrove extent and species composition with minimal manual tuning. GEE's built-in tools for cloud filtering, spectral index computation, and machine learning model training enable rapid and reproducible mapping pipelines without the need for local high-performance hardware. Several studies suggest using GEE to increase access to large-scale time series analyses for mangrove monitoring. The lessons learned from 2023 thus both reinforce our insights from 2022 and chart new avenues for future methodological development and application.

9. Future Directions

Some possible next steps might pave the path for more effective use of GEE in mangrove monitoring in the future. In this section, we highlight areas deserving special attention in potential future work on mangroves using GEE.

9.1. Fusing Data

Several authors demonstrated that GEE can be used for merging multisource remote sensing data to obtain better accuracy [86,88,94,96,102]. However, more research is needed in this area. Time series of imagery stored in GEE (e.g., Sentinel) allows detection of differences in mangrove phenological patterns, particularly in areas where they are irregular and patchy [80,91]. Tracking phenological patterns highlights changes in amplitude and phase of mangroves, leveraging the detection of leaf density and synchronizing timing shifts of mangrove canopy with environmental cues like temperature, rainfall, or tidal patterns.

9.2. Incorporating Spectral and Ecological Properties

Future research should concentrate on improving approaches and adding advanced modeling/algorithms/tools that account for spectral signatures and ecological properties of various mangrove typologies [113].

9.3. Mapping with Spatial Awareness

To improve the precision of mangrove classification, object-based techniques such as Simple Non-Iterative Clustering (SNIC) has the potential to improve classification accuracy and decrease problems in mangrove spatial separability [99,100]. Additionally, there is an urgency to include state-of-the-art deep learning classifiers like U-Net and Stacked AutoEncoder into GEE [91,92].

9.4. Incorporating More Environmental Data

Many authors have proposed utilizing climatic and atmospheric data, including pollution, sea temperature, soil moisture, nighttime lights, drought severity, sea surface temperature, and pests/diseases, to enhance our knowledge of mangrove dynamics [79,80]. Therefore, future research should leverage many of the environmental gridded data products already available on GEE [119].

9.5. Using Diverse Spectral Indicators

There is an apparent need for looking into more combinations of indices to improve the detection of mangroves [37,74]. There are already GEE-based algorithms and utilities developed by researchers that can automatically generate numerous spectral indices from various satellite data products, like Landsat, Sentinel-1, and -2 [122]. However, single-index approaches, like the NDVI, have been criticized for being too sensitive to climatic variables [85]. Prior works have highlighted the importance of exploring mangrove-specific indices [39]. However, for varying geographical settings, systematic evaluation of these emerging indices within the GEE environment is recommended. Furthermore, future studies could evaluate the effectiveness of the Tasselled Cap approach for identifying mangroves on tidal flats and monitoring cover changes [85].

Looking forward, researchers should conduct comparative studies across diverse biogeographic regions—tropical, subtropical, and temperate—to assess index robustness under varying tidal regimes, canopy densities, and seasonal conditions. Such work will refine best practices for index selection and open avenues for automated, scalable mangrove monitoring pipelines that adapt to local spectral characteristics.

9.6. Improving Reference Data Collection

As GEE offers interactive mapping facilities, there is potential to develop GEE-based automatic methods to collect and expand training samples for improved classification accuracy [91]. Also, there is a need for the development of refined and automated techniques using GEE to improve visual assessments [108]. While tools such as collect earth online (https://www.collect.earth/ (accessed on 1 March 2023)), earthmap (https://earthmap.org/ (accessed on 1 March 2023)), sepal (https://sepal.io/ (accessed on 9 January 2024)), and Sentinel Hub (https://apps.sentinel-hub.com/eo-browser/ (accessed on 1 March 2023)) offer non-coding approaches, advanced processing require additional payment. Novel ideas for visualization are needed for practical visualization outcomes.

9.7. Incorporating a Temporal Segmentation Algorithm

The use of temporal segmentation techniques in GEE was recently highlighted by Pasquarella et al. [123] as a way to assist the worldwide mapping, monitoring, and analysis of landscapes, with an emphasis on understanding the trends and patterns in their features. LandTrendr, CCDC, EWMACD, VCT, and VeRDET are just a few of the algorithms that use band and spectral-index-based time series to detect and monitor changes. They can provide useful insights for tracking the rate and pattern of erosion and colonization in mangrove habitats [85]. Despite the advantages, there are few studies that used a temporal segmentation method to analyze a mangrove forest [85,124]. To more correctly portray the dynamic processes and changing timeline of mangrove forests, future studies should emphasize investigating and applying the many temporal segmentation techniques accessible on GEE.

9.8. Detecting Degradation

Detecting mangrove degradation is key to anticipating new mangrove deforestation regimes or preventing the complete loss of mangroves. While degradation can be defined, conceptualized, and represented in different ways [125], few articles focus on tracking mangrove degradation using dense remote sensing time series. New approaches to disentangle mangrove loss from degraded mangrove are necessary. Some key examples are algorithms such as Continuous Degradation Detection (CODED) [126], which delineate ephemeral changes linked to degradation over tropical forests but poorly tested on mangrove ecosystems. Although global degradation datasets like Tropical Moist Forests [127] provide and understanding of land-cover transitions, field validation is still necessary to confirm mangrove degradation.

9.9. Expanding Potential Drivers

Some drivers of mangrove change are unusual but important. For instance, the effect of oil spills in the Nigerian delta over mangroves can be mapped through satellite imagery [128]. This is useful to inform authorities about inefficient oil extraction practices that gradually degrade mangrove and the role of locals for the establishment of community vanguards to develop a sustainable platform for mangrove conservation [129]. Another understudied driver is Nipa palm expansion that has documented the loss of 28,000 ha of mangroves [130]. Natural drivers such as hurricanes cause damage to mangrove forests depending on the storm level. However, understanding mangrove resurgence after hurricane impact remains uncertain. While some local studies suggest >5 years for canopy recovery, constant monitoring and mangrove resilience should be quantified, especially during repetitive hurricane events and the long-term effects of climate change. Salinity is a gradual driver of mangrove change. Salinity is rising due to climate change, and it significantly impedes forest growth and ecosystem functions for mangroves. Understanding how salinity influences mangrove structure is relevant at regional scales. Satellite imagery, mostly hyperspectral [131] can assist in identifying wide mangrove areas affected by salinity. The proliferation of invasive species is another driver of mangrove change that is poorly evaluated using satellite imagery. Monitoring and eradicating these invasive species are vital to preserving the ecosystem. Sentinel 2 and hyperspectral imagery show promising results in identifying the spatial variability of invasive species [132]. Data collection from high-resolution imagery available in GEE and empowering citizen science through GEE-free apps will increase the amount of training data to improve the classification of potential drivers of mangrove change.

9.10. Focusing on Carbon

GEE is used to derive diverse biophysical variables related to mangrove at a global scale. Height, biomass and total carbon estimates are now passively calculated using the GEDI's LIDAR system combined with optical/SAR imagery [133,134]. However, more regional and local biophysical variable estimates are needed to reduce uncertainty and improve accuracy. More reliable estimates can support blue carbon credits that may help tropical coastal countries for climate change goals [135].

10. Conclusions

A thorough review of 54 relevant studies published within the 2017–2022 period showed that GEE has been widely and increasingly used for mapping mangroves in many locations around the world. We categorized the studies based on the breadth of GEE integration, the primary motivations for doing so, and the subfields of mangrove remote sensing to better understand how GEE may be used in mangrove studies. We documented growing interest in GEE among mangrove remote sensing experts. GEE is particularly useful for data extraction, preprocessing, modeling, computational tasks, accuracy evaluations, and the development of interactive mapping interfaces, all of which are critical to mangrove research. However, some constraints prevent GEE from reaching its full potential in mangrove research. By identifying these limitations, we outlined clear directions and priorities for future studies.

An essential function of this present work, like any other review paper, is to outline challenges and propose potential avenues for future research [52]. The present work is an indispensable asset for researchers seeking to expand upon prior investigations. The current effort would bring together existing knowledge, techniques, methodology, and findings, offering a coherent and thorough summary of current trends in leveraging GEE for mangrove studies. As the first of its kind, this current study provides an important starting point for the emerging field of mangrove research using GEE, as well as a guidepost for new research endeavors looking to capitalize on the potential of this cloud-computing platform.

Author Contributions: K.M.A.I., conceptualization, investigation, data preparation, data curation, methodology, software, formal analysis, visualization, writing—original draft, and writing—review and editing; P.M.-S., formal analysis, writing—original draft, and writing—review and editing; E.B., writing—review and editing; R.K., supervision, writing—original draft, and writing—review and editing. All authors have read and agreed to the published version of the manuscript.

Funding: This work was partially supported by the NASA SERVIR program under grant 80NSSC23K0245.

Data Availability Statement: Data available on request.

Acknowledgments: The authors would like to extend their appreciation to the Graduate Writing Center of Oregon State University, USA, for their assistance in refining the manuscript. Additionally, the authors would like to express their gratitude to all the corresponding authors of the reviewed articles for their cordial assistance in resolving any questions that arose during the analysis period.

Conflicts of Interest: The authors declare no conflicts of interest.

Abbreviations

The following abbreviations are used in this manuscript:

CODED	Continuous Degradation Detection
EO-1	Earth Observing-1
EVI	Enhanced Vegetation Index
GEE	Google Earth Engine
LAI	Leaf Area Index
LANDSAT	Land Satellite
LSWI	Land Surface Water Index
MDPI	Multidisciplinary Digital Publishing Institute
MMRI	Modular Mangrove Recognition Index
mNDWI	Modified Normalized Difference Water Index
MVI	Mangrove Vegetation Index
NDVI	Normalized Difference Vegetation Index
RF	Random Forest
RS	Remote Sensing
SNIC	Non-Iterative Clustering

Appendix A

Table A1. Study areas and the country affiliation of researchers.

Article Title	Affiliated Country of First Author	Context/Study Area
10-m-resolution mangrove maps of China derived from multisource and multitemporal satellite observations	China	China
A cloud computing-based approach to mapping mangrove erosion and progradation: Case studies from the Sundarbans and French Guiana	Singapore	India, Bangladesh, French Guiana
A detailed mangrove map of China for 2019 derived from Sentinel-1 and-2 images and Google Earth images	China	China
A history of the rehabilitation of mangroves and an assessment of their diversity and structure using Landsat annual composites (1987–2019) and transect plot inventories	Thailand	Thailand
A mangrove forest map of China in 2015: Analysis of time series Landsat 7/8 and Sentinel-1A imagery in Google Earth Engine cloud computing platform	China	China
Advancing the Mapping of Mangrove Forests at National-Scale Using Sentinel-1 and Sentinel-2 Time Series Data with Google Earth Engine: A Case Study in China	China	China
An Improved Submerged Mangrove Recognition Index-Based Method for Mapping Mangrove Forests by Removing the Disturbance of Tidal Dynamics and S. alterniflora	China	China
Application of Artificial Neural Networks for Mangrove Mapping Using Multitemporal and Multisource Remote Sensing Imagery	Iran	Iran
Assessing cyclone disturbances (1988–2016) in the Sundarbans mangrove forests using Landsat and Google Earth Engine	Japan	India, Bangladesh
Brazilian Mangrove Status: Three Decades of Satellite Data Analysis	Brazil	Brazil
Characterizing Spatiotemporal Patterns of Mangrove Forests in Can Gio Biosphere Reserve Using Sentinel-2 Imagery	Vietnam	Vietnam
Cloud-Based Monitoring and Evaluation of the Spatial-Temporal Distribution of Southeast Asia's Mangroves Using Deep Learning	UK	Cambodia, Laos, Malaysia, Myanmar, Philippines, Indonesia, Thailand, Vietnam
Combing Sentinel-1 and Sentinel-2 image time series for invasive Spartina alterniflora mapping on Google Earth Engine: a case study in Zhangjiang Estuary	China	China
Development and application of a new mangrove vegetation index (MVI) for rapid and accurate mangrove mapping	Philippines	Philippines, Japan
Distribution and drivers of Vietnam mangrove deforestation from 1995 to 2019	Vietnam	Vietnam
Eleven Years of Mangrove-Mudflat Dynamics on the Mud Volcano-Induced Prograding Delta in East Java, Indonesia: Integrating UAV and Satellite Imagery	Netherlands	Indonesia
Elucidating the phenology of the Sundarbans mangrove forest using 18-year time series of MODIS vegetation indices	Japan	Bangladesh
Evaluating Combinations of Sentinel-2 Data and Machine-Learning Algorithms for Mangrove Mapping in West Africa	USA	Senegal, Gambia
Evaluating mangrove conservation and sustainability through spatiotemporal (1990–2020) mangrove cover change analysis in Pakistan	Pakistan	Pakistan
Evaluation of Spatiotemporal Dynamics of Guyana's Mangroves Using SAR and GEE	Guyana	Guyana
Extrapolating canopy phenology information using Sentinel-2 data and the Google Earth Engine platform to identify the optimal dates for remotely sensed image acquisition of semiarid mangroves	Mexico	Mexico
Gap-Free Monitoring of Annual Mangrove Forest Dynamics in Ca Mau Province, Vietnamese Mekong Delta, Using the Landsat-7–8 Archives and Post-Classification Temporal Optimization	Netherlands	Vietnam

Table A1. *Cont.*

Article Title	Affiliated Country of First Author	Context/Study Area
How to automate timely large-scale mangrove mapping with remote sensing	USA	USA, China
Identifying and forecasting potential biophysical risk areas within a tropical mangrove ecosystem using multisensor data	USA	India
Identifying large-area mangrove distribution based on remote sensing: A binary classification approach considering subclasses of non-mangroves	China	China
Improved estimates of mangrove cover and change reveal catastrophic deforestation in Myanmar	Singapore	Myanmar
Incorporating the Plant Phenological Trajectory into Mangrove Species Mapping with Dense Time Series Sentinel-2 Imagery and the Google Earth Engine Platform	China	China
Large-Scale High-Resolution Coastal Mangrove Forests Mapping Across West Africa With Machine Learning Ensemble and Satellite Big Data	USA	Senegal, Gambia, Guinea Bissau, Guinea, Sierra Leone, Liberia, Ivory Coast, Ghana, Togo, Benin, Nigeria
Mangrove Ecosystem Mapping Using Sentinel-1 and Sentinel-2 Satellite Images and Random Forest Algorithm in Google Earth Engine	Iran	Iran
Mangrove Forest Cover and Phenology with Landsat Dense Time Series in Central Queensland, Australia	Australia	Australia
Mapping and dynamic analysis of mangrove forest during 2009–2019 using landsat-5 and sentinel-2 satellite data along Odisha Coast	India	India
Mapping mangrove dynamics and colonization patterns at the Suriname coast using historic satellite data and the LandTrendr algorithm	Netherlands	Suriname
Mapping National Mangrove Cover for Belize Using Google Earth Engine and Sentinel-2 Imagery	USA	Belize
Previous Shoreline Dynamics Determine Future Susceptibility to Cyclone Impact in the Sundarban Mangrove Forest	Singapore	India, Bangladesh
Radar and optical remote sensing for near real-time assessments of cyclone impacts on coastal ecosystems	USA	India, Bangladesh
Remote Sensing-Based Mapping of Senescent Leaf C:N Ratio in the Sundarbans Reserved Forest Using Machine Learning Techniques	China	Bangladesh
Sub-continental-scale mapping of tidal wetland composition for East Asia: A novel algorithm integrating satellite tide-level and phenological features	China	China, North and South Korea
The Google Earth Engine Mangrove Mapping Methodology (GEEMMM)	UK	Myanmar
The Key Reason of False Positive Misclassification for Accurate Large-Area Mangrove Classifications	China	China
Turning the Tide on Mapping Marginal Mangroves with Multidimensional Space-Time Remote Sensing	Australia	Australia
Using multi-indices approach to quantify mangrove changes over the Western Arabian Gulf along Saudi Arabia coast	USA	Saudi Arabia
Annual Change Analysis of Mangrove Forests in China during 1986–2021 Based on Google Earth Engine	China	China

Table A1. *Cont.*

Article Title	Affiliated Country of First Author	Context/Study Area
Assessment of mangrove cover dynamics and its health status in the Gulf of Khambhat, Western India, using high-resolution multitemporal satellite data and Google Earth Engine	India	India
Decision surface optimization in mapping exotic mangrove species (Sonneratia apetala) across latitudinal coastal areas of China	China	China
Identification of Mangrove Changes in The Mahakam Delta in 2007–2017 using Alos/Palsar and Landsat	Indonesia	Indonesia
Landsat-8-based coastal ecosystem mapping in South Africa using Random Forest classification in Google Earth Engine	South Africa	South Africa
Mapping Multidecadal Mangrove Extent in the Northern Coast of Vietnam Using Landsat Time Series Data on Google Earth Engine Platform	Vietnam	Vietnam
Monitoring detailed mangrove hurricane damage and early recovery using multisource remote sensing data	Mexico	Mexico
Quantifying Mangrove Extent Using a Combination of Optical and Radar Images in a Wetland Complex, Western Region, Ghana	Ghana	Ghana
Remap: An online remote sensing application for land cover classification and monitoring	Australia	Australia
The environmental burdens of special economic zones on the coastal and marine environment: A remote sensing assessment in Myanmar	USA	Myanmar
The national nature reserves in China: Are they effective in conserving mangroves?	China	China
Tracking changes in extent and distribution of tropical coastal covers using simple semi-supervised classification	USA	Colombia
Tracking Deforestation, Drought, and Fire Occurrence in Kutai National Park, Indonesia	Canada	Indonesia

Table A2. Summary of selected 2023 remote sensing studies using Google Earth Engine (GEE) for mangrove monitoring.

Title	Publication Date	Total Citations	Average/Year	Summary of GEE Implementation: Methods, Advantages, and Future Directions
GWL FCS30: a global 30 m wetland map with a fine classification system using multi-sourced and time series remote sensing imagery in 2020 [136]	JAN 2023	78	26	Leveraged multisource, time series imagery in GEE; generated phenological composites, stratified random samples, and a Random Forest classifier; all preprocessing, classification, and accuracy assessment ran within GEE. Future work should integrate additional data sources to further refine map detail.
Identifying mangroves through knowledge extracted from trained Random Forest models: An interpretable mangrove mapping approach (IMMA) [137]	JUL 2023	36	12	Employed GEE to extract spectral-band thresholds from pretrained Random Forest models; GEE aided automation; interpretable delineation of mangrove extents without manual parameter tuning was possible.
Remote sensing for cost-effective blue carbon accounting [63]	MAR 2023	28	9.33	Demonstrated that cloud-computing platforms like GEE, combined with multisensor fusion, big-data integration, and AI algorithms, allow sophisticated blue-carbon estimation and have the potential to attain higher accuracy and precision without the need to download large volumes of observational data.
Long-Term Wetland Monitoring Using the Landsat Archive: A Review [138]	FEB 2023	17	5.67	Reviewed GEE-based workflows using intra-annual Landsat TM/ETM+/OLI-TIRS time series and efficient index calculations for regional to continental-scale change detection. Highlighted potential of GEE to reduce technology gaps between the Global North and South by democratizing access to large-scale time series analyses.
Collaborative multiple change detection methods for monitoring the spatiotemporal dynamics of mangroves in Beibu Gulf, China [139]	DEC 2023	13	4.33	Showcased GEE implementation of temporal-segmentation algorithms (e.g., CCDC) to deliver robust, automated mapping of mangrove extent and dynamics over time.
Mangrove species mapping through phenological analysis using Random Forest algorithm on Google Earth Engine [140]	APR 2023	13	4.33	Used GEE to access multisource, multitemporal imagery; applied cloud filtering, tidal-effect thresholding, and computed 21 spectral indices; generated max/mean/median composites; and ran Random Forest classification. Noted that rapidly mapping mangrove species was possible without requiring local training samples.
Monitoring of 35-Year Mangrove Wetland Change Dynamics and Agents in the Sundarbans Using Temporal Consistency Checking [141]	FEB 2023	13	4.33	Accessed the complete Landsat archive in GEE; applied cloud filtering, median composites, spectral-Tasseled-Cap indices, pixel unmixing, gap filling, and temporal segmentation algorithm; ran GEE-based Random Forest and preclassification workflows; compared age-structure map products; and computed accuracy while accounting for tidal effects.
Annual Mangrove Vegetation Cover Changes (2014–2020) in Indian Sundarbans National Park Using Landsat 8 and Google Earth Engine [142]	MAR 2023	12	4	Retrieved and preprocessed Landsat 8 in GEE; performed cloud/shadow masking, spectral-index computation, and median compositing; then conducted time series change detection. Authors demonstrated that the speed and efficiency of GEE were superior for large-scale mangrove monitoring.
An enhanced approach to mangrove forest analysis in the Colombian Pacific coast using optical and SAR data in Google Earth Engine [143]	APR 2023	10	3.33	Fused Landsat and SAR inputs in GEE with cloud/speckle (Refined Lee) filtering; computed spectral indices and GLCM texture metrics; generated image composites; and tuned Random Forest hyperparameters, which achieved improved classification accuracy.
Spatiotemporal monitoring of Qeshm mangrove forests through machine learning classification of SAR and optical images on Google Earth Engine [144]	OCT 2023	9	3	Employed GEE's free access to Landsat 8, Sentinel-1/2; fused optical and radar data; applied spectral indices; and trained SVM and Random Forest models, computing overall accuracy entirely within GEE's cloud environment. Noted that GEE gives researchers instant access to vast multisensor and multitemporal imagery, plus built-in tools for preprocessing, analysis, and machine learning. It enables fast, reproducible mapping without local hardware constraints.

Table A2. *Cont.*

Title	Publication Date	Total Citations	Average/Year	Summary of GEE Implementation: Methods, Advantages and, Future Directions
Satellite Data Reveal Concerns Regarding Mangrove Restoration Efforts in Southern China [145]	SEP 2023	7	2.33	Used GEE to access the Landsat archive; calculate spectral indices; apply user-defined thresholding; analyze time series. Tidal influence was considered. Authors used temporal segmentation algorithm for disturbance detection. Accuracy was assessed using F1-score metrics using GEE.
Comparison of the Applicability of J-M Distance Feature Selection Methods for Coastal Wetland Classification [146]	JUN 2023	7	2.33	Accessed Sentinel and DEM data in GEE; performed cloud masking, SAR noise removal, and terrain correction; fused datasets; computed spectral indices and GLCM metrics; and trained a Random Forest classifier; evaluated performance via confusion-matrix.
Time series (2001/2002–2021) analysis of Earth observation data using Google Earth Engine (GEE) for detecting changes in land use land cover (LULC) with specific reference to forest cover in East Godavari Region, Andhra Pradesh, India [147]	MAY 2023	7	2.33	This paper is not explicitly mangrove-focused. Authors harnessed GEE's cloud infrastructure to automate a 20-year LULC change-detection. The study accessed and preprocessed Landsat archives; applied cloud masking and normalization. GEE-based Hansen global forest change data was used. Classification and Regression Tree algorithm was utilized, and accuracy was assessed using GEE.
Comparison of vegetation indices and image classification methods for mangrove mapping at semi-detailed scale in southwest of Rio de Janeiro, Brazil [148]	APR 2023	7	2.33	Retrieved Landsat in GEE, applied topographic masking, median composites, and computed spectral indices. Ran supervised classifiers (CART, RF, Minimum Distance) and unsupervised methods (K-means, X-means, Cascade Simple K-means, LVQ, Cobweb), noting GEE's capacity to handle unprecedented data volumes. Authors highlighted the MVI-Cobweb method for accurately identifying various mangrove ecotypes and recommended testing newer algorithms in future studies.
Understanding the natural expansion of white mangrove (Laguncularia racemosa) in an ephemeral inlet based on geomorphological analysis and remote sensing data [149]	JUL 2023	7	2.33	Analyzed Sentinel-2 NDVI time series via the GEE plugin in QGIS. Authors successfully detected spatial expansion patterns of L. racemosa linked to geomorphological changes.
Understanding the states and dynamics of mangrove forests in land cover transitions of The Gambia using a Fourier transformation of Landsat and MODIS time series in Google Earth Engine [150]	FEB 23 2023	6	2	Accessed Landsat/MODIS in GEE; applied cloud masking, generated gap-free median mosaics, computed spectral indices, and performed harmonics-based phenological analysis with K-means clustering for sample design. Ran Random Forest classification, assessed accuracy, and recommended longer data records and CCDC temporal segmentation for future work.

References

1. Barbier, E.B. Natural barriers to natural disasters: Replanting mangroves after the tsunami. *Front. Ecol. Environ.* **2006**, *4*, 124–131. [CrossRef]
2. Spalding, M. *World Mangrove Atlas*; International Society for Mangrove Ecosystems: Okinawa, Japan, 1997.
3. Alongi, D.M. Present state and future of the world's mangrove forests. *Environ. Conserv.* **2002**, *29*, 331–349. [CrossRef]
4. Mitra, A. *Mangrove Forests in India: Exploring Ecosystem Services*; Springer International Publishing: Cham, Switzerland, 2020.
5. Craft, C. Mangroves. In *Creating and Restoring Wetlands*; Elsevier: Amsterdam, The Netherlands, 2016; pp. 233–263.
6. Costanza, R.; de Groot, R.; Sutton, P.; van der Ploeg, S.; Anderson, S.J.; Kubiszewski, I.; Farber, S.; Turner, R.K. Changes in the global value of ecosystem services. *Glob. Environ. Change* **2014**, *26*, 152–158. [CrossRef]
7. Held, A.; Ticehurst, C.; Lymburner, L.; Williams, N. High resolution mapping of tropical mangrove ecosystems using hyperspectral and radar remote sensing. *Int. J. Remote Sens.* **2003**, *24*, 2739–2759. [CrossRef]
8. Worthington, T.A.; Andradi-Brown, D.A.; Bhargava, R.; Buelow, C.; Bunting, P.; Duncan, C.; Fatoyinbo, L.; Friess, D.A.; Goldberg, L.; Hilarides, L.; et al. Harnessing Big Data to Support the Conservation and Rehabilitation of Mangrove Forests Globally. *One Earth* **2020**, *2*, 429–443. [CrossRef]
9. Davie, J.D.S.; Hegerl, E.J.; Saenger, P. *Global Status of Mangrove Ecosystems*; IUCN: Gland, Switzerland, 1983.
10. Clüsener-Godt, M.; Cárdenas Tomažič, M.R. 8—The Importance of Mangrove Ecosystems for Nature Protection and Food Productivity: Actions of UNESCO's Man and the Biosphere Programme. In *Halophytes for Food Security in Dry Lands*; Khan, M.A.; Ozturk, M.; Gul, B.; Ahmed, M.Z., Eds.; Academic Press: San Diego, CA, USA, 2016; pp. 125–140.
11. Fao. *The World's Mangroves 2000–2020*; FAO: Rome, Italy, 2023.
12. Kusmana, C. Distribution and Current Status of Mangrove Forests in Indonesia. In *Mangrove Ecosystems of Asia: Status, Challenges and Management Strategies*; Faridah-Hanum, I., Latiff, A., Hakeem, K.R., Ozturk, M., Eds.; Springer: New York, NY, USA, 2014; pp. 37–60.
13. Tangah, J.; Bajau, F.E.; Jilimin, W.; Baba, S.; Chan, H.T.; Kezuka, M. *Rehabilitation of Mangroves in Sabah—The SFD-ISME Collaboration (2011–2014)*; Technical Report; Sabah Forestry Department, International Society for Mangrove Ecosystems and Tokio Marine & Nichido Fire Insurance Co., Ltd.: Tokyo, Japan, 2015.
14. United Nations Development Programme; World Conservation Union. *Mangroves for the Future: A Strategy for Promoting Investment in Coastal Ecosystem Conservation*; Technical Report; United Nations Development Programme: New York, NY, USA; World Conservation Union: Gland, Switzerland; 2015.
15. Adame, M.F.; Kauffman, J.B.; Medina, I.; Gamboa, J.N.; Torres, O.; Caamal, J.P.; Reza, M.; Herrera-Silveira, J.A. Carbon Stocks of Tropical Coastal Wetlands within the Karstic Landscape of the Mexican Caribbean. *PLoS ONE* **2013**, *8*, e56569. [CrossRef]
16. Atwood, T.B.; Connolly, R.M.; Almahasheer, H.; Carnell, P.E.; Duarte, C.M.; Ewers Lewis, C.J.; Irigoien, X.; Kelleway, J.J.; Lavery, P.S.; Macreadie, P.I.; et al. Global patterns in mangrove soil carbon stocks and losses. *Nat. Clim. Change* **2017**, *7*, 523–528. [CrossRef]
17. Bouillon, S.; Borges, A.V.; Castañeda Moya, E.; Diele, K.; Dittmar, T.; Duke, N.C.; Kristensen, E.; Lee, S.Y.; Marchand, C.; Middelburg, J.J.; et al. Mangrove production and carbon sinks: A revision of global budget estimates. *Glob. Biogeochem. Cycles* **2008**, *22*, 3052. [CrossRef]
18. Kristensen, E.; Bouillon, S.; Dittmar, T.; Marchand, C. Organic carbon dynamics in mangrove ecosystems: A review. *Aquat. Bot.* **2008**, *89*, 201–219. [CrossRef]
19. Moraes, O. Blue carbon in area-based coastal and marine management schemes—A review. *J. Indian Ocean Reg.* **2019**, *15*, 193–212. [CrossRef]
20. Sanderman, J.; Hengl, T.; Fiske, G.; Solvik, K.; Adame, M.F.; Benson, L.; Bukoski, J.J.; Carnell, P.; Cifuentes-Jara, M.; Donato, D.; et al. A global map of mangrove forest soil carbon at 30 m spatial resolution. *Environ. Res. Lett.* **2018**, *13*, 055002. [CrossRef]
21. Vanderklift, M.A.; Marcos-Martinez, R.; Butler, J.R.A.; Coleman, M.; Lawrence, A.; Prislan, H.; Steven, A.D.L.; Thomas, S. Constraints and opportunities for market-based finance for the restoration and protection of blue carbon ecosystems. *Mar. Policy* **2019**, *107*, 103429. [CrossRef]
22. Dittmar, T.; Hertkorn, N.; Kattner, G.; Lara, R.J. Mangroves, a major source of dissolved organic carbon to the oceans. *Glob. Biogeochem. Cycles* **2006**, *20*, 2570. [CrossRef]
23. Diop, S.; Barusseau, J.P.; Descamps, C., Eds. *The Land/Ocean Interactions in the Coastal Zone of West and Central Africa*; Estuaries of the World; Springer International Publishing: Cham, Switzerland, 2014.
24. Donato, D.C.; Kauffman, J.B.; Murdiyarso, D.; Kurnianto, S.; Stidham, M.; Kanninen, M. Mangroves among the most carbon-rich forests in the tropics. *Nat. Geosci.* **2011**, *4*, 293–297. [CrossRef]
25. Mizanur Rahman, M.; Nabiul Islam Khan, M.; Fazlul Hoque, A.K.; Ahmed, I. Carbon stock in the Sundarbans mangrove forest: Spatial variations in vegetation types and salinity zones. *Wetl. Ecol. Manag.* **2015**, *23*, 269–283. [CrossRef]

26. Thomas, S.; Dargusch, P.; Harrison, S.; Herbohn, J. Why are there so few afforestation and reforestation Clean Development Mechanism projects? *Land Use Policy* **2010**, *27*, 880–887. [CrossRef]
27. Younes Cárdenas, N.; Joyce, K.E.; Maier, S.W. Monitoring mangrove forests: Are we taking full advantage of technology? *Int. J. Appl. Earth Obs. Geoinf.* **2017**, *63*, 1–14. [CrossRef]
28. Giri, C.; Ochieng, E.; Tieszen, L.L.; Zhu, Z.; Singh, A.; Loveland, T.; Masek, J.; Duke, N. Status and distribution of mangrove forests of the world using earth observation satellite data: Status and distributions of global mangroves. *Glob. Ecol. Biogeogr.* **2011**, *20*, 154–159. [CrossRef]
29. Kuenzer, C.; Bluemel, A.; Gebhardt, S.; Quoc, T.V.; Dech, S. Remote Sensing of Mangrove Ecosystems: A Review. *Remote Sens.* **2011**, *3*, 878–928. [CrossRef]
30. Purnamasayangsukasih, P.R.; Norizah, K.; Ismail, A.A.M.; Shamsudin, I. A review of uses of satellite imagery in monitoring mangrove forests. *IOP Conf. Ser. Earth Environ. Sci.* **2016**, *37*, 012034. [CrossRef]
31. Lewis, R.R.; Milbrandt, E.C.; Brown, B.; Krauss, K.W.; Rovai, A.S.; Beever, J.W.; Flynn, L.L. Stress in mangrove forests: Early detection and preemptive rehabilitation are essential for future successful worldwide mangrove forest management. *Mar. Pollut. Bull.* **2016**, *109*, 764–771. [CrossRef] [PubMed]
32. Lee, S.Y.; Primavera, J.H.; Dahdouh-Guebas, F.; McKee, K.; Bosire, J.O.; Cannicci, S.; Diele, K.; Fromard, F.; Koedam, N.; Marchand, C.; et al. Ecological role and services of tropical mangrove ecosystems: A reassessment. *Glob. Ecol. Biogeogr.* **2014**, *23*, 726–743. [CrossRef]
33. Farzanmanesh, R.; Khoshelham, K.; Thomas, S. Technological opportunities for measuring and monitoring blue carbon initiatives in mangrove ecosystems. *Remote Sens. Appl. Soc. Environ.* **2021**, *24*, 100612. [CrossRef]
34. Diniz, C.; Cortinhas, L.; Nerino, G.; Rodrigues, J.; Sadeck, L.; Adami, M.; Souza, P. Brazilian Mangrove Status: Three Decades of Satellite Data Analysis. *Remote Sens.* **2019**, *11*, 808. [CrossRef]
35. Zhao, C.; Qin, C.Z. A detailed mangrove map of China for 2019 derived from Sentinel-1 and-2 images and Google Earth images. *Geosci. Data J.* **2022**, *9*, 74–88. [CrossRef]
36. Hickey, S.; Radford, B. Turning the Tide on Mapping Marginal Mangroves with Multi-Dimensional Space-Time Remote Sensing. *Remote Sens.* **2022**, *14*, 3365. [CrossRef]
37. Chamberlain, D.; Phinn, S.; Possingham, H. Mangrove Forest Cover and Phenology with Landsat Dense Time Series in Central Queensland, Australia. *Remote Sens.* **2021**, *13*, 3032. [CrossRef]
38. Yancho, J.; Jones, T.; Gandhi, S.; Ferster, C.; Lin, A.; Glass, L. The Google Earth Engine Mangrove Mapping Methodology (GEEMMM). *Remote Sens.* **2020**, *12*, 3758. [CrossRef]
39. Tran, T.V.; Reef, R.; Zhu, X. A Review of Spectral Indices for Mangrove Remote Sensing. *Remote Sens.* **2022**, *14*, 4868. [CrossRef]
40. Pham, T.D.; Xia, J.; Ha, N.T.; Bui, D.T.; Le, N.N.; Tekeuchi, W. A Review of Remote Sensing Approaches for Monitoring Blue Carbon Ecosystems: Mangroves, Seagrassesand Salt Marshes during 2010–2018. *Sensors* **2019**, *19*, 1933. [CrossRef]
41. Schill, S.R.; McNulty, V.P.; Perez, D.; Shono, K.; Friedman, K. *Remote Sensing Techniques for Mapping and Monitoring Mangroves at Fine Scales*; FAO: Rome, Italy, 2024.
42. Ma, Y.; Wu, H.; Wang, L.; Huang, B.; Ranjan, R.; Zomaya, A.; Jie, W. Remote sensing big data computing: Challenges and opportunities. *Future Gener. Comput. Syst.* **2015**, *51*, 47–60. [CrossRef]
43. Giri, C.; Pengra, B.; Zhu, Z.; Singh, A.; Tieszen, L.L. Monitoring mangrove forest dynamics of the Sundarbans in Bangladesh and India using multi-temporal satellite data from 1973 to 2000. *Estuar. Coast. Shelf Sci.* **2007**, *73*, 91–100. [CrossRef]
44. Gorelick, N.; Hancher, M.; Dixon, M.; Ilyushchenko, S.; Thau, D.; Moore, R. Google Earth Engine: Planetary-scale geospatial analysis for everyone. *Remote Sens. Environ.* **2017**, *202*, 18–27. [CrossRef]
45. Alonso, A.; Muñoz Carpena, R.; Kennedy, R.E.; Murcia, C. Wetland Landscape Spatio-Temporal Degradation Dynamics Using the New Google Earth Engine Cloud-Based Platform: Opportunities for Non-Specialists in Remote Sensing. *Trans. ASABE* **2016**, *59*, 1331–1342.
46. Jahromi, M.N.; Jahromi, M.N.; Zolghadr-Asli, B.; Pourghasemi, H.R.; Alavipanah, S.K. Google Earth Engine and Its Application in Forest Sciences. In *Spatial Modeling in Forest Resources Management*; Shit, P.K., Pourghasemi, H.R., Das, P., Bhunia, G.S., Eds.; Springer International Publishing: Cham, Switzerland, 2021; pp. 629–649.
47. Chen, B.; Xiao, X.; Li, X.; Pan, L.; Doughty, R.; Ma, J.; Dong, J.; Qin, Y.; Zhao, B.; Wu, Z.; et al. A mangrove forest map of China in 2015: Analysis of time series Landsat 7/8 and Sentinel-1A imagery in Google Earth Engine cloud computing platform. *ISPRS J. Photogramm. Remote Sens.* **2017**, *131*, 104–120. [CrossRef]
48. Tinh, P.; MacKenzie, R.; Hung, T.; Hanh, N.; Hanh, N.; Manh, D.; Ha, H.; Tuan, M. Distribution and drivers of Vietnam mangrove deforestation from 1995 to 2019. *Mitig. Adapt. Strateg. Glob. Change* **2022**, *27*, 29. [CrossRef]
49. Bhargava, R.; Sarkar, D.; Friess, D. A cloud computing-based approach to mapping mangrove erosion and progradation: Case studies from the Sundarbans and French Guiana. *Estuar. Coast. Shelf Sci.* **2021**, *248*, 106798. [CrossRef]

50. Besely, S.; van der Wegen, M.; Grueters, U.; Reyns, J.; Dijkstra, J.; Roelvink, D. Eleven Years of Mangrove-Mudflat Dynamics on the Mud Volcano-Induced Prograding Delta in East Java, Indonesia: Integrating UAV and Satellite Imagery. *Remote Sens.* **2021**, *13*, 1084. [CrossRef]
51. Gilani, H.; Naz, H.; Arshad, M.; Nazim, K.; Akram, U.; Abrar, A.; Asif, M. Evaluating mangrove conservation and sustainability through spatiotemporal (1990–2020) mangrove cover change analysis in Pakistan. *Estuar. Coast. Shelf Sci.* **2021**, *249*, 107128. [CrossRef]
52. Palmatier, R.W.; Houston, M.B.; Hulland, J. Review articles: Purpose, process, and structure. *J. Acad. Mark. Sci.* **2018**, *46*, 1–5. [CrossRef]
53. Martín-Martín, A.; Orduna-Malea, E.; Thelwall, M.; Delgado López-Cózar, E. Google Scholar, Web of Science, and Scopus: A systematic comparison of citations in 252 subject categories. *J. Inf.* **2018**, *12*, 1160–1177. [CrossRef]
54. Pranckutė, R. Web of Science (WoS) and Scopus: The Titans of Bibliographic Information in Today's Academic World. *Publications* **2021**, *9*, 12. [CrossRef]
55. Clarivate. Web of Science Editorial Selection Process. 2023. Available online: https://clarivate.com/academia-government/scientific-and-academic-research/research-discovery-and-referencing/web-of-science/web-of-science-core-collection/editorial-selection-process/ (accessed on 31 March 2023).
56. Ibharim, N.; Mustapha, M.; Lihan, T.; Mazlan, A. Mapping mangrove changes in the Matang Mangrove Forest using multi temporal satellite imageries. *Ocean Coast. Manag.* **2015**, *114*, 64–76. [CrossRef]
57. Jean-Baptiste, N.; Jensen, J.R. Measurement of Mangrove Biophysical Characteristics in the Bocozelle Ecosystem in Haiti Using ASTER Multispectral Data. *Geocarto Int.* **2006**, *21*, 3–8. [CrossRef]
58. Quader, M.A.; Agrawal, S.; Kervyn, M. Multi-decadal land cover evolution in the Sundarban, the largest mangrove forest in the world. *Ocean Coast. Manag.* **2017**, *139*, 113–124. [CrossRef]
59. Giri, C. Observation and Monitoring of Mangrove Forests Using Remote Sensing: Opportunities and Challenges. *Remote Sens.* **2016**, *8*, 783. [CrossRef]
60. Granell, C.; Havlik, D.; Schade, S.; Sabeur, Z.; Delaney, C.; Pielorz, J.; Usländer, T.; Mazzetti, P.; Schleidt, K.; Kobernus, M.; et al. Future Internet technologies for environmental applications. *Environ. Model. Softw.* **2016**, *78*, 1–15. [CrossRef]
61. Wang, L.; Jia, M.; Yin, D.; Tian, J. A review of remote sensing for mangrove forests: 1956–2018. *Remote Sens. Environ.* **2019**, *231*, 111223. [CrossRef]
62. Bunting, P.; Rosenqvist, A.; Hilarides, L.; Lucas, R.M.; Thomas, N.; Tadono, T.; Worthington, T.A.; Spalding, M.; Murray, N.J.; Rebelo, L.M. Global Mangrove Extent Change 1996–2020: Global Mangrove Watch Version 3.0. *Remote Sens.* **2022**, *14*, 3657. [CrossRef]
63. Malerba, M.E.; Costa, M.D.d.P.; Friess, D.A.; Schuster, L.; Young, M.A.; Lagomasino, D.; Serrano, O.; Hickey, S.M.; York, P.H.; Rasheed, M.; et al. Remote sensing for cost-effective blue carbon accounting. *Earth-Sci. Rev.* **2023**, *238*, 104337. [CrossRef]
64. Breiman, L.; Friedman, J.H.; Olshen, R.A.; Stone, C.J. *Classification and Regression Trees*; Routledge: New York, NY, USA, 2017.
65. Friedman, J.H. Greedy Function Approximation: A Gradient Boosting Machine. *Ann. Stat.* **2001**, *29*, 1189–1232. [CrossRef]
66. Jin, X.; Han, J. K-Means Clustering. In *Encyclopedia of Machine Learning*; Sammut, C., Webb, G.I., Eds.; Springer: Boston, MA, USA, 2010; pp. 563–564.
67. Breiman, L. Random Forests. *Mach. Learn.* **2001**, *45*, 5–32. [CrossRef]
68. Shmilovici, A. Support Vector Machines. In *Data Mining and Knowledge Discovery Handbook*; Maimon, O., Rokach, L., Eds.; Springer: Boston, MA, USA, 2005; pp. 257–276.
69. Tax, D.M.; Duin, R.P. Support Vector Data Description. *Mach. Learn.* **2004**, *54*, 45–66. [CrossRef]
70. Bloomfield, P. *Fourier Analysis of Time Series: An Introduction*; John Wiley & Sons: Hoboken, NJ, USA, 2004.
71. Otsu, N. A Threshold Selection Method from Gray-Level Histograms. *IEEE Trans. Syst. Man Cybern.* **1979**, *9*, 62–66. [CrossRef]
72. Kennedy, R.E.; Yang, Z.; Cohen, W.B. Detecting trends in forest disturbance and recovery using yearly Landsat time series: 1. LandTrendr—Temporal segmentation algorithms. *Remote Sens. Environ.* **2010**, *114*, 2897–2910. [CrossRef]
73. Forthofer, R.N.; Lee, E.S.; Hernandez, M. 13 - Linear Regression. In *Biostatistics*, 2nd, ed.; Forthofer, R.N., Lee, E.S., Hernandez, M., Eds.; Academic Press: San Diego, CA, USA, 2007; pp. 349–386.
74. Pimple, U.; Simonetti, D.; Hinks, I.; Oszwald, J.; Berger, U.; Pungkul, S.; Leadprathom, K.; Pravinvongvuthi, T.; Maprasoap, P.; Gond, V. A history of the rehabilitation of mangroves and an assessment of their diversity and structure using Landsat annual composites (1987-2019) and transect plot inventories. *For. Ecol. Manag.* **2020**, *462*, 118007. [CrossRef]
75. Murray, N.J.; Keith, D.A.; Simpson, D.; Wilshire, J.H.; Lucas, R.M. Remap: An online remote sensing application for land cover classification and monitoring. *Methods Ecol. Evol.* **2018**, *9*, 2019–2027. [CrossRef]
76. Bhargava, R.; Friess, D. Previous Shoreline Dynamics Determine Future Susceptibility to Cyclone Impact in the Sundarban Mangrove Forest. *Front. Mar. Sci.* **2022**, *9*, 814577. [CrossRef]

77. Li, H.; Jia, M.; Zhang, R.; Ren, Y.; Wen, X. Incorporating the Plant Phenological Trajectory into Mangrove Species Mapping with Dense Time Series Sentinel-2 Imagery and the Google Earth Engine Platform. *Remote Sens.* **2019**, *11*, 2479. [CrossRef]
78. Nedd, G.; Oyedotun, T.; Simard, M. Evaluation of Spatio-Temporal Dynamics of Guyana's Mangroves Using SAR and GEE. *Earth Syst. Environ.* **2021**, *7*, 509–524. [CrossRef]
79. Shrestha, S.; Miranda, I.; Kumar, A.; Pardo, M.; Dahal, S.; Rashid, T.; Remillard, C.; Mishra, D. Identifying and forecasting potential biophysical risk areas within a tropical mangrove ecosystem using multi-sensor data. *Int. J. Appl. Earth Obs. Geoinf.* **2019**, *74*, 281–294. [CrossRef]
80. Mandal, M.; Kamruzzaman, M.; Hosaka, T. Elucidating the phenology of the Sundarbans mangrove forest using 18-year time series of MODIS vegetation indices. *Tropics* **2020**, *29*, 41–55. [CrossRef]
81. Valenti, V.L.; Carcelen, E.C.; Lange, K.; Russo, N.J.; Chapman, B. Leveraging Google Earth Engine User Interface for Semiautomated Wetland Classification in the Great Lakes Basin at 10 m With Optical and Radar Geospatial Datasets. *IEEE J. Sel. Top. Appl. Earth Obs. Remote Sens.* **2020**, *13*, 6008–6018. [CrossRef]
82. Scheip, C.M.; Wegmann, K.W. HazMapper: A global open-source natural hazard mapping application in Google Earth Engine. *Nat. Hazards Earth Syst. Sci.* **2021**, *21*, 1495–1511. [CrossRef]
83. Baloloy, A.; Blanco, A.; Ana, R.; Nadaoka, K. Development and application of a new mangrove vegetation index (MVI) for rapid and accurate mangrove mapping. *Isprs J. Photogramm. Remote Sens.* **2020**, *166*, 95–117. [CrossRef]
84. Cissell, J.; Canty, S.; Steinberg, M.; Simpson, L. Mapping National Mangrove Cover for Belize Using Google Earth Engine and Sentinel-2 Imagery. *Appl. Sci.* **2021**, *11*, 4258. [CrossRef]
85. de Jong, S.; Shen, Y.; de Vries, J.; Bijnaar, G.; van Maanen, B.; Augustinus, P.; Verweij, P. Mapping mangrove dynamics and colonization patterns at the Suriname coast using historic satellite data and the LandTrendr algorithm. *Int. J. Appl. Earth Obs. Geoinf.* **2021**, *97*, 102293. [CrossRef]
86. Ghorbanian, A.; Zaghian, S.; Asiyabi, R.M.; Amani, M.; Mohammadzadeh, A.; Jamali, S. Mangrove Ecosystem Mapping Using Sentinel-1 and Sentinel-2 Satellite Images and Random Forest Algorithm in Google Earth Engine. *Remote Sens.* **2021**, *13*, 2565. [CrossRef]
87. Guild, R.; Wang, X.; Russon, A.E. Tracking Deforestation, Drought, and Fire Occurrence in Kutai National Park, Indonesia. *Remote Sens.* **2022**, *14*, 5630. [CrossRef]
88. Parida, B.; Kumar, P. Mapping and dynamic analysis of mangrove forest during 2009–2019 using landsat-5 and sentinel-2 satellite data along Odisha Coast. *Trop. Ecol.* **2020**, *61*, 538–549. [CrossRef]
89. Rahman, M.; Zhang, X.; Ahmed, I.; Iqbal, Z.; Zeraatpisheh, M.; Kanzaki, M.; Xu, M. Remote Sensing-Based Mapping of Senescent Leaf C:N Ratio in the Sundarbans Reserved Forest Using Machine Learning Techniques. *Remote Sens.* **2020**, *12*, 1375. [CrossRef]
90. Vu, T.T.P.; Pham, T.D.; Saintilan, N.; Skidmore, A.; Luu, H.V.; Vu, Q.H.; Le, N.N.; Nguyen, H.Q.; Matsushita, B. Mapping Multi-Decadal Mangrove Extent in the Northern Coast of Vietnam Using Landsat Time-Series Data on Google Earth Engine Platform. *Remote Sens.* **2022**, *14*, 4664. [CrossRef]
91. Zhang, R.; Jia, M.; Wang, Z.; Zhou, Y.; Mao, D.; Ren, C.; Zhao, C.; Liu, X. Tracking annual dynamics of mangrove forests in mangrove National Nature Reserves of China based on time series Sentinel-2 imagery during 2016–2020. *Int. J. Appl. Earth Obs. Geoinf.* **2022**, *112*, 102918. [CrossRef]
92. Zhao, C.; Qin, C.Z. Identifying large-area mangrove distribution based on remote sensing: A binary classification approach considering subclasses of non-mangroves. *Int. J. Appl. Earth Obs. Geoinf.* **2022**, *108*, 102750. [CrossRef]
93. Johnston-González, R.; Coca, O.; Duque, D.; Ricaurte-Villota, C. Tracking changes in extent and distribution of tropical coastal covers using simple semi-supervised classification. *Remote Sens. Appl. Soc. Environ.* **2022**, *27*, 100822. [CrossRef]
94. Aja, D.; Miyittah, M.K.; Angnuureng, D.B. Quantifying Mangrove Extent Using a Combination of Optical and Radar Images in a Wetland Complex, Western Region, Ghana. *Sustainability* **2022**, *14*, 16687. [CrossRef]
95. De Alban, J.; Jamaludin, J.; de Wen, D.; Than, M.; Webb, E. Improved estimates of mangrove cover and change reveal catastrophic deforestation in Myanmar. *Environ. Res. Lett.* **2020**, *15*, 034034. [CrossRef]
96. Dong, D.; Wang, C.; Yan, J.; He, Q.; Zeng, J.; Wei, Z. Combing Sentinel-1 and Sentinel-2 image time series for invasive Spartina alterniflora mapping on Google Earth Engine: A case study in Zhangjiang Estuary. *J. Appl. Remote Sens.* **2020**, *14*, 044504.
97. Mondal, P.; Dutta, T.; Qadir, A.; Sharma, S. Radar and optical remote sensing for near real-time assessments of cyclone impacts on coastal ecosystems. *Remote Sens. Ecol. Conserv.* **2022**, *8*, 506–520. [CrossRef]
98. Solanki, J.B.; Lele, N.; Das, A.K.; Maurya, P.; Kumari, R. Assessment of mangrove cover dynamics and its health status in the Gulf of Khambhat, Western India, using high-resolution multi-temporal satellite data and Google Earth Engine. *Environ. Monit. Assess.* **2022**, *194*, 896. [CrossRef]
99. Xia, Q.; He, T.; Qin, C.; Xing, X.; Xiao, W. An Improved Submerged Mangrove Recognition Index-Based Method for Mapping Mangrove Forests by Removing the Disturbance of Tidal Dynamics and S. alterniflora. *Remote Sens.* **2022**, *14*, 3112. [CrossRef]

100. Bessinger, M.; Lück-Vogel, M.; Skowno, A.; Conrad, F. Landsat-8 based coastal ecosystem mapping in South Africa using random forest classification in Google Earth Engine. *S. Afr. J. Bot.* **2022**, *150*, 928–939. [CrossRef]
101. Hauser, L.; Binh, N.; Hoa, P.; Quan, N.; Timmermans, J. Gap-Free Monitoring of Annual Mangrove Forest Dynamics in Ca Mau Province, Vietnamese Mekong Delta, Using the Landsat-7-8 Archives and Post-Classification Temporal Optimization. *Remote Sens.* **2020**, *12*, 3729. [CrossRef]
102. Zhao, C.; Qin, C.Z. The Key Reason of False Positive Misclassification for Accurate Large-Area Mangrove Classifications. *Remote Sens.* **2021**, *13*, 2909. [CrossRef]
103. Aung, T.S.; Overland, I.; Vakulchuk, R.; Xie, Y. The environmental burdens of special economic zones on the coastal and marine environment: A remote sensing assessment in Myanmar. *Remote Sens. Appl. Soc. Environ.* **2022**, *28*, 100809. [CrossRef]
104. Hu, L.; Xu, N.; Liang, J.; Li, Z.; Chen, L.; Zhao, F. Advancing the Mapping of Mangrove Forests at National-Scale Using Sentinel-1 and Sentinel-2 Time-Series Data with Google Earth Engine: A Case Study in China. *Remote Sens.* **2020**, *12*, 3120. [CrossRef]
105. Liu, X.; Fatoyinbo, T.; Thomas, N.; Guan, W.; Zhan, Y.; Mondal, P.; Lagomasino, D.; Simard, M.; Trettin, C.; Deo, R.; et al. Large-Scale High-Resolution Coastal Mangrove Forests Mapping Across West Africa With Machine Learning Ensemble and Satellite Big Data. *Front. Earth Sci.* **2021**, *8*, 560933. [CrossRef]
106. Lu, Y.; Wang, L. How to automate timely large-scale mangrove mapping with remote sensing. *Remote Sens. Environ.* **2021**, *264*, 112585. [CrossRef]
107. Mondal, P.; Liu, X.; Fatoyinbo, T.; Lagomasino, D. Evaluating Combinations of Sentinel-2 Data and Machine-Learning Algorithms for Mangrove Mapping in West Africa. *Remote Sens.* **2019**, *11*, 2928. [CrossRef]
108. Zhao, C.; Qin, C.Z. 10-m-resolution mangrove maps of China derived from multi-source and multi-temporal satellite observations. *Isprs J. Photogramm. Remote Sens.* **2020**, *169*, 389–405. [CrossRef]
109. Le, H.; Tran, T.; Gyeltshen, S.; Nguyen, C.; Tran, D.; Luu, T.; Duong, M. Characterizing Spatiotemporal Patterns of Mangrove Forests in Can Gio Biosphere Reserve Using Sentinel-2 Imagery. *Appl. Sci.* **2020**, *10*, 4058. [CrossRef]
110. Valderrama-Landeros, L.; Flores-Verdugo, F.; Rodriguez-Sobreyra, R.; Kovacs, J.; Flores-de Santiago, F. Extrapolating canopy phenology information using Sentinel-2 data and the Google Earth Engine platform to identify the optimal dates for remotely sensed image acquisition of semiarid mangroves. *J. Environ. Manag.* **2021**, *279*, 111617. [CrossRef] [PubMed]
111. Google. *Sentinel Collections in Earth Engine: Earth Engine Data Catalog*; Google: Mountain View, CA, USA, 2023.
112. Ghorbanian, A.; Ahmadi, S.; Amani, M.; Mohammadzadeh, A.; Jamali, S. Application of Artificial Neural Networks for Mangrove Mapping Using Multi-Temporal and Multi-Source Remote Sensing Imagery. *Water* **2022**, *14*, 244. [CrossRef]
113. Lomeo, D.; Singh, M. Cloud-Based Monitoring and Evaluation of the Spatial-Temporal Distribution of Southeast Asia's Mangroves Using Deep Learning. *Remote Sens.* **2022**, *14*, 2291. [CrossRef]
114. Mullissa, A.; Vollrath, A.; Odongo-Braun, C.; Slagter, B.; Balling, J.; Gou, Y.; Gorelick, N.; Reiche, J. Sentinel-1 SAR Backscatter Analysis Ready Data Preparation in Google Earth Engine. *Remote Sens.* **2021**, *13*, 1954. [CrossRef]
115. Mentaschi, L.; Vousdoukas, M.I.; Pekel, J.F.; Voukouvalas, E.; Feyen, L. Global long-term observations of coastal erosion and accretion. *Sci. Rep.* **2018**, *8*, 12876. [CrossRef]
116. Google. *Earth Engine Quotas: Google Earth Engine*; Google: Mountain View, CA, USA, 2023.
117. Priyono, K.D.; Saifuddin, A.; Nugroho, F.S.; Jumadi, J. Identification of Mangrove Changes in The Mahakam Delta in 2007–2017 using Alos/Palsar and Landsat. *Int. J. GEOMATE* **2022**, *23*, 77–84. [CrossRef]
118. Google. *Machine Learning in Earth Engine: Google Earth Engine: Google for Developers*; Google: Mountain View, CA, USA, 2023.
119. Google. *Earth Engine Data Catalog: Google for Developers*; Google: Mountain View, CA, USA, 2023.
120. Li, W.; El-Askary, H.; Qurban, M.; Li, J.; ManiKandan, K.; Piechota, T. Using multi-indices approach to quantify mangrove changes over the Western Arabian Gulf along Saudi Arabia coast. *Ecol. Indic.* **2019**, *102*, 734–745. [CrossRef]
121. Hooper, S.; Kennedy, R.E. A spatial ensemble approach for broad-area mapping of land surface properties. *Remote Sens. Environ.* **2018**, *210*, 473–489. [CrossRef]
122. Montero, D.; Aybar, C.; Mahecha, M.D.; Wieneke, S. Spectral: Awesome Spectral Indices deployed via the Google Earth Engine JavaScript API. *Int. Arch. Photogramm. Remote Sens. Spat. Inf. Sci.* **2022**, *48*, 301–306. [CrossRef]
123. Pasquarella, V.J.; Arévalo, P.; Bratley, K.H.; Bullock, E.L.; Gorelick, N.; Yang, Z.; Kennedy, R.E. Demystifying LandTrendr and CCDC temporal segmentation. *Int. J. Appl. Earth Obs. Geoinf.* **2022**, *110*, 102806. [CrossRef]
124. Murillo-Sandoval, P.J.; Fatoyinbo, L.; Simard, M. Mangroves Cover Change Trajectories 1984–2020: The Gradual Decrease of Mangroves in Colombia. *Front. Mar. Sci.* **2022**, *9*, 892946. [CrossRef]
125. Lee, C.K.F.; Duncan, C.; Nicholson, E.; Fatoyinbo, T.E.; Lagomasino, D.; Thomas, N.; Worthington, T.A.; Murray, N.J. Mapping the Extent of Mangrove Ecosystem Degradation by Integrating an Ecological Conceptual Model with Satellite Data. *Remote Sens.* **2021**, *13*, 2047. [CrossRef]

126. Bullock, E.L.; Woodcock, C.E.; Olofsson, P. Monitoring tropical forest degradation using spectral unmixing and Landsat time series analysis. *Remote Sens. Environ.* **2020**, *238*, 110968. [CrossRef]
127. Vancutsem, C.; Achard, F.; Pekel, J.F.; Vieilledent, G.; Carboni, S.; Simonetti, D.; Gallego, J.; Aragão, L.E.O.C.; Nasi, R. Long-term (1990–2019) monitoring of forest cover changes in the humid tropics. *Sci. Adv.* **2021**, *7*, eabe1603. [CrossRef]
128. Obida, C.B.; Blackburn, G.A.; Whyatt, J.D.; Semple, K.T. Counting the cost of the Niger Delta's largest oil spills: Satellite remote sensing reveals extensive environmental damage with >1million people in the impact zone. *Sci. Total Environ.* **2021**, *775*, 145854. [CrossRef]
129. Onyena, A.P.; Sam, K. A review of the threat of oil exploitation to mangrove ecosystem: Insights from Niger Delta, Nigeria. *Glob. Ecol. Conserv.* **2020**, *22*, e00961. [CrossRef]
130. Barenblitt, A.; Fatoyinbo, L.; Thomas, N.; Stovall, A.; de Sousa, C.; Nwobi, C.; Duncanson, L. Invasion in the Niger Delta: Remote sensing of mangrove conversion to invasive Nypa fruticans from 2015 to 2020. *Remote Sens. Ecol. Conserv.* **2023**, *10*, 5–23. [CrossRef]
131. Lassalle, G.; Ferreira, M.P.; Cué La Rosa, L.E.; Del'Papa Moreira Scafutto, R.; de Souza Filho, C.R. Advances in multi- and hyperspectral remote sensing of mangrove species: A synthesis and study case on airborne and multisource spaceborne imagery. *ISPRS J. Photogramm. Remote Sens.* **2023**, *195*, 298–312. [CrossRef]
132. Arasumani, M.; Kumaresan, M.; Esakki, B. Mapping native and non-native vegetation communities in a coastal wetland complex using multi-seasonal Sentinel-2 time series. *Biol. Invasions* **2024**, *26*, 1105–1124. [CrossRef]
133. Lang, N.; Jetz, W.; Schindler, K.; Wegner, J.D. A high-resolution canopy height model of the Earth. *Nature Ecology & Evolution* **2023**, *7*, 1778–1789.
134. Simard, M.; Fatoyinbo, L.; Smetanka, C.; Rivera-Monroy, V.H.; Castañeda Moya, E.; Thomas, N.; Van der Stocken, T. Mangrove canopy height globally related to precipitation, temperature and cyclone frequency. *Nat. Geosci.* **2019**, *12*, 40–45. [CrossRef]
135. Zeng, Y.; Friess, D.A.; Sarira, T.V.; Siman, K.; Koh, L.P. Global potential and limits of mangrove blue carbon for climate change mitigation. *Curr. Biol.* **2021**, *31*, 1737–1743.e3. [CrossRef] [PubMed]
136. Zhang, X.; Liu, L.; Zhao, T.; Chen, X.; Lin, S.; Wang, J.; Mi, J.; Liu, W. GWLFCS30: A global 30 m wetland map with a fine classificationsystem using multi-sourced and time-series remote sensing imagery in 2020. *Earth Syst. Sci. Data* **2023**, *15*, 265–293. [CrossRef]
137. Zhao, C.; Jia, M.; Wang, Z.; Mao, D.; Wang, Y. Identifying mangroves through knowledge extracted from trained random forest models: An interpretable mangrove mapping approach (IMMA). *ISPRS J. Photogramm. Remote Sens.* **2023**, *201*, 209–225. [CrossRef]
138. Demarquet, Q.; Rapinel, S.; Dufour, S.; Hubert-Moy, L. Long-Term Wetland Monitoring Using the Landsat Archive: A Review. *Remote Sens.* **2023**, *15*, 820. [CrossRef]
139. Fu, B.; Yao, H.; Lan, F.; Li, S.; Liang, Y.; He, H.; Jia, M.; Wang, Y.; Fan, D. Collaborative multiple change detection methods for monitoring the spatio-temporal dynamics of mangroves in Beibu Gulf, China. *Giscience Remote Sens.* **2023**, *60*, 2202506. [CrossRef]
140. Aji, M.A.P.; Kamal, M.; Farda, N.M. Mangrove species mapping through phenological analysis using random forest algorithm on Google Earth Engine. *Remote Sens.-Appl.-Soc. Environ.* **2023**, *30*, 100978. [CrossRef]
141. Zhang, Z.; Ahmed, M.R.; Zhang, Q.; Li, Y.; Li, Y. Monitoring of 35-Year Mangrove Wetland Change Dynamics and Agents in the Sundarbans Using Temporal Consistency Checking. *Remote Sens.* **2023**, *15*, 625. [CrossRef]
142. Karsch, G.; Mukul, S.A.; Srivastava, S.K. Annual Mangrove Vegetation Cover Changes (2014-2020) in Indian Sundarbans National Park Using Landsat 8 and Google Earth Engine. *Sustainability* **2023**, *15*, 5592. [CrossRef]
143. Selvaraj, J.J.; Perez, B.E.G. An enhanced approach to mangrove forest analysis in the Colombian Pacific coast using optical and SAR data in Google Earth Engine. *Remote Sens.-Appl.-Soc. Environ.* **2023**, *30*, 100938. [CrossRef]
144. Mahdavifard, M.; Ahangar, S.K.; Feizizadeh, B.; Kamran, K.V.; Karimzadeh, S. Spatio-Temporal monitoring of Qeshm mangrove forests through machine learning classification of SAR and optical images on Google Earth Engine. *Int. J. Eng. Geosci.* **2023**, *8*, 239–250. [CrossRef]
145. Fan, C.; Hou, X.; Zhang, Y.; Li, D. Satellite Data Reveal Concerns Regarding Mangrove Restoration Efforts in Southern China. *Remote Sens.* **2023**, *15*, 4151. [CrossRef]
146. Zhang, X.; Lin, X.; Fu, D.; Wang, Y.; Sun, S.; Wang, F.; Wang, C.; Xiao, Z.; Shi, Y. Comparison of the Applicability of J-M Distance Feature Selection Methods for Coastal Wetland Classification. *Water* **2023**, *15*, 2212. [CrossRef]
147. Kaur, H.; Tyagi, S.; Mehta, M.; Singh, D. Time series (2001/2002–2021) analysis of Earth observation data using Google Earth Engine (GEE) for detecting changes in land use land cover (LULC) with specific reference to forest cover in East Godavari Region, Andhra Pradesh, India. *J. Earth Syst. Sci.* **2023**, *132*, 86. [CrossRef]

148. Rodrigues, F.H.; Cerri, R.I.; Kolya, A.d.A.; Veiga, V.M.; Reis, F.A.G.V. Comparison of vegetation indices and image classification methods for mangrove mapping at semi-detailed scale in southwest of Rio de Janeiro, Brazil. *Remote Sens.-Appl.-Soc. Environ.* **2023**, *30*, 100965. [CrossRef]
149. Flores-de Santiago, F.; Rodríguez-Sobreyra, R.; Álvarez Sánchez, L.F.; Valderrama-Landeros, L.; Amezcua, F.; Flores-Verdugo, F. Understanding the natural expansion of white mangrove (*Laguncularia racemosa*) in an ephemeral inlet based on geomorphological analysis and remote sensing data. *J. Environ. Manag.* **2023**, *338*, 117820. [CrossRef]
150. Harou, I.L.; Inyele, J.; Minang, P.; Duguma, L. Understanding the states and dynamics of mangrove forests in land cover transitions of The Gambia using a Fourier transformation of Landsat and MODIS time series in Google Earth Engine. *Front. For. Glob. Change* **2023**, *5*, 934019.

remote sensing

MDPI

Article

Can Synthetic Aperture Radar Enhance the Quality of Satellite-Based Mangrove Detection? A Focus on the Denpasar Region of Indonesia

Soohyun Kwon [1], Hyeon Kwon Ahn [2] and Chul-Hee Lim [1,3,*]

1 Department of Forestry, Environment, and Systems, Kookmin University, Seoul 02707, Republic of Korea; estheruno@kookmin.ac.kr

2 Department of Forest Resources, Kookmin University, Seoul 02707, Republic of Korea; postmist@kookmin.ac.kr

3 Forest Carbon Graduate School, Kookmin University, Seoul 02707, Republic of Korea

* Correspondence: clim@kookmin.ac.kr

Abstract: Mangrove forests are vital ecosystems with the highest global carbon absorption capacity, playing a crucial role in climate change mitigation. Therefore, their conservation and management are essential. However, as mangroves are primarily found in tropical regions, frequent cloud cover and limited accessibility pose significant challenges to effective monitoring using optical satellite imagery. In addition, many developing countries with extensive mangrove coverage face challenges in conducting precise monitoring due to limited technological infrastructure. To overcome these limitations, this study integrated open-access synthetic aperture radar (SAR) data with optical imagery to enhance the classification accuracy of mangrove forests in the Bali Denpasar–Badung region. The Sentinel-1 and Sentinel-2 datasets were used, and the U-Net deep learning model was employed for training and classification. A digital elevation model (DEM) from the Shuttle Radar Topography Mission (SRTM) was applied to exclude areas higher than 10 m above sea level, thereby improving the classification accuracy. Additionally, a time-series analysis was performed to assess changes in the mangrove distribution over the past decade, revealing a consistent increase in mangrove extent in the study area. The classification performance was evaluated using a confusion matrix, demonstrating that the combined SAR-optical model outperformed single-source models across all key metrics including precision, accuracy, recall, and F1-score. The findings highlight the effectiveness of integrating SAR and optical data for capturing the complex ecological and geographical characteristics of mangrove forests. Notably, SAR imagery, which is resistant to cloud cover, shows considerable potential for independent application in tropical mangrove monitoring, warranting further research to explore its capabilities in greater depth.

Keywords: mangrove; synthetic aperture radar (SAR); U-Net; optical imagery; Sentinel

1. Introduction

Mangrove forests are evergreen ecosystems that thrive at the interface between land and sea in tropical and subtropical regions and are distinguished by their high tolerance to salinity [1–4]. These ecosystems are among the most diverse and dynamic ecosystems in the world, playing a critical role in climate change mitigation and carbon storage because of their exceptional capacity for carbon sequestration [5]. However, since the

1980s, more than half of the global mangrove forests have experienced a rapid decline driven by commercial logging, fuelwood harvesting, charcoal production, mining activities, agricultural expansion, housing development, and the growth of aquaculture farms [6]. Asia, in particular, has recorded the highest net loss, with approximately 1.9 million hectares of mangrove forest disappearing since the 1980s [7]. This substantial decline highlights the urgent need for sustainable mangrove management and conservation efforts on a global scale.

Asia contains the world's largest mangrove forests, making it a critical region for the preservation and management of global mangrove ecosystems [8]. Indonesia, in particular, hosts the most extensive mangrove forest area and is widely recognized for its exceptional biodiversity, both in terms of total coverage and species diversity [9,10]. Over the past 30 years, the estimated total value of ecosystem services provided by Bali's mangrove forests has exceeded USD 2 million [11], underscoring their continuous contribution to ecological functions and ecosystem services. However, with its tourism-driven economy [12], Bali faces an ongoing challenge in balancing economic growth with environmental conservation efforts. Despite these challenges, the mangrove forests in Bali have remained relatively stable in size over the past decade, increasing slightly from 2122.35 hectares in 2010 to 2146.32 hectares in 2020 [13]. This stability represents a successful case of harmonizing local economic development with environmental preservation while maintaining the ecological and environmental value of mangrove forests. These characteristics position Bali's mangrove forests as a model for ecosystem preservation and the promotion of sustainable tourism. Bali's mangrove conservation efforts serve as a valuable reference for discussing sustainable management strategies at both regional and global levels. Previous studies on mangroves in the Bali region have focused primarily on estimating carbon storage, and research addressing mangrove classification has largely relied on medium-resolution satellite imagery. Accordingly, this study applied a high-resolution classification approach that integrates SAR and optical imagery, combined with deep learning techniques, focusing on the Badung–Denpasar area to improve classification accuracy.

Mangroves exhibit complex vegetation structures and geomorphological characteristics, including intertwined canopies, water bodies, and mudflats, which make it difficult to achieve precise classification using only spectral information [14,15]. Deep learning techniques are well-suited for mangrove classification as they can effectively learn such complex and abstract features and can also perform well with relatively limited labeled data. Moreover, mangroves typically occupy a small proportion of the overall area, leading to class imbalance issues that may reduce the classification accuracy. Therefore, satellite-based approaches offer an effective alternative, providing the foundation for accurate monitoring and conservation efforts. However, as most mangrove forests are located in tropical regions, their remoteness and frequent cloud cover often limit the utility of optical imagery alone. To overcome these challenges, the use of synthetic aperture radar (SAR), which can reliably collect data regardless of cloud cover, haze, or time of day, has gained attention. SAR imagery is particularly effective for mangrove classification due to its sensitivity to vegetation structure and moisture content [16–18].

This study aimed to improve the classification accuracy through the integration of optical and synthetic aperture radar (SAR) satellite imagery, with particular emphasis on the importance of generalizability. The complementary characteristics of SAR and optical imagery were leveraged to quantitatively assess the effectiveness of data integration. This approach is meaningful in that it enables the precise mapping of the mangrove distribution and area as well as the development of data-driven detection methods to support ecosystem conservation. However, many mangrove-rich countries lack the technical capacity to conduct high-resolution mapping, despite the increasing demand for accurate spatial information on mangrove distribution. Accordingly, this study proposes a classification framework that balances technical sophistication with broad accessibility, aiming to support sustainable mangrove management even in regions with limited technological infrastructure. Although previous studies have demonstrated the potential of SAR–optical data fusion, challenges remain in terms of scalability, consistency, and applicability across diverse environmental settings. In response to these challenges, the present study introduces a high-resolution classification framework that leverages open-access satellite imagery in conjunction with advanced deep learning algorithms. The findings of this study are expected to contribute not only to the advancement of mangrove classification techniques, but also to the development of policies and ecosystem-based conservation strategies.

2. Materials and Methods

2.1. Study Area

Bali is located in the southern region of Indonesia, with this study focusing on the Denpasar–Badung mangrove zone in southeastern Bali as the primary research area (Figure 1). Denpasar, the capital of Bali [19], is approximately centered at latitude −8.6667°S and longitude 115.217°E. The study area spans approximately 173.01 km^2 and holds significant ecological and economic importance, contributing over USD 5000 annually in coastal protection services [11].

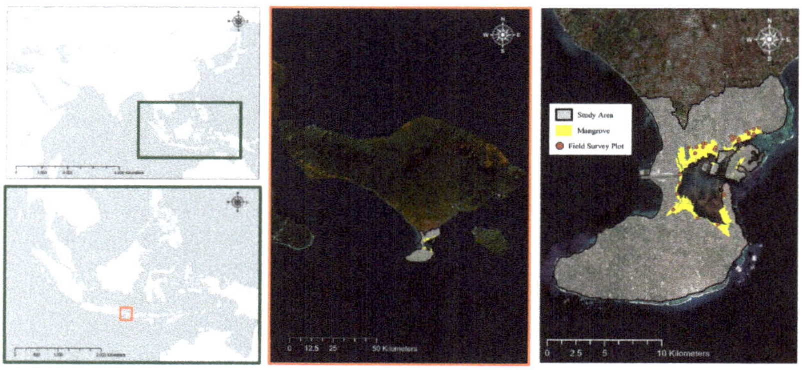

Figure 1. Study area in southeastern Bali, Indonesia, showing the distribution of mangrove forests in yellow and the field survey plot in red.

The region is located in a tropical monsoon climate characterized by the influence of seasonal winds that create distinct wet and dry seasons. The average annual temperature ranges between 25 and 26 °C, while the average annual precipitation is approximately 2576 mm [20]. Situated near Benoa Bay, the area is shaped by coastal conditions that contribute to the formation of saline soils with a high clay content. The Denpasar–Badung mangrove zone provides an optimal environment for mangrove ecosystems, featuring consistently high temperatures and humidity, elevated salinity levels, and muddy substrates that support mangrove growth [6,16,21].

To support classification and labeling efforts, field survey plots were jointly established in this region by Udayana University, providing reliable ground truth data for mangrove and non-mangrove delineation.

2.2. Data Acquisition and Preprocessing

This study utilized Google Earth Engine (GEE) for data processing, with all datasets acquired through the platform. The primary datasets included Sentinel-1 Ground Range Detected (GRD) C-Band SAR and Sentinel-2 Multispectral Instrument (MSI) imagery (Table 1). Additionally, mangrove labeling was conducted by extracting mangrove polygons from the European Space Agency (ESA) WorldCover 10 m v100 dataset. The ESA WorldCover dataset achieves an overall accuracy of 74.4% for global coverage, with the reported accuracy reaching approximately 80% for the Asian region [22]. To verify and improve the accuracy of the mangrove dataset, a visual comparison was conducted using high-resolution satellite imagery. Based on this assessment, we manually labeled and supplemented areas where gaps in mangrove coverage were identified. Subsequently, the final labels were refined with reference to the field survey data conducted by Udayana University. Sentinel-1 and Sentinel-2 images from 2020 and 2024 were extracted from GEE and processed using a 75 percentile value composite. The 75th percentile was selected as it effectively captures periods of stable vegetation growth while reducing the influence of seasonal fluctuations and outliers, particularly in tropical regions where optical image availability is limited.

Table 1. The data used in this study.

Satellite	Temporal Resolution (Day)	Collection Period	Number of Images	Mosaic Method
Sentinel-1	6	2020-01-01~2020-12-31	8	Percentile 75
Sentinel-2	5	2020-01-01~2020-12-31	41	Percentile 75
Sentinel-1	6	2024-01-01~2024-12-31	59	Percentile 75
Sentinel-2	5	2024-01-01~2024-12-31	36	Percentile 75

Sentinel-1 data were preprocessed using the Sentinel Toolbox following a series of standardized steps: (1) application of orbit files, (2) GRD border noise removal, (3) thermal noise removal, (4) radiometric calibration, and (5) terrain correction [23].

During post-processing, a digital elevation model (DEM) was applied to impose elevation constraints to ensure accurate terrain corrections. Elevation data were sourced from the Shuttle Radar Topography Mission (SRTM) DEM, which provides a vertical accuracy of approximately 2.39 m in lowland areas with elevations below 20 m [24,25].

Using this approach, classification was performed on both the 2020 and 2024 imagery to analyze temporal trends in mangrove distribution.

2.2.1. Band Composition

To construct the input layer for the U-Net model, various band and polarization combinations were synthesized. As the U-Net model accepts input layers consisting of either a single band or three bands, a multi-band dataset was generated by merging VV and VH polarizations from Sentinel-1. Additionally, a mangrove index derived from Sentinel-2 bands was incorporated to enhance the suitability of the dataset for model training (Table 2).

Table 2. The band used in this study.

Satellite	Polarization and Band	Derivation Formula	Source
Sentinel-1	VV		
	VV and VH	VV-VH	
Sentinel-2	Band 3, Band 8, and Band 11	EMI [1] = (NIR − SWIR)/(GREEN + NIR)	[26]

[1] EMI: enhanced mangrove index.

The VV polarization transmits and receives vertically polarized waves, while the VH polarization transmits vertically polarized waves and receives horizontally polarized waves [27]. VV polarization has been shown to achieve a higher accuracy in detecting buildings and vegetation than VH polarization, whereas VH polarization is more sensitive to surface roughness [28]. Based on these characteristics, it was hypothesized that mangroves, as high-moisture vegetation, would exhibit low VV-VH values, and the analysis was conducted accordingly.

The enhanced mangrove index (EMI) utilizes the reflectance contrast between near-infrared (NIR) and short wavelength infrared (SWIR) bands combined with the green band to distinguish mangroves from general vegetation more effectively than the existing mangrove indices [26].

To optimize vegetation classification, the input layer of the U-Net model was constructed by integrating the VV polarization, VV-VH polarization, and EMI index, all of which are highly responsive to vegetation characteristics.

2.2.2. Training Data Preparation

To improve the model's adaptability to various transformation scenarios, the training data were refined via data augmentation based on the labeled dataset [29]. In addition, techniques such as image rotation, scaling, and stride adjustments were applied to generate multiple variations of the training samples. This process enables the model to learn features from different perspectives and environmental conditions, even within the same class [29]. By incorporating these augmentations, the model achieved the improved detection of mangroves within images, while the increased diversity of the training dataset enhanced its generalization capabilities. The number of images extracted from each band used in model training is summarized in Table 3.

Table 3. Number of images extracted from each satellite band.

Polarization and Band	Image	Feature
VV	556	2328
VV-VH		
EMI		
Mixed band		

2.3. *U-Net-Based Classification Model Training and Post-Processing*

This study used the U-Net deep learning model, which was initially introduced in 2015 for biomedical segmentation applications. The U-Net is specifically designed for effective training with a limited number of images [30,31]. U-Net comprises two primary components: a contracting path and an expanding path. The contracting path includes a convolutional neural network (CNN) architecture to analyze images, extract key features, and reduce the resolution during training. The expanding path then restores the spatial details via upsampling, ultimately producing a refined segmentation output [31]. The U-shaped architecture of U-Net enables accurate image segmentation, even on relatively small datasets [31]. Recent research on satellite image classification has shown that U-Net achieves high classification accuracy and is frequently employed in the field [32–34]. Given the moderate size of the study area (approximately 173.01 km^2), the U-Net model was selected for its proven ability to achieve high-precision results with limited data availability.

The key hyperparameters used for model training are summarized as follows. All processes were executed in ArcGIS Pro. The batch size was set to 8, meaning that eight training samples were processed simultaneously. This configuration was selected to enhance convergence speed while ensuring efficient memory use. A validation ratio of 20% was applied to reserve 20% of the training dataset for validation to mitigate the risk of overfitting. The chip size for training was set to 128 pixels to ensure consistency in image processing and feature learning.

ResNet-50 was employed as the backbone model for U-Net. ResNet-50 is a 50-layer CNN architecture that leverages residual learning to address the vanishing gradient problem commonly encountered in deep neural networks. This feature facilitates efficient training and improves model performance [35]. To further optimize the training process, an early stopping technique was implemented, where training was stopped when the validation accuracy failed to improve for a specified number of consecutive epochs [36]. The validation loss curves for each band are shown in Figure 2.

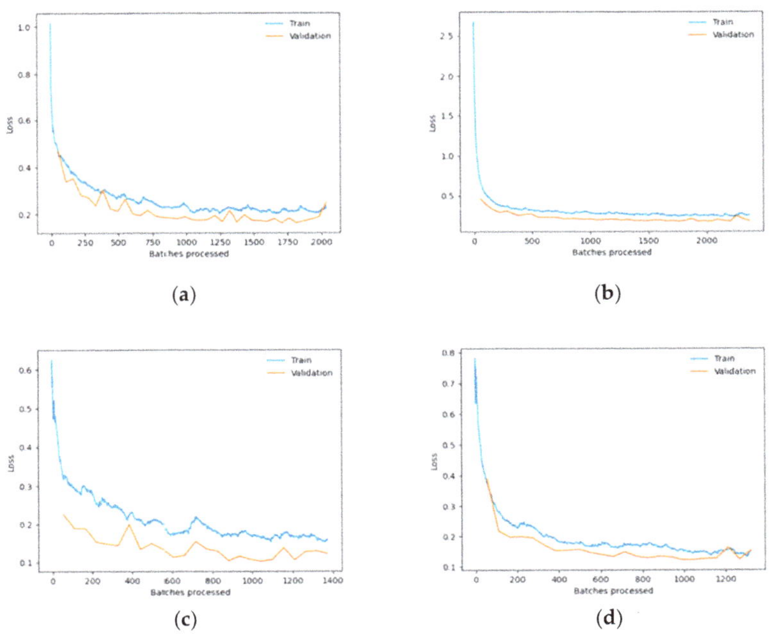

Figure 2. Validation loss graph. (**a**) VV graph. (**b**) VV-VH graph. (**c**) EMI graph. (**d**) Mixed band graph.

Once trained, the model was applied to classify the 2020 imagery and subsequently used to classify the 2024 data. The classification results revealed misclassification in inland areas. To address this issue, post-processing was conducted using a DEM to exclude areas with elevations exceeding 10 m above sea level, which improved the classification accuracy.

2.4. Accuracy Assessment

To evaluate the performance of the model in mangrove classification, a confusion matrix was used. The confusion matrix is a widely used tool for assessing classification accuracy by quantifying the ability of the model to distinguish between mangrove and non-mangrove regions [37]. For the accuracy assessment, 100 random points were sampled from the mangrove areas, and 400 points were sampled from the non-mangrove areas. The predicted classifications were then compared with the ground truth data, and the results were summarized in a confusion matrix. The classification performance of the model was evaluated using key metrics derived from the confusion matrix including the accuracy, precision, recall, and F1-score (Equations (1)–(4)). These metrics provided insights into the agreement between the predicted and actual values.

The accuracy reflects the overall classification performance [38], while precision and recall evaluate the correctness and completeness of mangrove detection, respectively [39]. The F1-score combines both as a balanced metric [40].

$$Accuracy = (tp + tn)/(tp + fn + fp + tn) \quad (1)$$

$$Precision = tp/(tp + fp) \quad (2)$$

$$Recall = tp/(tp + fn) \quad (3)$$

$$F1\text{-}score = 2 \times (Recall \times Precision)/(Recall + Precision) \quad (4)$$

Accuracy is a fundamental metric for evaluating the overall performance of a model because it assesses the predictive ability across both mangrove and non-mangrove areas, providing a comprehensive measure of classification accuracy [39]. Precision, which evaluates the reliability of mangrove predictions, is particularly crucial for minimizing the misclassification of non-mangrove areas as mangroves [41]. In mangrove management and conservation, such misclassifications could lead to unnecessary resource allocation, making high precision essential for ensuring the reliability of the prediction outcomes. Recall measures the effectiveness of the model in correctly identifying mangroves, ensuring that the actual mangrove areas are not omitted [42]. A high recall indicates the capability of the model to capture all of the mangrove areas, making it a key metric for conservation efforts. The F1-score, calculated as the harmonic mean of the precision and recall, is a balanced evaluation metric when both the precision and recall are equally crucial in assessing the model performance [40]. This study underscores the importance of accurate mangrove classification for effective management, conservation, and protection. The reliability and completeness of the classification model play a critical role in these efforts. Therefore, the performance of the model was evaluated using four metrics: accuracy, precision, recall, and F1-score.

3. Results

3.1. Classification of Mangroves by Composition

The labeled dataset used in this study represented approximately 11.23 km^2 of mangrove areas. The model was trained using this dataset, and classification was performed for 2020 and 2024. For the 2020 classification results, the VV polarization estimated the mangrove distribution to be approximately 12.67 km^2, while the VV-VH polarization yielded an estimated area of around 13.24 km^2. The EMI index indicated a mangrove area of approximately 11.74 km^2, whereas the mixed band approach identified a distribution of about 11.94 km^2. The results suggest that the EMI band provided the most accurate estimate of the actual mangrove distribution area. These classification outcomes are visualized in Figure 3, providing spatial comparisons across band combinations. Furthermore, the analysis indicates that EMI achieved the highest area estimation accuracy, whereas the mixed band approach demonstrated the strongest spatial correlation with the labeled dataset. These findings highlight that EMI is optimal for precise area estimation, whereas the mixed band method offers a more accurate spatial representation of the mangrove distribution.

The 2024 classification results in Figure 4 indicate that the estimated mangrove distribution area was approximately 12.66 km^2 based on VV polarization. The VV-VH polarization yielded an estimated area of 14.61 km^2, while the EMI index was 12.41 km^2. The mixed band approach identified a mangrove distribution of approximately 12.11 km^2. Comparing 2024 to 2020, the highest mangrove area expansion rate was observed in the VV-VH polarization, with a 10.37% increase. This trend was followed by the EMI index (5.71%) and the mixed band approach (1.42%). In contrast, VV polarization showed a slight decline of 0.08%, making it the only category that exhibited a reduction in mangrove extent.

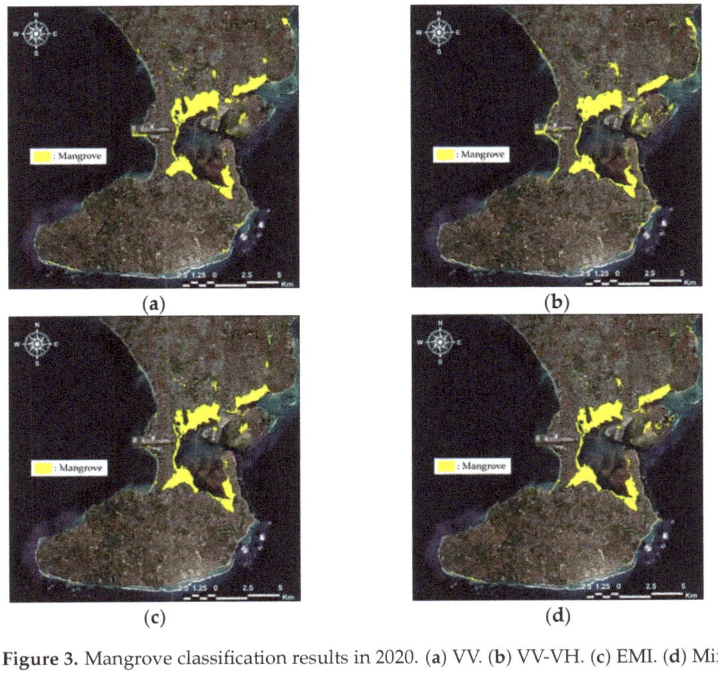

Figure 3. Mangrove classification results in 2020. (**a**) VV. (**b**) VV-VH. (**c**) EMI. (**d**) Mixed band.

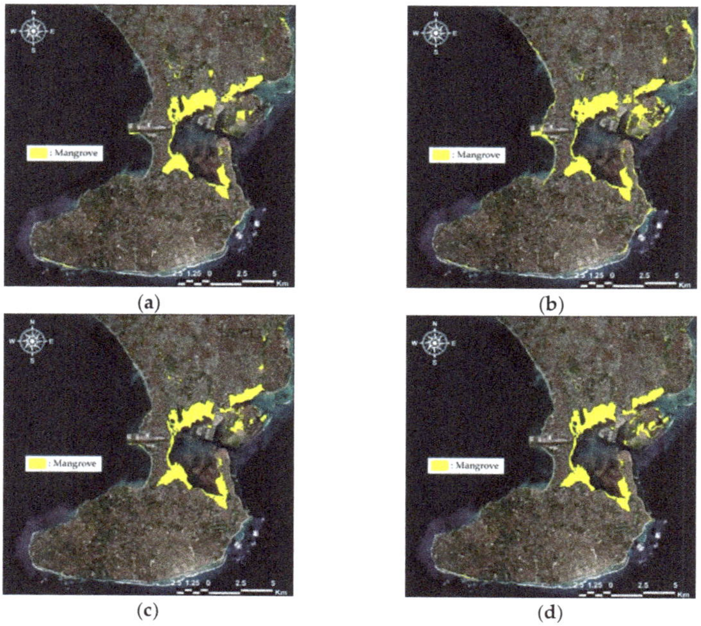

Figure 4. Mangrove classification results in 2024. (**a**) VV. (**b**) VV-VH. (**c**) EMI. (**d**) Mixed band.

3.2. Evaluation of Classification Performance

3.2.1. Accuracy Metrics and Validation

The evaluation results indicate that the mixed band approach achieved the highest classification accuracy among the considered methods. The classification accuracy for mangrove forests was in the following order: mixed band > EMI > VV > VV-VH. However, the recall values displayed a different ranking, with the mixed band approach achieving the highest recall, followed by VV, EMI, and VV-VH. The F1 score followed a similar pattern, ranked as mixed band > EMI > VV > VV-VV. Overall, the mixed band approach demonstrated superior classification accuracy across all evaluation metrics, followed by EMI, VV, and VV-VH in descending order.

Figure 5 shows the confusion matrices for each input type in the 2020 classification, allowing a visual comparison of classification performance across bands. Table 4 presents the accuracy metrics used to assess the performance of the model in mangrove classification. The mixed band approach exhibited the highest performance across all key metrics, with the accuracy (0.982), precision (0.93), recall (0.978947), and F1-score (0.953846) consistently achieving high values. Although the VV and VV-VH polarizations demonstrated a high overall accuracy, their precision values were relatively lower, at 0.84 and 0.76, respectively. Notably, VV-VH exhibited a lower recall (0.863636) and F1-score (0.808511), indicating that although it achieved a reasonable classification accuracy, its performance in other key metrics was suboptimal. These results suggest that the mixed band approach provided a more stable and superior classification performance for mangroves, while the VV band also demonstrated well-balanced accuracy across different evaluation metrics.

	VV	Predicted Class: Mangrove	Predicted Class: Non-Mangrove	
Actual Class	Mangrove	84	4	88
	Non-Mangrove	16	396	412
		100	400	500

(**a**)

	VV - VH	Predicted Class: Mangrove	Predicted Class: Non-Mangrove	
Actual Class	Mangrove	76	12	88
	Non-Mangrove	24	388	412
		100	400	500

(**b**)

	EMI	Predicted Class: Mangrove	Predicted Class: Non-Mangrove	
Actual Class	Mangrove	86	5	91
	Non-Mangrove	14	395	409
		100	400	500

(**c**)

	Mixed Band	Predicted Class: Mangrove	Predicted Class: Non-Mangrove	
Actual Class	Mangrove	93	2	95
	Non-Mangrove	7	398	405
		100	400	500

(**d**)

Figure 5. Confusion matrix for 2020. (**a**) VV. (**b**) VV-VH. (**c**) EMI. (**d**) Mixed band.

Table 4. Comparison of VV, VV-VH, EMI, and mixed bands for accuracy in 2020.

Polarization and Band	Accuracy	Precision	Recall	F1-Score
VV	0.959839	0.84	0.954545	0.893617
VV-VH	0.928	0.76	0.863636	0.808511
EMI	0.962	0.86	0.945055	0.900524
Mixed band	0.982	0.93	0.978947	0.953846

Figure 6 shows the confusion matrices for each input type in the 2024 classification, allowing a visual assessment of classification accuracy across different band combinations. In the 2024 classification results, the mixed band approach exhibited improvements in all accuracy metrics compared with 2020, achieving the highest values among all methods. This was followed by EMI, VV, and VV-VH, which maintained a performance pattern similar to their 2020 accuracy metrics. These findings indicate that the classification results were consistent and demonstrated statistical significance.

	VV	Predicted Class: Mangrove	Predicted Class: Non-Mangrove	
Actual Class	Mangrove	81	4	85
	Non-Mangrove	19	396	415
		100	400	500

(**a**)

	VV - VH	Predicted Class: Mangrove	Predicted Class: Non-Mangrove	
Actual Class	Mangrove	77	8	85
	Non-Mangrove	23	392	415
		100	400	500

(**b**)

	EMI	Predicted Class: Mangrove	Predicted Class: Non-Mangrove	
Actual Class	Mangrove	92	2	94
	Non-Mangrove	8	398	406
		100	400	500

(**c**)

	Mixed Band	Predicted Class: Mangrove	Predicted Class: Non-Mangrove	
Actual Class	Mangrove	95	1	96
	Non-Mangrove	5	399	404
		100	400	500

(**d**)

Figure 6. Confusion matrix for 2024. (**a**) VV. (**b**) VV-VH. (**c**) EMI. (**d**) Mixed band.

Table 5 summarizes the classification accuracy metrics for 2024. The mixed band approach achieved the highest performance across all metrics, with an accuracy of 0.954, precision of 0.95, recall of 0.99583, and an F1-score of 0.969388. Compared with 2020, the VV polarization showed a slight decrease in overall performance, whereas the VV-VH polarization exhibited improvements across all metrics. However, despite this increase, the precision value of VV-VH remained low compared with the other methods. This suggests that VV-VH misclassified a significant portion of non-mangrove areas as mangroves, likely due to the water surface reflections causing water bodies to be incorrectly classified as mangroves. The misclassification patterns observed in coastal areas further indicate that VV-VH polarization may be less effective for differentiating between mangroves and water bodies.

Table 5. Comparison of VV, VV-VH, EMI, and mixed bands for accuracy in 2024.

Polarization and Band	Accuracy	Precision	Recall	F1-Score
VV	0.954	0.81	0.952941	0.875676
VV-VH	0.938	0.77	0.905882	0.832432
EMI	0.98	0.92	0.978723	0.948454
Mixed band	0.988	0.95	0.989583	0.969388

3.2.2. Spatial Intercomparison with Existing Datasets

To evaluate the performance of the proposed classification approach, a comparative analysis was conducted against the Global Mangrove Watch (GMW) dataset (Figure 7). The comparison focused on the year 2020, which is the latest year for which GMW data are available. Developed as a global open-access dataset, GMW provides mangrove distribution maps generated using various satellite datasets to support global mangrove monitoring efforts [13].

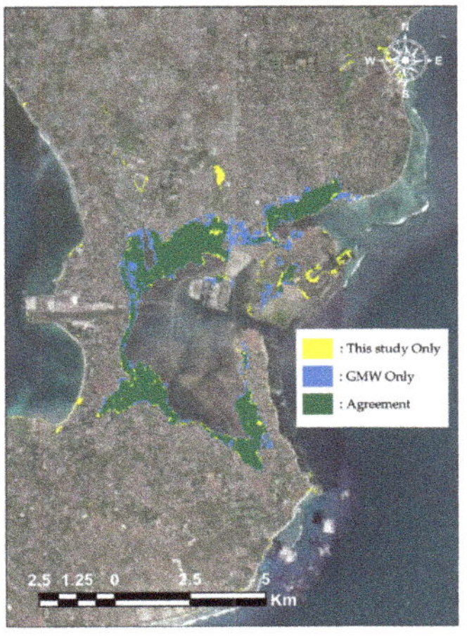

	Global Mangrove Watch	Predicted Class: Mangrove	Predicted Class: Non-Mangrove	
Actual Class	Mangrove	84	2	86
	Non-Mangrove	16	398	414
		100	400	500

Figure 7. Spatial agreement and disagreement between this study and the Global Mangrove Watch (GMW) classification in the southeastern Bali region (**left**), and the confusion matrix comparing GMW-derived mangrove labels to the reference data (**right**). This comparison was based on the year 2020, for which GMW data were available.

Quantitative evaluation using classification metrics such as accuracy, precision, recall, and F1-score revealed that the GMW 2020 dataset achieved an accuracy of 0.964, precision of 0.84, recall of 0.9767, and F1-score of 0.9032. In contrast, the mixed band classification proposed in this study outperformed GMW with an accuracy of 0.982, precision of 0.93, recall of 0.9789, and F1-score of 0.9538 (Table 6).

Although GMW exhibited fewer misclassifications in inland areas, it frequently misidentified coastal paddy fields or urban regions as mangroves. This was likely due to limitations in capturing the fine-scale environmental heterogeneity of specific coastal zones. In contrast, the proposed method integrated both SAR and optical features through a tailored band combination, resulting in enhanced spatial coherence and improved classification precision in complex coastal environments.

Table 6. GMW for accuracy in 2020.

Dataset	Accuracy	Precision	Recall	F1-Score
GMW 2020	0.964	0.84	0.976744	0.903225
Mixed band	0.982	0.93	0.978947	0.953846

These results suggest that the proposed approach provides a viable alternative for generating high-accuracy mangrove maps, particularly when recent or localized data are unavailable. Furthermore, it highlights the potential of the framework to complement existing global products such as GMW, especially in areas requiring fine-scale monitoring or updated assessments.

3.3. Site-Specific Detection Performance Evaluation

The study area contains a substantial waterway, which is essential to assess the classification performance of the model in these regions. To achieve a more precise evaluation, the classification results before the removal of areas above 10 m in elevation (based on the DEM) were analyzed. Figure 8 shows the mangrove classification results around the waterway in 2020. The findings revealed that the mixed band approach produced the most precise and distinct waterway classification, while the EMI band also demonstrated strong accuracy in identifying mangrove areas near the waterway. In contrast, SAR-based polarization (VV and VV-VH) was less effective in accurately classifying these areas.

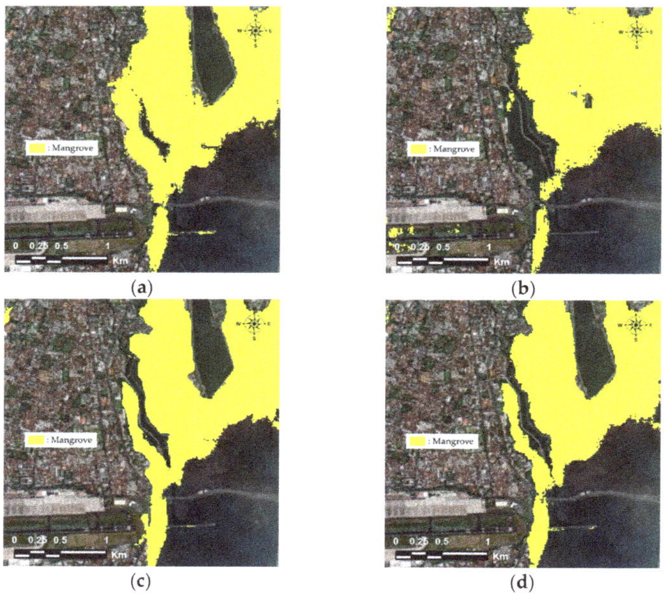

Figure 8. Classification results of mangroves around the channel in 2020. (**a**) VV. (**b**) VV-VH. (**c**) EMI. (**d**) Mixed band.

The classification accuracy assessment for waterways in 2024 showed notable differences compared with 2020 (Figure 9). While the mixed band approach exhibited the highest classification accuracy in 2020, EMI achieved the most accurate results in 2024, followed by the mixed band, VV, and VV-VH in descending order of performance.

Notably, despite expectations that SAR-based polarizations (VV and VV-VH) would be highly sensitive to water influence, they generally tended to be less effective in classifying mangroves near waterways. These findings suggest that SAR imagery may have struggled to accurately capture the complex backscattering characteristics of waterway-adjacent areas or may have caused confusion in delineating the boundaries between mangroves and waterways. Additionally, it was observed that none of the classification approaches could clearly distinguish the lower part of the waterway. This indicates that accurate mangrove classification becomes increasingly challenging in areas with intricately intertwined water bodies and mangrove vegetation.

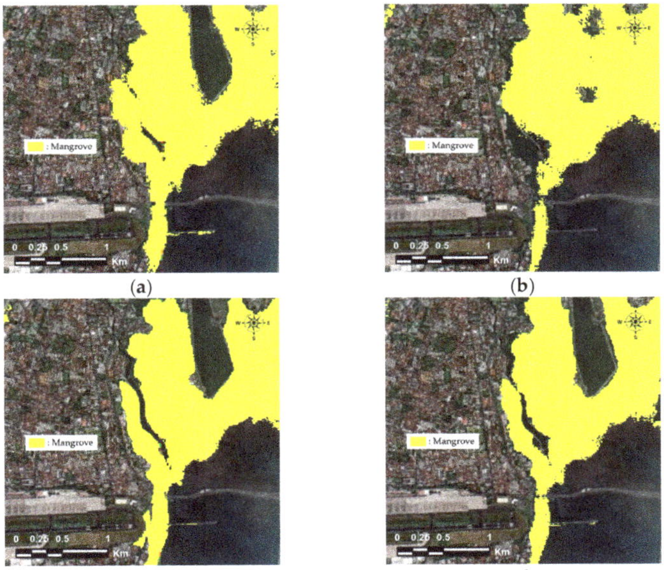

Figure 9. The classification results of mangroves around the channel in 2024. (**a**) VV. (**b**) VV-VH. (**c**) EMI. (**d**) Mixed band.

3.4. Time Series Assessment of Four Approaches

Figure 10 visually represents the areas of mangrove expansion and reduction in the 2024 classification results based on a comparison with the 2020 classification data. Overall, a decrease in mangrove coverage was observed in inland areas, suggesting that the 2020 classification contained many false positives in these regions. In contrast, the 2024 classification showed a notable reduction in such misclassifications, indicating improved accuracy. Mangrove coverage was observed to increase in the Serangan area in eastern Bali. An increase in NDVI was detected in this region; however, further analysis revealed that this change resulted from the planting of non-mangrove tree species rather than actual mangrove afforestation. This finding suggests that the model struggled to accurately differentiate between vegetation types, likely because VV and VV-VH polarizations are more responsive to detecting water bodies than distinguishing mangrove vegetation.

A time-series analysis of mangrove distribution in the study area revealed a consistent increase in mangrove extent over time. Specifically, the estimated area increased by 0.01 km^2 for VV, 1.37 km^2 for VV-VH, 0.67 km^2 for EMI, and 0.17 km^2 for the mixed band approach (Table 7). Among these, VV-VH exhibited the largest increase, while VV showed the slightest change, with an increase of only 0.01 km^2. An analysis of changes in the mangrove area from 2016 to 2020 indicated a total increase of 0.12 km^2 over four years. Considering this trend, the mixed band approach appeared to most accurately reflect the overall increase in mangrove distribution, suggesting its reliability for the long-term monitoring of mangrove changes. Unlike the other bands that displayed a consistent increasing trend, the VV polarization exhibited a slight decrease in the mangrove area, suggesting possible false positives and detection limitations. This result indicates that VV polarization may be relatively weak when differentiating between mangrove and non-mangrove areas.

Table 7. Comparison of changes in mangrove area.

	2020 (km^2)	2024 (km^2)	Increase (km^2)	Percentage Change
VV	12.67	12.66	−0.01	−0.08%
VV-VH	13.24	14.61	1.37	10.37%
EMI	11.74	12.41	0.67	5.71%
Mixed band	11.94	12.11	0.17	1.42%

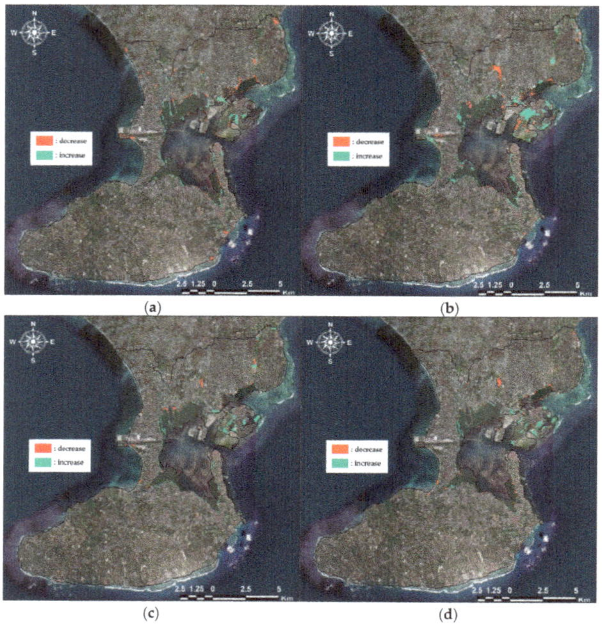

Figure 10. Mangrove change detection (2020–2024). Increase and decrease areas: (**a**) VV, (**b**) VV-VH, (**c**) EMI, and (**d**) mixed band.

Mangrove afforestation was observed in the northern part of the Denpasar mangrove forest, leading to an increase in the mangrove area (Figure 11). Based on this observation, an additional time-series analysis was conducted to further investigate this trend.

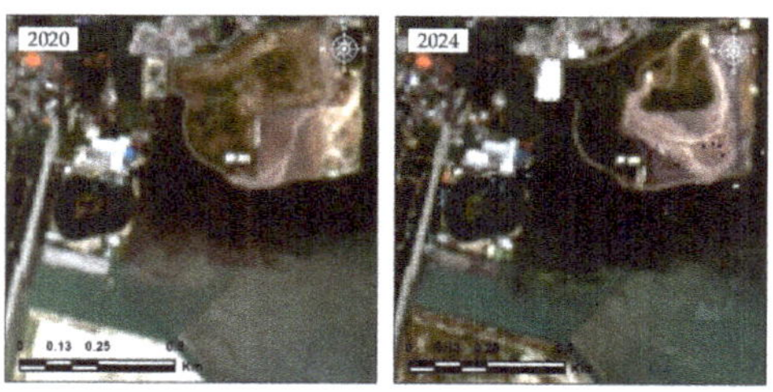

Figure 11. Sentinel-2 imagery of the Denpasar mangrove afforestation area (2020 vs. 2024) [43].

Based on this observation, an additional time-series analysis was conducted to further investigate this trend (Figure 12). Both the EMI and mixed band approaches exhibited a steady increase in classified mangrove areas over time. Notably, EMI demonstrated high sensitivity to water-rich or moist areas, which made it effective in detecting newly afforested mangrove regions. However, this sensitivity also led to a tendency to overclassify water bodies, resulting in potential misclassifications. In contrast, the mixed band approach, which integrates SAR imagery, provided more stable and accurate classification in areas with high moisture content, delivering more reliable results than EMI. By combining the strengths of both the optical and SAR data, the mixed band method proved to be particularly suitable for the long-term monitoring of mangrove changes in environments such as afforestation zones with high moisture levels.

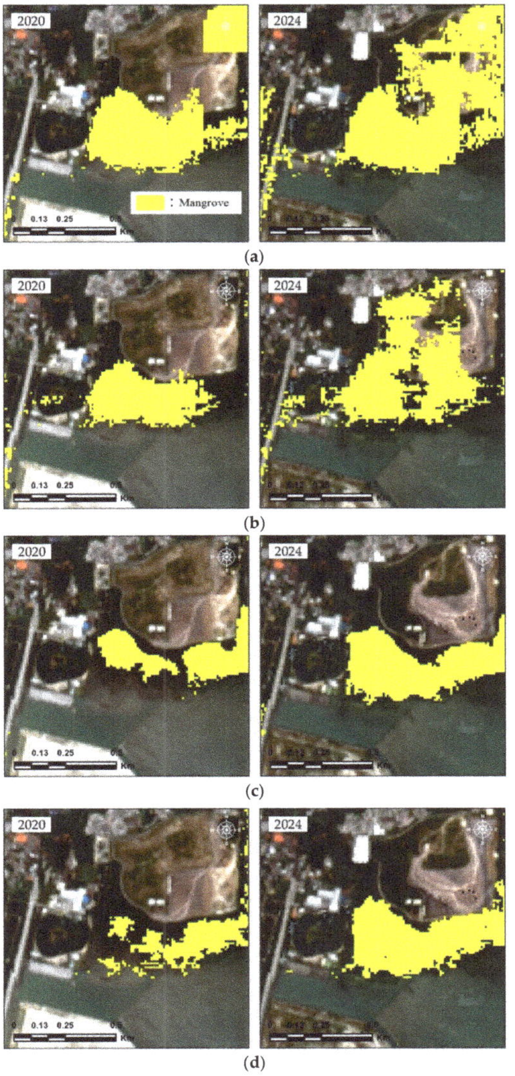

Figure 12. Time-series comparison between 2020 and 2024. (**a**) VV. (**b**) VV-VH. (**c**) EMI. (**d**) Mixed band.

4. Discussion

This study went beyond simple data fusion by analyzing the respective advantages and limitations of SAR and optical imagery and designing a combined band configuration that reflected these characteristics, thereby distinguishing itself from previous studies. The classification was conducted using only publicly available satellite data—Sentinel-1, Sentinel-2, and ESA WorldCover v100—demonstrating a practical and scalable approach, particularly in regions with limited technical infrastructure. While some earlier studies relied heavily on field-based reference data or high-resolution satellite imagery to enhance classification accuracy [44], the present study primarily utilized ESA WorldCover and Sentinel imagery, with high-resolution data and field information used only as supplementary inputs. Similarly, although Xie et al. (2024) [14] achieved high accuracy by comparing multiple deep learning models, their approach as based on Chinese proprietary satellite data, which may limit its generalizability. In contrast, our method was grounded entirely in open-access imagery, offering enhanced scalability and transferability across different regions and conditions. This approach thus presents a framework that balances accessibility, practicality, and classification accuracy. In particular, by conducting a temporal comparison of the classification results and demonstrating that the combined bands improved both the spatial consistency and precision, the study validated the practical utility of SAR–optical integration. The results yielded a high classification accuracy within the study area and suggest the potential applicability of this framework in other environmental contexts.

Notably, due to the lack of official mangrove label data for 2024, high-resolution satellite imagery was visually interpreted to verify the classification results. The model trained on 2020 data was successfully applied to the 2024 imagery, demonstrating its temporal transferability. This result further highlights the model's potential for practical use in near-real-time mangrove monitoring, especially when up-to-date reference data are unavailable.

The results of this study suggest that the classification performance varies depending on how different band combinations reflect the mangrove vegetation characteristics and their interactions with other environmental factors. To better capture the complex relationship between aquatic mangrove vegetation and environmental indicators, this study integrated optical imagery with SAR data. Since optical and SAR imagery are acquired through different physical mechanisms, combining both data types allows for a more comprehensive representation of the ecological and geographical characteristics, which a single data source may fail to capture. This integration aims to overcome the limitations of single-source data and enhance classification accuracy in complex environments such as mangrove ecosystems.

Additionally, the analysis demonstrated a significant improvement in classification accuracy when mixed bands were used compared with using optical or SAR data alone. This improvement can be attributed to the complementary characteristics of optical imagery and SAR data, which better captured the intricate properties of mangrove vegetation. The water body detection capability of SAR imagery combined with the spectral information from optical bands helped overcome the limitations of using individual data types.

However, when classification was restricted to areas with elevations below 10 m using DEM data, some actual mangrove regions were erroneously excluded, leading to omission errors. In lowland areas with similar spectral and textural characteristics, confusion still remains, especially in agricultural or high-moisture forest zones. The results suggest that the elevation criteria used in the DEM failed to fully capture the diversity and complexity of mangrove habitats, potentially restricting the accuracy of the classification data necessary for conservation and management efforts. To address these limitations, further research is needed to explore additional band combinations and develop new indices that extend beyond the current mixed band approach. In addition, strategies such as incorporating auxiliary datasets or developing region-specific post-processing rules may also be considered to better reflect the local characteristics and environmental conditions.

One of the primary challenges of optical imagery in tropical regions is persistent cloud cover, which limits data acquisition. In contrast, SAR data can operate independently of weather conditions, providing a significant advantage for mangrove classification in such environments. Time-series analysis revealed that VV-VH polarization-based classification provided a stable representation of mangrove area expansion, whereas VV polarization showed a minor decline in mangrove coverage. Despite this, the VV polarization still exhibited high-accuracy coverage, indicating that SAR data alone can be a reliable and effective classification method in cloud-dense conditions. The all-weather observation capability of SAR data makes it a practical alternative for mangrove classification in tropical regions.

However, overestimation was observed when classifying imagery from different years using the trained dataset. This issue is presumed to have resulted from changes in the temporal characteristics of the satellite imagery or discrepancies in the dataset. To ensure accurate time-series analysis, it is essential to develop an optimized methodology that minimizes both overestimation and underestimation by accounting for variations between datasets. To address these limitations, future studies should focus on refining the correction processes to further improve the classification accuracy.

5. Conclusions

This study adopted a hybrid approach that combined open-access SAR and optical imagery to improve the accuracy of mangrove classification and develop a model accessible to developing countries. The proposed framework prioritizes both methodological robustness and operational accessibility, enabling its deployment in areas with insufficient technological infrastructure and supporting evidence-based strategies for sustainable mangrove conservation and management in tropical regions. The U-Net-based SAR and optical imagery analysis demonstrated that the mixed band approach achieved the highest performance. While using SAR data alone resulted in lower recall values, its combination with optical data significantly improved the classification accuracy. The strong performance of the mixed band approach is likely due to the ability of SAR images to capture water-related differences between mangroves and other vegetation. These findings indicate that integrating SAR and optical data by leveraging their complementary characteristics plays a crucial role in enhancing the classification accuracy of mangrove forests. However, DEM failed to fully capture the complexity of mangrove habitats and exhibited a tendency toward overestimation in time-series classification. To mitigate these limitations, investigating novel band combinations and incorporating analysis techniques specifically optimized for time-series classification could significantly improve the classification accuracy.

Author Contributions: Conceptualization, S.K., H.K.A., C.-H.L.; Methodology, S.K., H.K.A., and C.-H.L.; Validation, S.K.; Formal analysis, S.K. and H.K.A.; Investigation, S.K.; Data curation, S.K.; Writing—original draft preparation, S.K.; Writing—review and editing, S.K. and C.-H.L.; Visualization, S.K.; Supervision, C.-H.L. All authors have read and agreed to the published version of the manuscript.

Funding: This research was funded by a National Research Foundation of Korea grant provided by the Ministry of Science and ICT (No. 2022R1C1C1008489) and a Kookmin University grant.

Data Availability Statement: The classified mangrove raster data for the Denpasar region, resulting from this study, is openly available at the Kookmin University CLIM Lab website: https://cms.kookmin.ac.kr/clim/research/researchdata1.do?mode=view&articleNo=5927333&article.offset=0&articleLimit=100, accessed on 1 March 2025.

Acknowledgments: We appreciate the cooperation of Udayana University in Bali, Indonesia, who provided the regional information and field campaign opportunity. We also appreciate Chang Bae Lee and Yong Ju Lee in the BEF lab at Kookmin University for the cooperating field dataset.

Conflicts of Interest: The authors declare no conflicts of interest.

Abbreviations

The following abbreviations are used in this manuscript:

CNN	Convolutional neural network
EMI	Enhanced mangrove index
ESA	European Space Agency
FAO	Food and Agriculture Organization
GEE	Google Earth Engine
MSI	Multispectral instrument
NIR	Near-infrared
SAR	Synthetic aperture radar
SWIR	Short wavelength infrared

References

1. Alongi, D.M. Present state and future of the world's mangrove forests. *Environ. Conserv.* **2002**, *29*, 331–349. [CrossRef]
2. IUCN. Environmental and Socio-Economic Value of Mangroves in Tsunami Affected Areas: Rapid Mangrove Valuation Study, Panama Village in South Eastern Coast of Sri Lanka. IUCN Report 2007. Available online: https://www.iwlearn.net/files/pdfs/IUCN%202007_Mangroves%20in%20Tsunami%20Affected%20Areas.pdf. (accessed on 14 January 2025).
3. Sandilyan, S.; Kathiresan, K. Mangrove conservation: A global perspective. *Biodivers. Conserv.* **2012**, *21*, 3523–3542. [CrossRef]
4. Hasan, M.; Hassan, L.; Abdullah Al, M.; Kamal, A.H.M.; Idris, M.H.; Hoque, M.Z.; Ali, A. Human intervention caused massive destruction of the second largest mangrove forest, Chakaria Sundarbans, Bangladesh. *Environ. Sci. Pollut. Res.* **2024**, *31*, 25329–25341. [CrossRef]
5. Zeng, Y.; Friess, D.A.; Sarira, T.V.; Siman, K.; Koh, L.P. Global potential and limits of mangrove blue carbon for climate change mitigation. *Curr. Biol.* **2021**, *31*, 1737–1743. [CrossRef] [PubMed]
6. Giesen, W.; Wulffraat, S.; Zieren, M.; Scholten, L. *Mangrove Guidebook for Southeast Asia*; FAO: Bangkok, Thailand, 2006; RAP Publication 2006/07.
7. Food and Agriculture Organization of the United Nations. *The World's Mangroves 1980–2005*; FAO Forestry Paper 153; FAO: Rome, Italy, 2007; ISBN 978-92-5-105856-5.
8. Food and Agriculture Organization of the United Nations. *Global Forest Resources Assessment 2020*; FAO: Rome, Italy, 2020; ISBN 978-92-5-132974-0.
9. Malik, A.; Mertz, O.; Fensholt, R. Mangrove forest decline: Consequences for livelihoods and environment in South Sulawesi. *Reg. Environ. Change* **2017**, *17*, 157–169. [CrossRef]
10. Spalding, M.D.; Blasco, E.; Field, C.D. *World Mangrove Atlas*; The International Society for Mangrove Ecosystems: Okinawa, Japan, 1997.
11. World Bank. *The Economics of Large-Scale Mangrove Conservation and Restoration in Indonesia*; World Bank: Washington, DC, USA, 2021.
12. Berger, A.A. *Bali Tourism*; Routledge: London, UK, 2013.
13. Bunting, P.; Rosenqvist, A.; Lucas, R.M.; Rebelo, L.-M.; Thomas, N.; Finlayson, C.M. Global Mangrove Extent Change 1996–2020: Global Mangrove Watch Version 3.0. *Remote Sens.* **2022**, *14*, 3657. [CrossRef]
14. Xie, Y.; Rui, X.; Zou, Y.; Tang, H.; Ouyang, N. Mangrove monitoring and extraction based on multi-source remote sensing data: A deep learning method based on SAR and optical image fusion. *Acta Oceanol. Sin.* **2024**, *43*, 110–121. [CrossRef]
15. Tran, T.V.; Reef, R.; Zhu, X.; Gunn, A. Characterising the distribution of mangroves along the southern coast of Vietnam using multi-spectral indices and a deep learning model. *Sci. Total Environ.* **2024**, *923*, 171367. [CrossRef]

16. Kathiresan, K.; Bingham, B.L. Biology of mangroves and mangrove ecosystems. *Adv. Mar. Biol.* **2001**, *40*, 81–251.
17. Chan, Y.K.; Koo, V. An introduction to synthetic aperture radar (SAR). *Prog. Electromagn. Res. B* **2008**, *2*, 27–60. [CrossRef]
18. Ulaby, F.T.; Moore, R.K.; Fung, A.K. *Microwave Remote Sensing: Active and Passive (Volume 1)*; Artech House: Norwood, MA, USA, 1981.
19. Widantara, K.W.; Mutaqin, B.W. Multi-hazard assessment in the coastal tourism city of Denpasar, Bali, Indonesia. *Nat. Hazards* **2024**, *120*, 7105–7138. [CrossRef]
20. World Bank. Climate Data: Historical Bali Watershed (EAP000769). Available online: https://climateknowledgeportal.worldbank.org/watershed/eap000769/climate-data-historical (accessed on 9 January 2025).
21. Ball, M.C. Comparative Ecophysiology of Mangrove Forest and Tropical Lowland Moist Forest. In *Tropical Forest Plant Ecophysiology*; Mulkey, S.S., Chazdon, R.L., Smith, A.O., Eds.; Chapman and Hall: London, UK, 1996; pp. 461–469.
22. European Space Agency. ESA World10m 2020 v100 Product Validation Report v1.1. ESA. 2020. Available online: https://worldcover2020.esa.int/data/docs/WorldCover_PVR_V1.1.pdf (accessed on 13 May 2025).
23. Google. Sentinel-1 Algorithms | Google Earth Engine. Google for Developers. 2022. Available online: https://developers.google.com/earth-engine/guides/sentinel1 (accessed on 16 January 2025).
24. Du, X.P.; Guo, H.D.; Fan, X.T.; Zhu, J.J.; Yan, Z.Z.; Zhan, Q. Vertical accuracy assessment of SRTM and ASTER GDEM over typical regions of China using ICESat/GLAS. *Earth Sci. J.* **2013**, *38*, 887–897.
25. Zhang, X.; Treitz, P.M.; Chen, D.; Quan, C.; Shi, L.; Li, X. Mapping mangrove forests using multi-tidal remotely-sensed data and a decision-tree-based procedure. *Int. J. Appl. Earth Obs. Geoinf.* **2017**, *62*, 201–214. [CrossRef]
26. Prayudha, B.; Ulumuddin, Y.I.; Siregar, V.; Agus, S.B.; Prasetyo, L.B.; Avianto, P.; Ramadhani, M.R. Enhanced mangrove index: A spectral index for discrimination understorey, nypa, and mangrove trees. *MethodsX* **2024**, *12*, 102778. [CrossRef] [PubMed]
27. Natural Resources Canada. Radar Polarimetry. Available online: https://natural-resources.canada.ca/maps-tools-and-publications/satellite-imagery-elevation-data-and-air-photos/tutorial-fundamentals-remote-sensing/microwave-remote-sensing/radar-polarimetry/9275 (accessed on 16 January 2025).
28. Ghasemi, M.; Karimzadeh, S.; Feizizadeh, B. Urban classification using preserved information of high dimensional textural features of Sentinel-1 images in Tabriz, Iran. *Earth Sci. Inform.* **2021**, *14*, 1745–1762. [CrossRef]
29. Shorten, C.; Khoshgoftaar, T.M. A survey on image data augmentation for deep learning. *J. Big Data* **2019**, *6*, 60. [CrossRef]
30. Siddique, N.; Paheding, S.; Elkin, C.P.; Devabhaktuni, V. U-net and its variants for medical image segmentation: A review of theory and applications. *IEEE Access* **2021**, *9*, 82031–82057. [CrossRef]
31. Ronneberger, O.; Fischer, P.; Brox, T. U-net: Convolutional Networks for Biomedical Image Segmentation. In *Proceedings of the Medical Image Computing and Computer-Assisted Intervention–MICCAI 2015, 18th International Conference, Munich, Germany, 5–9 October 2015, Proceedings, Part III*; Springer International Publishing: Munich, Germany, 2015; pp. 234–241.
32. Kim, J.; Lim, C.H.; Jo, H.W.; Lee, W.K. Phenological classification using deep learning and the Sentinel-2 satellite to identify priority afforestation sites in North Korea. *Remote Sens.* **2021**, *13*, 2946. [CrossRef]
33. Cho, W.; Lim, C.H. Simplified and high accessibility approach for the rapid assessment of deforestation in developing countries: A case of Timor-Leste. *Remote Sens.* **2023**, *15*, 4636. [CrossRef]
34. Chang, B.; Ahn, H.K.; Lim, C.H.; Cho, W. Pioneering approach for mangrove forest monitoring: Enhancing accessibility and assessment methodology in developing countries. *Heliyon*, 2025; *in press*.
35. Koonce, B. ResNet 50. In *Convolutional Neural Networks with Swift for TensorFlow: Image Recognition and Dataset Categorization*; Apress: Berkeley, CA, USA, 2021; pp. 63–72.
36. Yang, L.; Shami, A. On hyperparameter optimization of machine learning algorithms: Theory and practice. *Neurocomputing* **2020**, *415*, 295–316. [CrossRef]
37. Olofsson, P.; Foody, G.M.; Herold, M.; Stehman, S.V.; Woodcock, C.E.; Wulder, M.A. Good practices for estimating area and assessing accuracy of land change. *Remote Sens. Environ.* **2014**, *148*, 42–57. [CrossRef]
38. Moen, E.; Bannon, D.; Kudo, T.; Graf, W.; Covert, M.; Van Valen, D. Deep learning for cellular image analysis. *Nat. Methods* **2019**, *16*, 1233–1246. [CrossRef]
39. Sokolova, M.; Lapalme, G. A systematic analysis of performance measures for classification tasks. *Inf. Process. Manag.* **2009**, *45*, 427–437. [CrossRef]
40. Powers, D.M. Evaluation: From precision, recall and F-measure to ROC, informedness, markedness and correlation. *arXiv* **2020**, arXiv:2010.16061.
41. Saito, T.; Rehmsmeier, M. The precision-recall plot is more informative than the ROC plot when evaluating binary classifiers on imbalanced datasets. *PLoS ONE* **2015**, *10*, e0118432. [CrossRef] [PubMed]
42. Li, J.; Chen, W.; Ying, T.; Yang, L. Double-Exposure Algorithm: A Powerful Approach to Address the Accuracy Issues of Fractional Vegetation Extraction under Shadow Conditions. *Appl. Sci.* **2024**, *14*, 7719. [CrossRef]

43. European Space Agency (ESA). Sentinel-2 | Copernicus. Available online: https://www.esa.int/Applications/Observing_the_Earth/Copernicus/Sentinel-2 (accessed on 16 January 2025).
44. Sudiana, D.; Nisa, J.; Rizkinia, M.; Dimyati, R.D.; Anggraini, N.; Riyanto, I.; Prabuwono, A.S.; Sumantyo, J.T.S. Monitoring the Distribution of Mangrove Area Using Synthetic Aperture Radar (SAR) and Optic Remote Sensing Data Fusion Based on Deep Learning in Kotabaru Regency, Indonesia. *Evergreen* **2024**, *11*, 536–546. [CrossRef]

remote sensing

MDPI

Article

Mapping Tropical Forested Wetlands Biomass with LiDAR: A Machine Learning Comparison

Jonathan V. Solórzano [1,2], Candelario Peralta-Carreta [3] and J. Alberto Gallardo-Cruz [4,*]

[1] Centro de Investigaciones en Geografía Ambiental, Universidad Nacional Autónoma de México, Antigua Carretera a Pátzcuaro 8701, Ex-Hacienda de San José de la Huerta, Morelia 58058, Mexico; jsolorzano@ciga.unam.mx
[2] Comisión Nacional para el Conocimiento y Uso de la Biodiversidad, Periférico-Insurgentes Sur 4903, Parques del Pedregal, Mexico City 14010, Mexico
[3] Centro del Cambio Global y la Sustentabilidad, A.C., Centenario Instituto Juárez S/N, Reforma, Villahermosa 86080, Mexico; peralta.carreta@ccgs.mx
[4] Centro Transdisciplinar Universitario para la Sustentabilidad, Universidad Iberoamericana, Prolongación Paseo de la Reforma 880, Lomas de Santa Fe, Mexico City 01219, Mexico
* Correspondence: jose.gallardo@ibero.mx; Tel.: +52-5559504000 (ext. 7824)

Abstract: Mangroves and tropical swamp forests are ecosystems that play a critical role in carbon sequestration, coastal protection, and biodiversity support. Accurately estimating aboveground biomass (AGB) in these forests is crucial for global carbon management and conservation efforts. This study evaluates the potential of LiDAR-derived metrics to model the AGB of an area with mangroves and tropical swamp forests in Southeast Mexico. The study area, located in the Pantanos de Centla Protected Area, encompasses a gradient of seasonal waterlogged conditions, from saline to freshwater. Data were collected from 25 1250-m^2 plots, and three modeling approaches—linear regression, random forest, and XGBoost—were employed to estimate the AGB. The data were divided into training and test sets using an 80:20 ratio. The results indicate that the random forest model outperformed the others, achieving the lowest root mean squared error ($RMSE$ = 20.25 Mg/ha, $rRMSE$ = 12.25%, R^2 = 0.88). The most influential variables in this model were mean height (*zmean*), the 35th percentile of height (*zq35*), and the fourth percentile of returns (*p4th*), all positively correlated with the AGB. The model's robustness and uncertainty were evaluated through bootstrapping and spatial prediction across the study area, with higher AGB values concentrated near the main water channels. This study underscores the effectiveness of LiDAR-derived metrics for AGB estimation in complex forested environments.

Keywords: blue carbon; point cloud metrics; aboveground biomass (AGB); mangroves; tropical swamp forest; random forest; LiDAR

1. Introduction

Mangroves and tropical swamp forests are forested wetlands primarily found in coastal and riverine environments of Earth's tropical and subtropical latitudes; mangroves usually thrive under saline and brackish waters, while tropical swamp forests do so under freshwater conditions [1–4]. These forests play a crucial role in providing ecosystem services such as coastal protection against storms, carbon sequestration, habitat provision, and water purification [1,5–7].

Although these forested wetlands cover a small fraction of the world's forests, they are indispensable in the fight against climate change due to their ability to sequester large quantities of carbon, both in aboveground biomass (AGB) and in their extensive belowground root systems and soils [5,8–13]. For example, mangroves have been shown to sequester carbon at rates two to four times greater than terrestrial tropical forests, and they can store up to five times more carbon per hectare [1]. Given the critical role of mangroves and tropical swamp forests in global carbon cycles and climate change mitigation, the accurate monitoring of their AGB has become a priority for conservation efforts, carbon accounting, and the implementation of programs like Reducing Emissions from Deforestation and Forest Degradation (REDD+) [13–18].

Different methods have been designed to monitor the carbon stored in these ecosystems that include field sampling and remote sensing techniques. Although field-based methods are essential for monitoring AGB with accuracy, they often require a large investment in economic and human resources [6,19]. Furthermore, allometric models are usually used as the most common non-destructive method to convert field measures to AGB estimates [20–23]. Consequently, remote sensing technologies represent crucial tools to relate field-based biomass estimates with particular proxies and extrapolate this relation to larger areas [5,10,24,25]. This ability not only optimizes AGB monitoring and assessment but also enables obtaining a spatially explicit AGB characterization [26,27]. Therefore, the integration of remote sensing and field-based methods is one of the most used techniques to monitor AGB in forests and forested wetlands [24,28,29]. Nonetheless, each step in obtaining AGB estimates for a particular area involves its own uncertainties and errors [23,30,31].

In the last decades, Light Detection and Ranging (LiDAR) has attracted significant attention for modeling the aboveground biomass (AGB) component of forested wetlands due to the relatively small error associated with its AGB predictions [32–36]. This is attributable to the ability of LiDAR point clouds to extract metrics that describe the forests' three-dimensional structure with high spatial resolution [37,38]. Therefore, LiDAR is usually considered among the best remote sensing data to model AGB [39–41], especially in forests with a complex vertical structure, where traditional optical remote sensing methods (e.g., [42–45]) or even certain synthetic aperture radar (SAR) approaches may fall short [27,29,46].

On the other hand, different modeling techniques have been used to predict AGB from LiDAR metrics, including parametric and non-parametric methods [25,47–49]. Among these, machine learning algorithms usually outperform other parametric approaches, thanks to their lack of data a priori distribution (e.g., normal distribution) and ability to capture different types of non-linear relationships [50,51]. Thus, some of the most frequently used approaches include random forest, support vector machines, and artificial neural networks, among others [36,48,50,52,53]. Nevertheless, the potential of machine learning algorithms is frequently limited by the available sample size [54–56].

In this regard, the objective of this study is to evaluate the potential of LiDAR metrics to model a forested wetland AGB in Mexico, which includes areas of mangroves and tropical swamp forests. A comparison among three different methods (linear regression, random forest, and XGBoost) will be made to identify the best-performing approach for estimating AGB (i.e., smaller errors associated with AGB predictions). In addition, the variables included in the best-performing models will be identified and discussed to obtain further insights. Finally, the uncertainty associated with different steps of the AGB estimation (e.g., field sampling and modeling) will be characterized and propagated to obtain a wall-to-wall AGB estimation with its corresponding uncertainty.

2. Materials and Methods

The analysis workflow included two main sections (Figure 1): (1) finding and fitting the model that achieved the smallest error and (2) evaluating the uncertainty associated with AGB predictions both on the test set and the AGB prediction of the complete study area. The first part included per-plot AGB calculation, LiDAR point metrics extraction, selection of the model that achieved the smallest error (*RMSE*) using a cross-validation approach, and training the final model and evaluation on the test set. In addition, the correlation between AGB and LiDAR metrics included in the model that achieved the lowest error were analyzed. The second part included obtaining the uncertainty of three estimates: field AGB measures, AGB predictions by the model, and AGB predictions on the complete study area. In the last two types of predictions, the uncertainty associated with field measures was added to the uncertainty resulting from the modeling process. The complete analysis was performed in R 4.4.1 [57] using the following packages: BIOMASS to calculate AGB estimates [58]; lidR to process LiDAR data [59]; and tidymodels [60], randomForest [61], xgboost [62], yardstick [63], and modeltime [64] to train and evaluate the models. The scripts used in this study are available at https://github.com/JonathanVSV/AGBLiDAR (accessed on 10 March 2025).

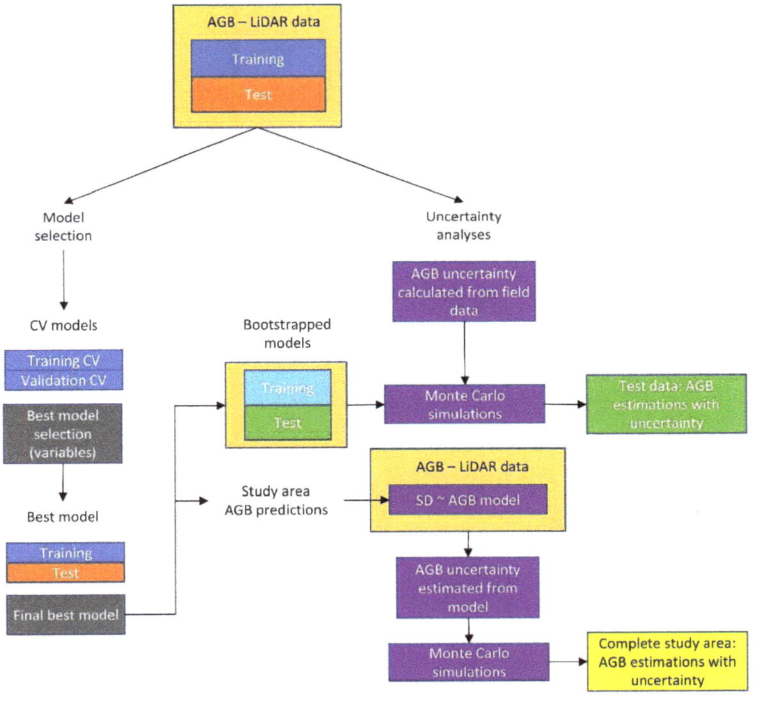

Figure 1. Schematic representation of all the steps followed in the study.

2.1. Study Site

The study was conducted in the surrounding forests of El Cometa Lagoon, which comprise a transition from mangrove to tropical swamp forest under a spectrum of brackish to freshwater flooding conditions. This lagoon is located inside the Pantanos de Centla Biosphere Reserve, Mexico (18°28′5″N, 92°27′15″W; Figure 2) [65], one of the largest wetlands protected areas in the country and Mesoamerica, covering roughly 3027 km^2. The tropical forested wetland around the lagoon is highly conserved, and no human settlement or human-induced covers can be found (i.e., the closest human settlement is approximately 12 km from the lagoon's center). The lagoon's main water channel flows approximately 25 km into the San Pedro–San Pablo River and, ultimately, into the Gulf of Mexico. The area experiences an annual mean temperature of 27 °C and a yearly mean precipitation of 1693 mm, with most rainfall occurring between June and October [9]. The dominant woody species in the study site are *Rhizophora mangle*, *Bucida buceras*, and *Pachira aquatica*. The distribution of these species responds mainly to proxies of salinity and waterlogging conditions [65].

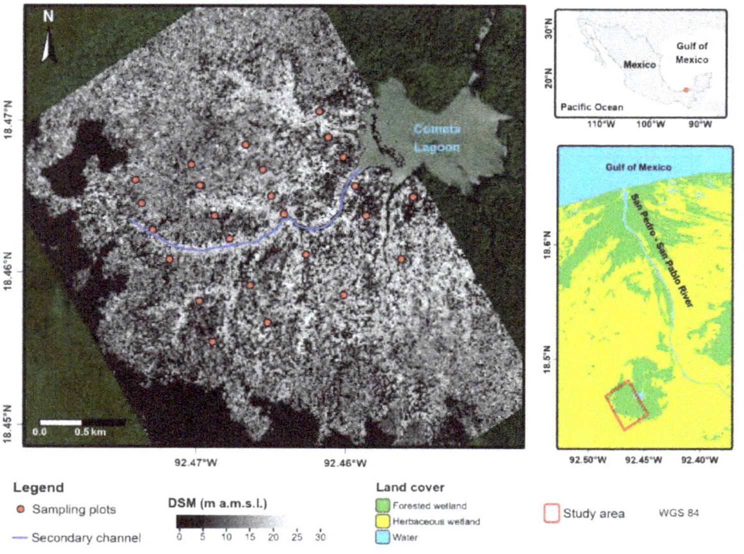

Figure 2. Location of the study site and sampled plots. In addition, a digital surface model (DSM) calculated from the LiDAR data is shown in grayscale.

2.2. Data

2.2.1. Sampling Method

A total of 3.37 ha was sampled, divided into 25 1250-m^2 plots (25 × 50 m). These plots were separated between 150 and 350 m following two gradients: distance to the lagoon and distance to the widest secondary channel (Figure 2). This sampling design assumed that these two gradients could serve as proxies of other environmental variables that determine species composition and AGB, such as salinity and flooding conditions. All woody plants with a diameter at breast height (DBH, at 1.3 m) $\geq$ 10 cm were registered between November 2016 and March 2017. The information recorded for each individual was species identity, DBH (cm), and height (H, measured in m). Due to the presence of aerial roots in *R. mangle* individuals, its DBH was measured at 10 cm above the highest root. In addition, the central coordinates of each plot were registered with a high-accuracy GNSS—GPS (i.e., Trimble Geo 7X) that allowed the differential correction of each coordinate with an average horizontal true error of 1 m.

2.2.2. Biomass Calculation

The AGB for each tree was calculated using the following Equation (1), developed for tropical forests [20]:

$$AGB = 0.0673\left(\rho DBH^2 H\right)^{0.976} \quad (1)$$

where ρ is the species mean wood density, extracted from the global wood density database [66], *DBH* stands for diameter at breast height, and *H* equals individual height. The mean and standard deviation of each species' wood density were calculated using global data and its taxonomic identity. The taxonomic identity of four individuals remained unknown; therefore, for these individuals, its wood density was assumed to be the average of the other species found in the same plot [20]. Once the AGB for each tree was obtained, the plot's AGB was calculated as the sum of all the trees' AGB and extrapolated to 1 ha (see Table S1).

To obtain the confidence intervals for the AGB predictions, a Monte Carlo simulation was implemented using the uncertainty in each species' wood density and an assumed error in diameter measures similar to [58,67]. From this procedure, the standard deviation (i.e., a measure of uncertainty) for each plot's AGB was obtained and extrapolated to 1 ha (see Table S1).

2.2.3. LiDAR Data

An airborne discrete return LiDAR point cloud acquired in March 2014 with a Riegl LMS-Q780 sensor (Riegl, Horn, Austria) was used to extract several metrics characterizing the vertical structure of the forest. The point cloud had a total of 192,252,873 points with a point density of 8 points/m^2 and up to seven returns per pulse. The LiDAR scan angle ranged from $-15°$ to $15°$, and the footprint encompassed approximately 9.5 km^2 (see Figure 2). The point cloud was acquired and classified by the vendor (CartoData SA de CV) into two classes: ground and unclassified and reported a precision of 0.001 m in the x-, y-, and z-axes. The first class had a z-value range of -1.58 to 2.9 m a.m.s.l., while the unclassified had a range of -1.37 to 31.5 m a.m.s.l. From the total number of points, 47.7% comprised of first returns, 29.2% to second returns, 15.18% to third, 5.95% to fourth, 1.67% to fifth, 0.33% to sixth, and 0.05% to seventh returns.

Ground-classified points were utilized to generate a digital terrain model (DTM) via universal kriging with 20 neighbors and 0.5 m resolution. This DTM served to normalize the point vegetation point cloud. Subsequently, the extent of each plot within the LiDAR data was delineated, and metrics summarizing the forest's vertical structure and the return intensity were computed. These metrics included vertical height (z) and intensity (i) statistical metrics such as mean, maximum (*max*), standard deviation (*sd*), skewness (*skew*), kurtosis (*kurt*), and percentiles in 5% steps (*q5–95*). Additionally, the cumulative percentage of return in the ith layer was measured according to [68] (*pcum1–9*), as well as height percentage of returns above the mean and 2 m (*pzabovezmean* and *pzabove2*), percentage of xth returns (*p1–5th*), number of points (n), the approximate area covered by each plot (*area*), and average absolute scan angle (*angle*). These variables have been previously reported as useful for modeling AGB, especially percentile metrics [31,32,35,36,46,49,52,53,69–72].

2.3. Modeling

2.3.1. Datasets

The complete dataset consisted of 25 plots, with AGB values as the dependent variable and 50 predictive variables. To reduce the size of the predictor variables and avoid redundant variables that may lead to overfitting and unnecessary computational complexity [72], highly correlated predictors were removed (Pearson correlation ≥ 0.8), resulting in a total of 27 predictive variables.

2.3.2. Algorithms Training

Three types of algorithms were used to train the models: linear regression and two machine learning algorithms: random forest and XGBoost. Given the different assumptions and characteristics of these models, linear regression was considered the simplest model and least flexible (as it only considers linear relations and assumes normally distributed data), while random forest and XGBoost corresponded to two machine learning algorithms capable of considering non-linear relations and working with data with non-parametric distribution. Although the latter two are based on decision trees to perform regressions, random forest outputs a prediction by averaging the results obtained by random trees, while XGBoost uses boosting to iteratively augment and build random trees [73]. Based on previous studies, we hypothesized that one of the two machine learning algorithms could result in the best modeling procedure; however, due to the relatively small sample size, we hypothesized that they could overfit the training data, obtaining poor results on the test data. Thus, the linear model represented the model with the lowest probability of overfitting, while XGBoost represented the one with the highest odds.

In order to reduce the chance of overfitting due to the large number of predictors (27) in relation to the sample size (25), all the combinations of these 27 variables, in groups of three, were evaluated to obtain the model with the smallest error (i.e., *RMSE*). The error of the models was evaluated as the root mean squared error (*RMSE*; Equation (2)):

$$RMSE = \sqrt{\sum_{i=1}^{n} \frac{(\hat{y}_i - y_i)}{n}} \tag{2}$$

where $\hat{y}$ stands for the predicted AGB value and y for the observed AGB value of each plot (i), while n stands for the number of sampled plots. The model with the lowest average *RMSE* obtained from the cross-validation procedure was regarded as the best model and selected to be fitted using all the training data and evaluated on the test set. Additionally, the *RMSE* was expressed in relative terms (*rRMSE*; Equation (3)), while the mean absolute error (*MAE*; Equation (4)) and mean absolute percentage error (*MAPE*; Equation (5)) were also calculated:

$$rRMSE = \frac{RMSE}{\frac{1}{n}\sum_{i=1}^{n} AGB_i} \tag{3}$$

$$MAE = \frac{1}{n}\sum_{i=1}^{n} |y_i - \hat{y}_i| \tag{4}$$

$$MAPE = \frac{100}{n}\sum_{i=1}^{n} \frac{|y_i - \hat{y}_i|}{y_i} \tag{5}$$

where *RMSE* stands for the *RMSE* obtained in Equation (2), n for the total number of sampled plots (i.e., 25), *AGB* for the AGB obtained in each plot (i), $\hat{y}$ stands for the predicted AGB value, and y stands for the observed AGB value of each plot (i).

First, cross-validation models (CV models) were developed to identify the optimal combination of predictive variables. Once this model was identified, the final model was fitted once using the training dataset and evaluated on the test set. These two types of models were performed on the data without considering the uncertainty of each plot's AGB value.

From the complete dataset, 20 randomly selected plots were selected as the training dataset (80%) and the remaining five as the test set (20%). In addition, for the CV models, the training dataset was divided into a 6-fold cross-validation set with two repeats to obtain 12 training/validation sets at 80–85%/15–20% (16–17 plots/3–4 plots). Thus, for each combination of three predictive variables, 12 metrics were obtained and then averaged to obtain the mean error for each model (Figure 1). Once the model that achieved the lowest mean error on the validation sets was identified, the final model was fitted once on the training data and evaluated on the test set. Then, using this final model, the importance of each predictive variable was calculated to identify the most influential variables. These importance values were scaled so the highest value corresponded to 100.

Afterward, the uncertainty of the predictions made by the final model was evaluated by calculating the standard deviation for the AGB predictions using a bootstrapping approach with 1000 randomizations. In each iteration, the complete dataset was randomly split, with 60–70% used for training and 30–40% reserved for testing. This procedure enabled the estimation of the prediction variability for each plot by altering the composition of the training set. In addition, the uncertainty of each AGB value resulting from the field data was propagated using a similar approach to the one suggested in [31,58]. In this process, a random number following a normal distribution with its corresponding standard deviation values (i.e., obtained from the field sampling calculation) was added to each model's AGB prediction. Ultimately, between 339 and 388 predictions were generated per plot, and the standard deviation was derived from this distribution of predictions.

2.3.3. Correlation Analyses

To gain a deeper understanding of the relationships between the variables included in the final model (i.e., the model with the lowest *RMSE* on the test set), a Pearson's correlation analysis was conducted to characterize the relationship between each predictive variable and AGB.

2.3.4. Prediction of the Complete Study Area

Using the model that achieved the lowest *RMSE*, the AGB was estimated across the entire LiDAR footprint, including areas that were not sampled. To exclude predictions for herbaceous wetlands, a mask was applied, retaining only areas where the mean height (*zqmean*) was greater than or equal to 1.5 m. This mask retained only areas covered by forested wetlands (i.e., mangroves or tropical swamp forests). The metrics for this model were extracted from the LiDAR point cloud at a resolution of 25 m, and the AGB predictions were based on these variables. The final model was retrained 1000 times with bootstrapped training data, and AGB predictions were made using each bootstrap-trained model.

The uncertainty in AGB estimates from field sampling was propagated using a similar approach as the one for calculating the standard deviation with the final model. Thus, a random error was simulated and summed to the AGB prediction to obtain variation in the predictions. Since, no a priori standard deviation value was known for the complete range of AGB values, a simple linear regression was fitted to predict the average standard deviation as a function of the AGB values (see Figure 3). This model was then used to simulate the error term for each pixel in the predicted AGB image. The mean and coefficient of variation (*CoV*) of these predictions were calculated for each pixel based on the outputs of the 1000 models. Finally, the total AGB for the study area was computed using the mean AGB values.

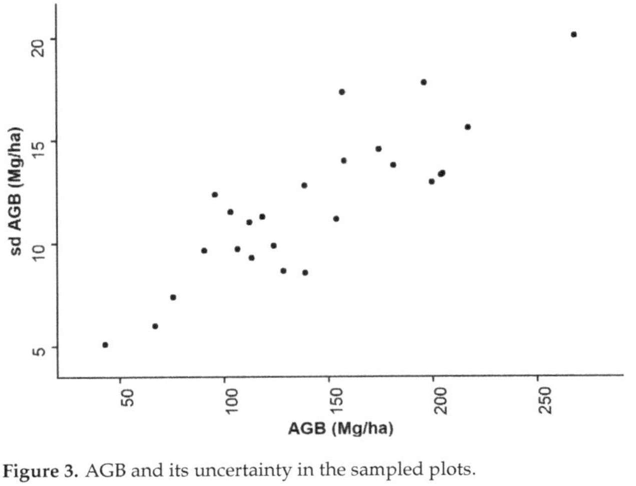

Figure 3. AGB and its uncertainty in the sampled plots.

3. Results

3.1. Forest Structure

The AGB data of the mangrove and tropical swamp forest plots showed a mean AGB value of 142.8 Mg/ha and a standard deviation of 53.62 Mg/ha. Additionally, the data reveals a positive relation between the mean height and AGB. As the mean height increases, there is a noticeable increase in AGB, with the lowest AGB being 42.98 Mg/ha at a mean height of 8.19 m and the highest reaching 268.17 Mg/ha at a mean height of 16.6 m (Table S1).

The uncertainty of each plot's AGB followed a general positive relation between the AGB uncertainty (i.e., sdAGB) and AGB mean values (Figure 3). This translates into plots with larger AGB also having more uncertainty. Overall, the highest AGB values are found in plots with mean heights exceeding 13 m, with the maximum AGB recorded at the highest mean height of 16.6 m (Figure 3).

3.2. Models

The model that achieved the lowest error on the test set was the random forest (*RMSE* = 20.24 Mg/ha, *rRMSE* = 12.25%, R^2 = 0.88, *MAE* = 14.25 Mg/ha, and *MAPE* = 9.19%), followed by the linear regression (*RMSE* = 31.80 Mg/ha, *rRMSE* = 19.24%, R^2 = 0.63, *MAE* = 23.34 Mg/ha, and *MAPE* = 13.63%) and XGBoost (*RMSE* = 36.22 Mg/ha, *rRMSE* = 21.91%, R^2 = 0.46, *MAE* = 31.46 Mg/ha, and *MAPE* = 21.83%) (Table 1 and Figure 4). All the models achieved a smaller error on the training set, while they differed mostly in their performance on the test set (Table 1 and Figure 4).

Table 1. Models that achieved the lowest *RMSE* on the test data for the random forest, XGBoost, and linear models.

Dataset	Model	Var 1	Var 2	Var 3	*RMSE*	*rRMSE* (%)	R^2	*MAE*	*MAPE*
Training	Random forest	zmean	zq35	p4th	14.31	10.43	0.97	10.58	9.91
	XGBoost	zq55	p4th	zq95	5.80	4.23	0.99	2.94	1.76
	Linear	zmean	p5th	p2th	19.60	14.29	0.87	14.02	12.54
Test	Random forest	zmean	zq35	p4th	20.24	12.25	0.88	14.25	9.19
	XGBoost	zq55	p4th	zq95	36.22	21.91	0.46	31.46	21.83
	Linear	zmean	p5th	p2th	31.80	19.24	0.63	23.34	13.63

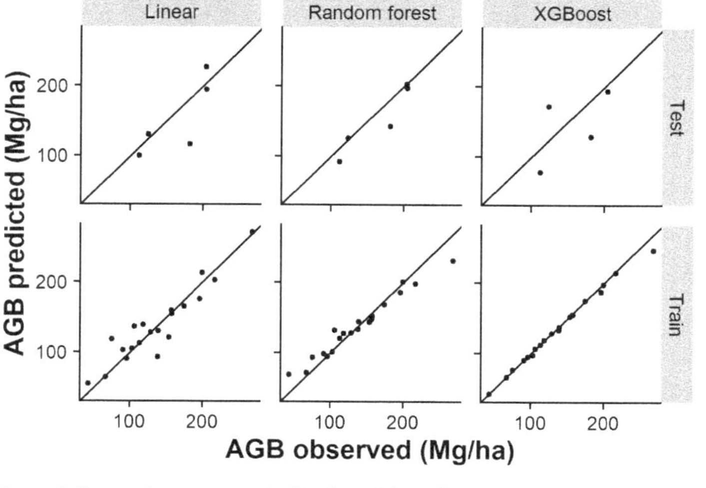

Figure 4. Comparisons among the fitted models on the training and test sets.

The variables incorporated in the final best model were *zmean*, *zq35*, and *p4th*, in order of relative importance (100, 93.75, and 73.67, respectively). On the other hand, the correlation analyses showed that the three variables included in the model with the lowest error showed a significant positive relation with the AGB (Table 2).

Table 2. Correlation coefficients between the LiDAR metrics included in the model that achieved the lowest *RMSE* on the test set and AGB.

Var 1	Var 2	Variable Importance	Pearson Coeff	*p*
AGB	zmean	100	0.85	<0.001
AGB	zq35	93.7	0.87	<0.001
AGB	p4th	73.7	0.53	0.006

3.3. Uncertainty Analyses

The bootstrap analyses showed that the uncertainty in the complete gradient of observed AGB values was similar to the pattern shown in the per-plot AGB calculations (Figure 5). Therefore, a positive relationship was observed between the standard deviation and AGB values.

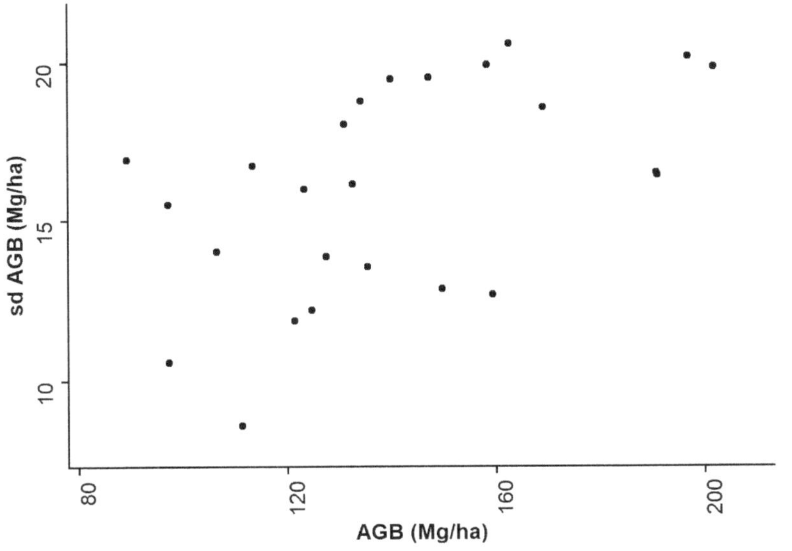

Figure 5. Uncertainty of the AGB predictions made by the random forest trained and tested on the bootstrapped datasets.

3.4. AGB Prediction for the Complete Study Area

The forested wetland's AGB in the LiDAR footprint was a total of 96,096.71 ± 16,447.72 Mg, distributed across approximately 7.61 km^2 of forested wetlands. The spatial distribution of the AGB revealed elevated values along the secondary water channels of the lagoon and lower values towards the peripheries of the forest, particularly at the southwestern edge (Figure 6). The *CoV* indicated lower variability in regions with higher AGB and greater variability in areas with lower AGB. Despite this pattern, the overall variation across the entire study area was relatively small (i.e., less than 12%; Figure 6).

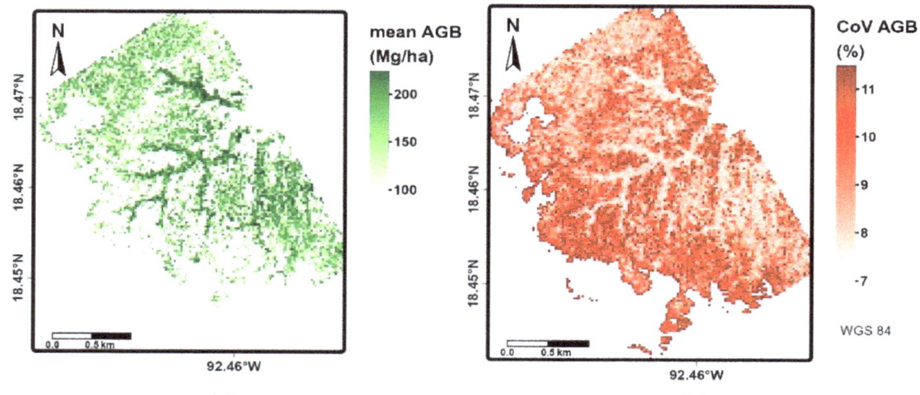

(a) (b)

Figure 6. (**a**) Mean AGB prediction of the study area. (**b**) Coefficient of variation (CoV) of the AGB of the study area.

4. Discussion

This study demonstrates that LiDAR-derived metrics can effectively model AGB in mangroves and tropical swamp forests with a relatively small error ($RMSE$ = 20.25 Mg/ha, $rRMSE$ = 12.25%, R^2 = 0.88, MAE = 14.25 Mg/ha, and $MAPE$ = 9.19%). Among the three modeling techniques evaluated—linear regression, random forest, and XGBoost—the random forest algorithm emerged as the most effective, outperforming both linear regression and XGBoost in terms of the observed error on the test set. The model with the lowest error included the *zq35*, *zmean*, and *p4th* metrics. Furthermore, the robustness of these predictions was confirmed through the randomization of both the training and test sets, as well as across the entire study area, indicating the model's reliability and generalization in diverse conditions.

4.1. Models

Comparing the $RMSE$ achieved on the training and test sets of the three best-performing models, it is evident the XGBoost overfitted on the training dataset, since it showed the lowest $RMSE$ on the training set and the highest $RMSE$ on the test set. In turn, the linear regression model underfitted on the training set but performed intermediately on the test set. Consequently, the random forest model represented a mid-point between the over- and underfitting of the other two models on the training set.

The linear models probably performed worse than the random forest ones due to their limited capacity to capture non-linear relationships between the LiDAR metrics and AGB. In contrast, machine learning techniques, such as random forest and XGBoost, are capable of working with non-parametric data and capturing non-linear interactions [73,74]. Nonetheless, due to the small sample size available in this study, the boosting mechanism in XGBoost caused an overfitting on the training data. Similar to previous reports, our study confirms that random forest is a robust modeling method capable of obtaining relatively low errors even when working with small sample sizes [48,50,52,53]. Future studies could benefit from using non-parametric modeling techniques to predict AGB based on LiDAR metrics [72].

The model that achieved the lowest $RMSE$ included *zmean*, *zq35*, and *p4th* metrics. The first two variables are associated with the vertical structure of the forest, and no variable characterizing the intensity of the laser's return was included. These findings agree with previous reports, since only vertical structure metrics were included in the best models [36,52,53].

Previous attempts to model AGB with LiDAR-derived metrics have reported that the maximum canopy height or quantiles above 90% are among the best height metrics for modeling AGB [32,35,36,49]. Our findings indicate that the model with the lowest *RMSE* did not include the total height (*zmax*); however, the mean height was highly correlated with the latter (*zmean*; $r = 0.81$). In fact, *zmax* was removed from the modeling process to avoid highly collinear variables. Although *zmax* was not tested in our models, it surely represents a variable with a high potential for predicting AGB. Thus, our results confirm the utility of LiDAR-derived percentile height metrics to model AGB, although not necessarily using single predictor models and the most frequent percentiles (i.e., max height) [31,32,35,36,46,49,52,53,69–72].

We infer that the positive relation between AGB and higher *zq35* and *zmean* might be associated with communities characterized by higher and denser canopies. Previous studies have found similar results, in which taller stands usually contain larger carbon stocks [19,26,32,36,48–50]. In addition, we conjecture that the positive relation between AGB and the *p4th* variable is related to the canopy's density. Since the studied forest has waterlogged areas, we hypothesize that forests with sparser canopies might have a lower percentage of fourth returns due to the absorption by water.

Compared to a previous attempt to model the same forest's AGB using gray-level co-occurrence metrics (GLCM) texture metrics [43] from a very high-resolution multispectral image, the current study achieved a lower *RMSE* and higher R^2 (*RMSE* = 20.25 Mg/ha; $R^2 = 0.88$ vs. *RMSE* = 26.62 Mg/ha; $R^2 = 0.65$). This result underscores the greater potential of LiDAR-derived metrics in comparison with texture-derived from multispectral imagery.

In the context of previous reports, particular plots' AGB correspond to some of the largest AGB values reported in mangroves or tropical swamp forests in Mexico, but most of the reported AGB values fall into the AGB range reported for the region [8,9,75–79]. The similarity in AGB estimates of these previous reports supports the accuracy of our AGB estimates. In turn, comparing these values with other forested wetlands worldwide, our AGB values fall in the smaller to intermediate AGB values [5,26,80]. Therefore, our AGB values are similar to other tropical swamp forests in Indonesia [48] and certain mangroves in Brazil [36,49], while smaller than certain tropical swamp forests of Indonesia and Brunei [19,50] and particular mangroves found in Gabon and Mozambique [32,81]. Thus, our AGB estimates in general align with worldwide AGB estimates for neotropical mangroves and forested wetlands in Mexico.

4.2. Spatial Patterns of AGB

Higher AGB values were concentrated in the areas closer to the main water channels (AGB $\geq$ 200 Mg/ha) and the southern part of the study area. These areas have an important presence of *Rhizophora mangle*, since these are commonly flooded and under higher salinity regimes, which coincides with previous reports of the relation of the inland presence of *R. mangle* with higher AGB values [22,79,82].

The total AGB in the forested cover of the study area (approximately 7.61 km^2) was 96,096.71 $\pm$ 16,447.72 Mg, which represents a considerable carbon sink in the region. Certainly, the LiDAR point cloud did not cover the complete forest that surrounds El Cometa Lagoon or other forested areas in the region, but previous studies reported similar or slightly lower AGB values in other forested areas inside the biosphere reserve and the neighboring Laguna de Términos protected area [9,75,77]. Moreover, other authors reported enormous carbon stocks in the belowground component of the forest in Pantanos de Centla Biosphere Reserve and neighboring protected areas (i.e., Laguna de Términos), especially in the swamp forests and mixed forests [8,77]. Thus, the reserve represents large carbon reservoirs that should continue to be protected. Additionally, this information can be vital to promote the involvement of these areas as beneficiaries of programs such as payment for environmental services or carbon bonds.

4.3. Uncertainty Analyses

Several aspects contribute to the uncertainty in AGB estimation, including the size and shape of the plots, biomass model selection, scale differences between the plots and remote sensing data, date differences between the field sampling and LiDAR data, and field coordinates precision, among others [30,67,83,84]. Although all these aspects contribute to AGB uncertainty and interact among them, frequently, its effect can only be quantified for a few of them.

Since AGB uncertainty is calculated as a sum of the individual trees' AGB, a larger uncertainty will usually be associated with plots with larger individuals or more trees. This exact pattern was observable in our results (Figure 3). Nonetheless, certain plots showed some deviation from this general pattern, which can be explained by either a higher presence of individuals with large DAP or height or species not identified to the species level. In the first case, it corresponds to particularly large individuals either in height or DAP that were present in the field, but no action can be performed to reduce the AGB uncertainty in these plots.

In the second case, previous studies have shown that including species identity for AGB calculation decreases the uncertainty in AGB values and errors associated with AGB modeling from LiDAR metrics [36,49]. Although we identified most individuals up to the species level, there were four morphospecies that we were unable to identify taxonomically. These species showed a higher uncertainty in their wood density values, since they were calculated from the identified species in the same plot. Although the occurrence of these species certainly increased the uncertainty in the AGB estimates, due to their rarity, their contribution to the AGB uncertainty was modest.

We assumed that the height metrics had no error in the Monte Carlo simulations, since they were directly measured in the field (although some previous studies have assumed a 0.5 m error [31]). This is obviously optimistic and unrealistic; however, there was no available field information to characterize this error. Although this aspect was not characterized in the field, we expect that the actual uncertainty in AGB values is slightly larger than the one presented here.

It is worth noting that, even though the size of our sampling plots could be regarded as small, based on previous studies [19,48,84,85], we were able to obtain predictions with relatively low uncertainty. This might be caused by the correspondence of scales between the LiDAR and field data [30] and the high precision used to capture the plots' coordinates.

Furthermore, when analyzing the predictions made on the entire study area, the coefficient of variation values ranged from 0 to 12%, which shows relatively low uncertainty in the predictions made by the model (Figure 6). Interestingly, most of the uncertainty related to AGB predictions in the study area can be attributed to the field sampling uncertainty, since the one derived exclusively from the bootstrapped models corresponded only to 0–3% variation. This finding underscores the relevance of propagating AGB uncertainties into the AGB predictions, since it is very likely that overoptimistic error estimates can be obtained without including this process [30,31].

In the AGB prediction of the complete study area, a larger uncertainty can be associated with larger AGB values. This is a direct consequence of propagating the uncertainty obtained in the AGB calculation from field data into the modeling uncertainty. Without this propagation, the modeled AGB uncertainties did not show any particular pattern. Finally, the predictions of the model made on the observed AGB values range should be regarded as trustworthy, which agrees with previous findings [32,36,43].

4.4. Potential Limitations of the Study

Due to the relatively small sample size and study area, the generalization of our results to a larger area might lead to AGB estimations with larger uncertainty than the one shown in this study. Although we used a bootstrapping process to measure the variability of AGB predictions using different training data over the complete study area, admittedly, a more robust generalization could be obtained with a larger sample that covers a larger area [54,55]. Nonetheless, due to financial and time restrictions, it was impossible to achieve this. Future studies should prioritize data sharing to increase the available sample size, although its generalization might be still hindered by the lack of LiDAR data over large areas.

Seasonal variations can alter the information registered from LiDAR sensors, especially in deciduous communities or forests with dead individuals [70,71,86]. Since the forested wetland studied in this work consisted of a perennial community, the effects of seasonality on the LiDAR metrics should be small and should consist of variations caused by leaf growth/non-growth cycles [71,87]. Thus, we expect that our AGB estimates and LiDAR metrics could vary with the season of data acquisition; however, this variation should be small. Conversely, we expect that the highest seasonal variation could be expected in the LiDAR metrics in response to waterlogging conditions [88]. Since the LiDAR data were registered in the dry season, we expected less interference of water with the NIR laser; however, a higher water level could make the extraction of the DTM difficult and consequently, the vegetation normalization process. Therefore, LiDAR data acquisition in forested wetlands should be prioritized during the period with the lowest water level to facilitate forest height normalization.

We acknowledge that a 3-year time lag between the LiDAR data acquisition and the field data registration could have added noise to our models. Since the area corresponds to a protected area with a small presence of human activity (except fishing), no major disturbances were expected in this period. We confirmed this scenario by visually interpreting Sentinel-2 and Google Earth images. Better models could have been obtained with a smaller time gap between the LiDAR and field data acquisitions; however, due to financial issues, this was not possible.

Previous studies suggest that locally developed equations might be better for decreasing the error for calculating the AGB [89,90], while others suggest the opposite [91,92]. In addition, in other available allometric equations, other frequent companion species in our study, such as *Pachira aquatica* and *Bucida buceras*, were not included, neither in locally developed equations [93] nor global [22]. Thus, we chose the pantropical equation developed by [20] to calculate the AGB for each species using its corresponding wood-specific gravity. Nonetheless, we acknowledge that using different allometric models could potentially alter the AGB estimates by a relatively moderate magnitude.

4.5. Future Monitoring Proposals

The significance of this research extends beyond the immediate study area. As LiDAR technology becomes more widely available and its application in forested ecosystems continues to expand, the approach developed in this study could serve as scientific evidence to support AGB estimations in other mangroves or tropical swamp forests. Notably, Mexico, with its extensive forested wetland surface, stands out as a key area for applying these methodologies [5,10,26].

Previous studies have reported that one of the best AGB predictors is the maximum canopy height [32,36,49]. Although this variable can be obtained using other remote sensing techniques (e.g., Interferometric Synthetic Aperture Radar (InSAR) [29,46]), the use of LiDAR has the advantage of providing other metrics that describe the 3D structure of the forest (e.g., height percentiles or percentages of xth returns). As in this study, future AGB monitoring proposals could decrease the error in AGB predictions by including other cloud metrics derived from LiDAR in its models and using models that include more than one predictor.

Since 2019, Global Ecosystem Dynamics Investigation (GEDI) has democratized the use of LiDAR data to monitor forests' AGB over large extents, since it is the first LiDAR data with worldwide cover and free access [94]. Although this mission is a huge advance in the development of tools to monitor AGB using remote sensors, its acquisition nature can prevent overlapping with field-acquired data to produce wall-to-wall AGB predictions [35]. Although the current study did not use GEDI data, our results provide a valuable link between LiDAR and field data that could afterward be used as a link between LiDAR and GEDI data to inform future attempts to model AGB over larger areas. For example, the most effective model in this study included *zq35* and *zmean*, two metrics that can easily be obtained from GEDI products. Therefore, future monitoring methods can build upon these results to model tropical wetlands' AGB with GEDI data and ultimately combine them with other types of images (e.g., multispectral and SAR) to enable wall-to-wall AGB mapping [95–99].

5. Conclusions

This study underscores the efficacy of LiDAR-derived metrics in modeling AGB in mangroves and tropical swamp forests, achieving a *RMSE* of 20.25 Mg/ha (*rRMSE* = 12.25%), with random forest emerging as the superior modeling technique. The model's success was attributed to the inclusion of vertical structure metrics, such as *zmean* and *zq35*, which were significantly correlated with the AGB values. The findings highlight the robustness and reliability of the random forest models, even with the constraints of a relatively small sample size. Compared to other methods, including linear regression and XGBoost, random forest demonstrated a balanced performance between overfitting and underfitting, validating its effectiveness in handling LiDAR data for biomass estimations. Moreover, the study reveals that the spatial distribution of AGB correlates closely with the forest's proximity to water channels and its vertical structure, affirming the importance of canopy metrics in biomass modeling. While the analysis acknowledged limitations related to sample size, the model's predictions remained relatively robust. This research contributes valuable insights into the application of LiDAR technology for AGB estimations and emphasizes the potential for scaling such methodologies using additional remote sensing tools. Future research should explore integrating SAR and multispectral data to further enhance AGB modeling and leverage freely available datasets like GEDI to extend these techniques to broader areas.

Supplementary Materials: The following supporting information can be downloaded at https://www.mdpi.com/article/10.3390/rs17061076/s1: Table S1: Structural attributes of the sampled plots.

Author Contributions: Conceptualization, supervision, and writing—review and editing, J.V.S., J.A.G.-C. and C.P.-C.; methodology, software, validation, formal analysis, data curation, visualization, investigation, and writing—original draft preparation, J.V.S.; resources, funding acquisition, and project administration, J.A.G.-C. All authors have read and agreed to the published version of the manuscript.

Funding: This research was funded by LANRESC (Laboratorio Nacional de Resiliencia Costera) (Grant No. 271544, 2016), CONACyT-FORDECyT (Grant No. 273646), and Universidad Iberoamericana (División de Investigación—Convocatoria 17).

Data Availability Statement: The data used in this study can be found at https://github.com/JonathanVSV/AGBLiDAR (accessed on 10 March 2025).

Acknowledgments: We are grateful to the people of Ribera baja de San Francisco, Campeche, for their hospitality. We also wish to thank Derio A. Jiménez-López, Matías Hernández-Gómez Ana Fernández-Montes de Oca, Luis Gerardo Cervantes-Jiménez, Nikolay M. Luna, Miguelina Sánchez-González, Marco A. Domínguez, Rubi E. Muñoz Vázquez, and Jorge E. Navarro Ramos for their help in the fieldwork.

Conflicts of Interest: The authors declare no conflicts of interest.

Abbreviations

The following abbreviations are used in this manuscript:

LiDAR	Light Detection and Ranging
AGB	Aboveground Biomass
DBH	Diameter at Breast Height
RMSE	Root Mean Squared Error
SD	Standard Deviation
CoV	Coefficient of Variation
SAR	Synthetic Aperture Radar
InSAR	Interferometric Synthetic Aperture Radar
GEDI	Global Ecosystem Dynamics Investigation

References

1. Donato, D.; Kauffman, J.B.; Murdiyarso, D.; Kurnianto, S.; Stidham, M.; Kanninen, M. Mangroves Among the Most Carbon-Rich Forests in the Tropics. *Nat. Geosci.* **2011**, *4*, 293–297. [CrossRef]
2. Alongi, D.M. Carbon Cycling and Storage in Mangrove Forests. *Annu. Rev. Mar. Sci.* **2014**, *6*, 195–219. [CrossRef] [PubMed]
3. Lugo, A.E.; Brown, S.; Brinson, M. *Ecosystems of the World, 15. Forested Wetlands*; Elsevier Science Publishers: Amsterdam, The Netherlands, 1990; Volume 2, p. 527. ISBN 0444428127.
4. Ainslie, W.B. Forested Wetlands. In *Southern Forest Resource Assessment*; Wear, D.N., Greis, J.G., Eds.; General Technical Report; U.S. Department of Agriculture, Forest Service: Washington, DC, USA, 2002; pp. 479–499.
5. Hutchison, J.; Manica, A.; Swetnam, R.; Balmford, A.; Spalding, M. Predicting Global Patterns in Mangrove Forest Biomass. *Conserv. Lett.* **2014**, *7*, 233–240. [CrossRef]
6. FAO. *The World's Mangroves 2000–2020*; FAO: Rome, Italy, 2023.
7. Murdiyarso, D.; Purbopuspito, J.; Kauffman, J.B.; Warren, M.W.; Sasmito, S.D.; Donato, D.C.; Manuri, S.; Krisnawati, H.; Taberima, S.; Kurnianto, S. The Potential of Indonesian Mangrove Forests for Global Climate Change Mitigation. *Nat. Clim. Change* **2015**, *5*, 1089–1092. [CrossRef]
8. Sjögersten, S.; Barreda-Bautista, B.d.l.; Brown, C.; Boyd, D.; Lopez-Rosas, H.; Hernandez, M.E.; Monroy, R.; Rincon, M.; Vane, C.; Moss-Hayes, V.; et al. Coastal Wetland Ecosystems Deliver Large Carbon Stocks in Tropical Mexico. *Geoderma* **2021**, *403*, 115173. [CrossRef]
9. Kauffman, J.B.; Trejo, H.H.; Carmen Jesus Garcia, M.d.; Heider, C.; Contreras, W.M. Carbon Stocks of Mangroves and Losses Arising from Their Conversion to Cattle Pastures in the Pantanos de Centla, Mexico. *Wetl. Ecol. Manag.* **2016**, *24*, 203–216. [CrossRef]
10. Giri, C. Chandra Observation and Monitoring of Mangrove Forests Using Remote Sensing: Opportunities and Challenges. *Remote Sens.* **2016**, *8*, 783. [CrossRef]
11. Gao, G.; Beardall, J.; Jin, P.; Gao, L.; Xie, S.; Gao, K. A Review of Existing and Potential Blue Carbon Contributions to Climate Change Mitigation in the Anthropocene. *J. Appl. Ecol.* **2022**, *59*, 1686–1699. [CrossRef]
12. Ribeiro, K.; Pacheco, F.S.; Ferreira, J.W.; Sousa-Neto, E.R.d.; Hastie, A.; Krieger Filho, G.C.; Alvalá, P.C.; Forti, M.C.; Ometto, J.P. Tropical Peatlands and Their Contribution to the Global Carbon Cycle and Climate Change. *Glob. Change Biol.* **2021**, *27*, 489–505. [CrossRef]

13. Hilmi, N.; Chami, R.; Sutherland, M.D.; Hall-Spencer, J.M.; Lebleu, L.; Benitez, M.B.; Levin, L.A. The Role of Blue Carbon in Climate Change Mitigation and Carbon Stock Conservation. *Front. Clim.* **2021**, *3*, 710546. [CrossRef]
14. Larrazábal, A.; McCall, M.K.; Mwampamba, T.H.; Skutsch, M. The Role of Community Carbon Monitoring for REDD+: A Review of Experiences. *Curr. Opin. Environ. Sustain.* **2012**, *4*, 707–716. [CrossRef]
15. Theilade, I.; Rutishauser, E.; Poulsen, M.K. Community Assessment of Tropical Tree Biomass: Challenges and Opportunities for REDD+. *Carbon Balance Manag.* **2015**, *10*, 17. [CrossRef] [PubMed]
16. Tokola, T. Remote Sensing Concepts and Their Applicability in REDD+ Monitoring. *Curr. For. Rep.* **2015**, *1*, 252–260. [CrossRef]
17. Choudhary, B.; Dhar, V.; Pawase, A.S. Blue Carbon and the Role of Mangroves in Carbon Sequestration: Its Mechanisms, Estimation, Human Impacts and Conservation Strategies for Economic Incentives. *J. Sea Res.* **2024**, *199*, 102504. [CrossRef]
18. Rahman; Lokollo, F.F.; Manuputty, G.D.; Hukubun, R.D.; Krisye; Maryono; Wawo, M.; Wardiatno, Y. A Review on the Biodiversity and Conservation of Mangrove Ecosystems in Indonesia. *Biodivers. Conserv.* **2024**, *33*, 875–903. [CrossRef]
19. Manuri, S.; Andersen, H.-E.; McGaughey, R.J.; Brack, C. Assessing the Influence of Return Density on Estimation of Lidar-Based Aboveground Biomass in Tropical Peat Swamp Forests of Kalimantan, Indonesia. *Int. J. Appl. Earth Obs. Geoinf.* **2017**, *56*, 24–35. [CrossRef]
20. Chave, J.; Réjou-Méchain, M.; Búrquez, A.; Chidumayo, E.; Colgan, M.S.; Delitti, W.B.C.; Duque, A.; Eid, T.; Fearnside, P.M.; Goodman, R.C.; et al. Improved Allometric Models to Estimate the Aboveground Biomass of Tropical Trees. *Glob. Change Biol.* **2014**, *20*, 3177–3190. [CrossRef]
21. Feldpausch, T.R.; Banin, L.; Phillips, O.L.; Baker, T.R.; Lewis, S.L.; Quesada, C.A.; Affum-Baffoe, K.; Arets, E.J.M.M.; Berry, N.J.; Bird, M.; et al. Height-Diameter Allometry of Tropical Forest Trees. *Biogeosciences* **2011**, *8*, 1081–1106. [CrossRef]
22. Komiyama, A.; Ong, J.E.; Poungparn, S. Allometry, Biomass, and Productivity of Mangrove Forests: A Review. *Aquat. Bot.* **2008**, *89*, 128–137. [CrossRef]
23. Vorster, A.G.; Evangelista, P.H.; Stovall, A.E.L.; Ex, S. Variability and Uncertainty in Forest Biomass Estimates from the Tree to Landscape Scale: The Role of Allometric Equations. *Carbon Balance Manag.* **2020**, *15*, 8. [CrossRef]
24. Kuenzer, C.; Bluemel, A.; Gebhardt, S.; Quoc, T.V.; Dech, S. Remote Sensing of Mangrove Ecosystems: A Review. *Remote Sens.* **2011**, *3*, 878–928. [CrossRef]
25. Lu, D.; Chen, Q.; Wang, G.; Liu, L.; Li, G.; Moran, E. A Survey of Remote Sensing-Based Aboveground Biomass Estimation Methods in Forest Ecosystems. *Int. J. Digit. Earth* **2016**, *9*, 63–105. [CrossRef]
26. Simard, M.; Fatoyinbo, L.; Smetanka, C.; Rivera-Monroy, V.H.; Castañeda-Moya, E.; Thomas, N.; Van der Stocken, T. Mangrove Canopy Height Globally Related to Precipitation, Temperature and Cyclone Frequency. *Nat. Geosci.* **2019**, *12*, 40–45. [CrossRef]
27. Saatchi, S. SAR Methods for Mapping and Monitoring Forest Biomass. In *The Synthetic Aperture Radar (SAR) Handbook: Comprehensive Methodologies for Forest Monitoring and Biomass Estimation*; Flores-Anderson, A.I., Herndon, K.E., Thapa, R.B., Cherrington, E., Eds.; NASA: Washington, DC, USA, 2019; pp. 207–247.
28. Feyisa, S.; Soromessa, T.; Bekele, T.; Bereta, A.; Hailemariam, F. Above Ground Biomass Estimation Methods and Challenges: A Review. *Int. J. Energy Technol. Policy* **2020**, *9*, 12–25. [CrossRef]
29. Singh, A.; Mahajan, S. A Review of SAR Image Processing for Mangrove Above Ground Biomass Estimation. *J. Plant Sci. Res.* **2023**, *38*, 779–789. [CrossRef]
30. Réjou-Méchain, M.; Barbier, N.; Couteron, P.; Ploton, P.; Vincent, G.; Herold, M.; Mermoz, S.; Saatchi, S.; Chave, J.; de Boissieu, F.; et al. Upscaling Forest Biomass from Field to Satellite Measurements: Sources of Errors and Ways to Reduce Them. *Surv. Geophys.* **2019**, *40*, 881–911. [CrossRef]
31. Rodda, S.R.; Fararoda, R.; Gopalakrishnan, R.; Jha, N.; Réjou-Méchain, M.; Couteron, P.; Barbier, N.; Alfonso, A.; Bako, O.; Bassama, P.; et al. LiDAR-Based Reference Aboveground Biomass Maps for Tropical Forests of South Asia and Central Africa. *Sci. Data* **2024**, *11*, 334. [CrossRef]
32. Fatoyinbo, T.; Feliciano, E.A.; Lagomasino, D.; Lee, S.K.; Trettin, C. Estimating Mangrove Aboveground Biomass from Airborne LiDAR Data: A Case Study from the Zambezi River Delta. *Environ. Res. Lett.* **2018**, *13*, 025012. [CrossRef]
33. Stovall, A.E.L.; Fatoyinbo, T.; Thomas, N.M.; Armston, J.; Ebanega, M.O.; Simard, M.; Trettin, C.; Zogo, R.V.O.; Aken, I.A.; Debina, M.; et al. Comprehensive Comparison of Airborne and Spaceborne SAR and LiDAR Estimates of Forest Structure in the Tallest Mangrove Forest on Earth. *Sci. Remote Sens.* **2021**, *4*, 100034. [CrossRef]
34. Hernández-Stefanoni, J.L.; Castillo-Santiago, M.Á.; Mas, J.F.; Wheeler, C.E.; Andres-Mauricio, J.; Tun-Dzul, F.; George-Chacón, S.P.; Reyes-Palomeque, G.; Castellanos-Basto, B.; Vaca, R.; et al. Improving Aboveground Biomass Maps of Tropical Dry Forests by Integrating LiDAR, ALOS PALSAR, Climate and Field Data. *Carbon Balance Manag.* **2020**, *15*, 15. [CrossRef]
35. Duncanson, L.; Kellner, J.R.; Armston, J.; Dubayah, R.; Minor, D.M.; Hancock, S.; Healey, S.P.; Patterson, P.L.; Saarela, S.; Marselis, S.; et al. Aboveground Biomass Density Models for NASA's Global Ecosystem Dynamics Investigation (GEDI) Lidar Mission. *Remote Sens. Environ.* **2022**, *270*, 112845. [CrossRef]
36. Souza Pereira, F.R.d.; Kampel, M.; Soares, M.L.G.; Estrada, G.C.D.; Bentz, C.; Vincent, G. Reducing Uncertainty in Mapping of Mangrove Aboveground Biomass Using Airborne Discrete Return Lidar Data. *Remote Sens.* **2018**, *10*, 637. [CrossRef]

37. Heumann, B.W. Satellite Remote Sensing of Mangrove Forests: Recent Advances and Future Opportunities. *Prog. Phys. Geogr.* **2011**, *35*, 87–108. [CrossRef]
38. Luo, S.; Wang, C.; Xi, X.; Nie, S.; Fan, X.; Chen, H.; Ma, D.; Liu, J.; Zou, J.; Lin, Y.; et al. Estimating Forest Aboveground Biomass Using Small-Footprint Full-Waveform Airborne LiDAR Data. *Int. J. Appl. Earth Obs. Geoinf.* **2019**, *83*, 101922. [CrossRef]
39. Garcia, M.; Saatchi, S.; Ferraz, A.; Silva, C.A.; Ustin, S.; Koltunov, A.; Balzter, H. Impact of Data Model and Point Density on Aboveground Forest Biomass Estimation from Airborne LiDAR. *Carbon Balance Manag.* **2017**, *12*, 4. [CrossRef] [PubMed]
40. Zaki, N.A.M.; Latif, Z.A. Carbon Sinks and Tropical Forest Biomass Estimation: A Review on Role of Remote Sensing in Aboveground-Biomass Modelling. *Geocarto Int.* **2017**, *32*, 701–716. [CrossRef]
41. Abbas, S.; Wong, M.S.; Wu, J.; Shahzad, N.; Irteza, S.M. Approaches of Satellite Remote Sensing for the Assessment of Above-Ground Biomass Across Tropical Forests: Pan-Tropical to National Scales. *Remote Sens.* **2020**, *12*, 3351. [CrossRef]
42. Tran, T.V.; Reef, R.; Zhu, X. A Review of Spectral Indices for Mangrove Remote Sensing. *Remote Sens.* **2022**, *14*, 4868. [CrossRef]
43. Solórzano, J.V.; Gallardo-Cruz, J.A.; González, E.J.; Peralta-Carreta, C.; Hernández-Gómez, M.; Oca, A.F.-M.d.; Cervantes-Jiménez, L.G. Contrasting the Potential of Fourier Transformed Ordination and Gray Level Co-Occurrence Matrix Textures to Model a Tropical Swamp Forest's Structural and Diversity Attributes. *J. Appl. Remote Sens.* **2018**, *12*, 036006. [CrossRef]
44. Bastin, J.-F.F.; Barbier, N.; Couteron, P.; Adams, B.; Shapiro, A.; Bogaert, J.; Cannière, C.D. Aboveground Biomass Mapping of African Forest Mosaics Using Canopy Texture Analysis: Towards a Regional Approach. *Ecol. Appl.* **2014**, *24*, 1984–2001. [CrossRef]
45. Ballhorn, U.; Jubanski, J.; Kronseder, K.; Siegert, F. Airborne LiDAR Measurements to Estimate Tropical Peat Swamp Forest Above Ground Biomass. In Proceedings of the 2012 IEEE International Geoscience and Remote Sensing Symposium, Munich, Germany, 22–27 July 2012; pp. 1660–1663.
46. Schlund, M.; Erasmi, S.; Scipal, K. Comparison of Aboveground Biomass Estimation from InSAR and LiDAR Canopy Height Models in Tropical Forests. *IEEE Geo-Sci. Remote Sens. Lett.* **2020**, *17*, 367–371. [CrossRef]
47. Kumar, L.; Mutanga, O. Remote Sensing of Above-Ground Biomass. *Remote Sens.* **2017**, *9*, 935. [CrossRef]
48. Englhart, S.; Jubanski, J.; Siegert, F. Quantifying Dynamics in Tropical Peat Swamp Forest Biomass with Multi-Temporal LiDAR Datasets. *Remote Sens.* **2013**, *5*, 2368–2388. [CrossRef]
49. Salum, R.B.; Souza-Filho, P.W.M.; Simard, M.; Silva, C.A.; Fernandes, M.E.B.; Cougo, M.F.; Nascimento, W.d.; Rogers, K. Improving Mangrove Above-Ground Biomass Estimates Using LiDAR. *Estuar. Coast. Shelf Sci.* **2020**, *236*, 106585. [CrossRef]
50. Zadbagher, E.; Marangoz, A.M.; Becek, K. Estimation of Above-Ground Biomass Using Machine Learning Approaches with InSAR and LiDAR Data in Tropical Peat Swamp Forest of Brunei Darussalam. *IForest* **2024**, *17*, 172–179. [CrossRef]
51. Lary, D.J.; Alavi, A.H.; Gandomi, A.H.; Walker, A.L. Machine Learning in Geosciences and Remote Sensing. *Geosci. Front.* **2016**, *7*, 3–10. [CrossRef]
52. Johnson, L.K.; Mahoney, M.J.; Bevilacqua, E.; Stehman, S.V.; Domke, G.M.; Beier, C.M. Fine-Resolution Landscape-Scale Biomass Mapping Using a Spatiotemporal Patchwork of LiDAR Coverages. *Int. J. Appl. Earth Obs. Geoinf.* **2022**, *114*, 103059. [CrossRef]
53. Tian, Y.; Zhang, Q.; Huang, H.; Huang, Y.; Tao, J.; Zhou, G.; Zhang, Y.; Yang, Y.; Lin, J. Aboveground Biomass of Typical Invasive Mangroves and Its Distribution Patterns Using UAV-LiDAR Data in a Subtropical Estuary: Maoling River Estuary, Guangxi, China. *Ecol. Indic.* **2022**, *136*, 108694. [CrossRef]
54. Chen, M.; Qiu, X.; Zeng, W.; Peng, D. Combining Sample Plot Stratification and Machine Learning Algorithms to Improve Forest Aboveground Carbon Density Estimation in Northeast China Using Airborne LiDAR Data. *Remote Sens.* **2022**, *14*, 1477. [CrossRef]
55. Schuh, M.S.; Favarin, J.A.S.; Marchesan, J.; Alba, E.; Berra, E.F.; Pereira, R.S. Machine Learning and Generalized Linear Model Techniques to Predict Aboveground Biomass in Amazon Rainforest Using LiDAR Data. *J. Appl. Remote Sens.* **2020**, *14*, 034518. [CrossRef]
56. Rajput, D.; Wang, W.J.; Chen, C.C. Evaluation of a Decided Sample Size in Machine Learning Applications. *BMC Bioinform.* **2023**, *24*, 48. [CrossRef] [PubMed]
57. R Core Team. *R: A Language and Environment for Statistical Computing*; R Foundation for Statistical Computing: Vienna, Austria, 2024.
58. Réjou-Méchain, M.; Tanguy, A.; Piponiot, C.; Chave, J.; Hérault, B. Biomass: An r Package for Estimating Above-Ground Biomass and Its Uncertainty in Tropical Forests. *Methods Ecol. Evol.* **2017**, *8*, 1163–1167. [CrossRef]
59. Roussel, J.-R.; Auty, D.; Coops, N.C.; Tompalski, P.; Goodbody, T.R.H.; Meador, A.S.; Bourdon, J.-F.; de Boissieu, F.; Achim, A. lidR: An r Package for Analysis of Airborne Laser Scanning (ALS) Data. *Remote Sens. Environ.* **2020**, *251*, 112061. [CrossRef]
60. Kuhn, M.; Wickham, H. Tidymodels: A Collection of Packages for Modeling and Machine Learning Using Tidyverse Principles. 2020. Available online: https://cran.r-project.org/web/packages/tidymodels/index.html (accessed on 10 March 2025).
61. Liaw, A.; Wiener, M. Classification and Regression by randomForest. *R News* **2002**, *2*, 18–22.
62. Chen, T.; He, T.; Benesty, M.; Khotilovich, V.; Tang, Y.; Cho, H.; Chen, K.; Mitchell, R.; Cano, I.; Zhou, T.; et al. Xgboost: Extreme Gradient Boosting. 2024. Available online: https://cran.r-project.org/web/packages/xgboost/index.html (accessed on 10 March 2025).

63. Kuhn, M.; Vaughan, D.; Hvitfeldt, E. Yardstick: Tidy Characterizations of Model Performance. 2024. Available online: https://cran.r-project.org/web/packages/yardstick/index.html (accessed on 10 March 2025).
64. Dancho, M. Modeltime: The Tidymodels Extension for Time Series Modeling. 2024. Available online: https://cran.r-project.org/web/packages/modeltime/index.html (accessed on 10 March 2025).
65. Solórzano, J.V.; Gallardo-Cruz, J.A.; Peralta-Carreta, C.; Martínez-Camilo, R.; Oca, A.F.-M. de Plant Community Composition Patterns in Relation to Microtopography and Distance to Water Bodies in a Tropical Forested Wetland. *Aquat. Bot.* **2020**, *167*, 103295. [CrossRef]
66. Zanne, A.E.; Lopez-Gonzalez, G.; Coomes, D.A.; Ilic, J.; Jansen, S.; Lewis, S.L.; Miller, R.B.; Swenson, N.G.; Wiemann, M.C.; Chave, J. Towards a Worldwide Wood Economics Spectrum. *Ecol. Lett.* **2009**, *12*, 351–366. [CrossRef]
67. Chave, J.; Condit, R.; Aguilar, S.; Hernandez, A.; Lao, S.; Perez, R. Error Propagation and Scaling for Tropical Forest Biomass Estimates. Philosophical transactions of the Royal Society of London. *Ser. B Biol. Sci.* **2004**, *359*, 409–420. [CrossRef]
68. Woods, M.; Lim, K.; Treitz, P. Predicting Forest Stand Variables from LiDAR Data in the Great Lakes—St. *Lawrence Forest of Ontario. For. Chron.* **2008**, *84*, 827–839. [CrossRef]
69. Ma, J.; Zhang, W.; Ji, Y.; Huang, J.; Huang, G.; Wang, L. Total and Component Forest Aboveground Biomass Inversion via LiDAR-Derived Features and Machine Learning Algorithms. *Front. Plant Sci.* **2023**, *14*, 1258521. [CrossRef]
70. Salas, E.A.L. Waveform LiDAR Concepts and Applications for Potential Vegetation Phenology Monitoring and Modeling: A Comprehensive Review. *Geo-Spat. Inf. Sci.* **2021**, *24*, 179–200. [CrossRef]
71. Cushman, K.C.; Armston, J.; Dubayah, R.; Duncanson, L.; Hancock, S.; Janík, D.; Král, K.; Krůček, M.; Minor, D.M.; Tang, H.; et al. Impact of Leaf Phenology on Estimates of Aboveground Biomass Density in a Deciduous Broadleaf Forest from Simulated GEDI Lidar. *Environ. Res. Lett.* **2023**, *18*, 065009. [CrossRef]
72. Borsah, A.A.; Nazeer, M.; Wong, M.S. LIDAR-Based Forest Biomass Remote Sensing: A Review of Metrics, Methods, and Assessment Criteria for the Selection of Allometric Equations. *Forests* **2023**, *14*, 2095. [CrossRef]
73. Chen, T.; Guestrin, C. XGBoost: A Scalable Tree Boosting System. In Proceedings of the 22nd ACM SIGKDD International Conference on Knowledge Discovery and Data Mining, ACM, San Francisco, CA, USA, 13–17 August 2016.
74. Breiman, L. Random Forests. *Mach. Learn.* **2001**, *45*, 5–32. [CrossRef]
75. Korpela, I.; Polvivaara, A.; Hovi, A.; Junttila, S.; Holopainen, M. Influence of Phenology on Waveform Features in Deciduous and Coniferous Trees in Airborne LiDAR. *Remote Sens. Environ.* **2023**, *293*, 113618. [CrossRef]
76. Simonson, W.; Allen, H.; Coomes, D. Effect of Tree Phenology on Lidar Measurement of Mediterranean Forest Structure. *Remote Sens.* **2018**, *10*, 659. [CrossRef]
77. Shan, J.; Toth, C.K. *Topographic Laser Ranging and Scanning: Principles and Processing*; CRC Press: Boca Raton, FL, USA, 2018.
78. Guerra-Santos, J.J.; Cerón-Bretón, R.M.; Cerón-Bretón, J.G.; Damián-Hernández, D.L.; Sánchez-Junco, R.C.; Carmen Guevara Carrió, E. del Estimation of the Carbon Pool in Soil and Above-Ground Biomass Within Mangrove Forests in Southeast Mexico Using Allometric Equations. *J. For. Res.* **2014**, *25*, 129–134. [CrossRef]
79. Adame, M.F.; Gamboa, J.N.; Torres, O.; Kauffman, J.B.; Reza, M.; Caamal, J.P.; Herrera-Silveira, J.A.; Medina, I. Carbon Stocks of Tropical Coastal Wetlands Within the Karstic Landscape of the Mexican Caribbean. *PLoS ONE* **2013**, *8*, e56569. [CrossRef]
80. Troche-Souza, C.; Priego-Santander, A.; Equihua, J.; Vázquez-Balderas, B. Spatial Distribution of Carbon Stocks Along Protected and Non-Protected Coastal Wetland Ecosystems in the Gulf of Mexico. *Ecosystems* **2024**, *72*, 724–738. [CrossRef]
81. Ávila-Acosta, C.R.; Domínguez-Domínguez, M.; Vázquez-Navarrete, C.J.; Acosta-Pech, R.G.; Martínez-Zurimendi, P. Aboveground Biomass and Carbon Storage in Mangrove Forests in Southeastern Mexico. *Resources* **2024**, *13*, 41. [CrossRef]
82. Velázquez-Pérez, C.; Hernández, C.; Romero-Berny, E.; Jesús-Navarrete, A. Estructura del manglar y su influencia en el almacén de carbono en la Reserva La Encrucijada, Chiapas, México. *Madera Bosques* **2023**, *25*, e2531885. [CrossRef]
83. Giri, C.; Ochieng, E.; Tieszen, L.L.; Zhu, Z.; Singh, A.; Loveland, T.; Masek, J.; Duke, N.C. Status and Distribution of Mangrove Forests of the World Using Earth Observation Satellite Data. *Glob. Ecol. Biogeogr.* **2011**, *20*, 154–159. [CrossRef]
84. Trettin, C.C.; Dai, Z.; Tang, W.; Lagomasino, D.; Thomas, N.; Lee, S.K.; Simard, M.; Ebanega, M.O.; Stoval, A.; Fatoyinbo, T.E. Mangrove Carbon Stocks in Pongara National Park, Gabon. *Estuar. Coast. Shelf Sci.* **2021**, *259*, 107432. [CrossRef]
85. Rahman, M.S.; Donoghue, D.N.M.; Bracken, L.J.; Mahmood, H. Biomass Estimation in Mangrove Forests: A Comparison of Allometric Models Incorporating Species and Structural Information. *Environ. Res. Lett.* **2021**, *16*, 124002. [CrossRef]
86. Zhao, F.; Guo, Q.; Kelly, M. Allometric Equation Choice Impacts Lidar-Based Forest Biomass Estimates: A Case Study from the Sierra National Forest, CA. *Agric. For. Meteorol.* **2012**, *165*, 64–72. [CrossRef]
87. Price, C.A.; Branoff, B.; Cummins, K.; Kakaï, R.G.; Ogurcak, D.; Papeș, M.; Ross, M.; Whelan, K.R.T.; Schroeder, T.A. Global Data Compilation Across Climate Gradients Supports the Use of Common Allometric Equations for Three Transatlantic Mangrove Species. *Ecol. Evol.* **2024**, *14*, e70577. [CrossRef]
88. Rutishauser, E.; Noor'an, F.; Laumonier, Y.; Halperin, J.; Rufi'ie; Hergoualch, K.; Verchot, L. Generic Allometric Models Including Height Best Estimate Forest Biomass and Carbon Stocks in Indonesia. *For. Ecol. Manag.* **2013**, *307*, 219–225. [CrossRef]

89. Day, J.W.; Conner, W.H.; Ley-Lou, F.; Day, R.H.; Navarro, A.M. The Productivity and Composition of Mangrove Forests, Laguna de Términos, Mexico. *Aquat. Bot.* **1987**, *27*, 267–284. [CrossRef]
90. Chavez, S.; Wdowinski, S.; Lagomasino, D.; Castañeda-Moya, E.; Fatoyinbo, T.; Moyer, R.P.; Smoak, J.M. Estimating Structural Damage to Mangrove Forests Using Airborne Lidar Imagery: Case Study of Damage Induced by the 2017 Hurricane Irma to Mangroves in the Florida Everglades, USA. *Sensors* **2023**, *23*, 6669. [CrossRef]
91. Picard, N.; Bosela, F.B.; Rossi, V. Reducing the Error in Biomass Estimates Strongly Depends on Model Selection. *Ann. For. Sci.* **2015**, *72*, 811–823. [CrossRef]
92. Réjou-Méchain, M.; Muller-Landau, H.C.; Detto, M.; Thomas, S.C.; Toan, T.L.; Saatchi, S.S.; Barreto-Silva, J.S.; Bourg, N.A.; Bunyavejchewin, S.; Butt, N.; et al. Local Spatial Structure of Forest Biomass and Its Consequences for Remote Sensing of Carbon Stocks. *Biogeosciences* **2014**, *11*, 6827–6840. [CrossRef]
93. Hernández-Stefanoni, J.L.; Reyes-Palomeque, G.; Castillo-Santiago, M.Á.; George-Chacón, S.P.; Huechacona-Ruiz, A.H.; Tun-Dzul, F.; Rondon-Rivera, D.; Dupuy, J.M. Effects of Sample Plot Size and GPS Location Errors on Aboveground Biomass Estimates from LiDAR in Tropical Dry Forests. *Remote Sens.* **2018**, *10*, 1586. [CrossRef]
94. Dubayah, R.; Armston, J.; Healey, S.P.; Bruening, J.M.; Patterson, P.L.; Kellner, J.R.; Duncanson, L.; Saarela, S.; Ståhl, G.; Yang, Z.; et al. GEDI Launches a New Era of Biomass Inference from Space. *Environ. Res. Lett.* **2022**, *17*, 095001. [CrossRef]
95. Saarela, S.; Holm, S.; Healey, S.P.; Patterson, P.L.; Yang, Z.; Andersen, H.-E.; Dubayah, R.O.; Qi, W.; Duncanson, L.I.; Armston, J.D.; et al. Comparing Frameworks for Biomass Prediction for the Global Ecosystem Dynamics Investigation. *Remote Sens. Environ.* **2022**, *278*, 113074. [CrossRef]
96. Tommaso, S.D.; Wang, S.; Lobell, D.B. Combining GEDI and Sentinel-2 for Wall-to-Wall Mapping of Tall and Short Crops. *Environ. Res. Lett.* **2021**, *16*, 125002. [CrossRef]
97. Adrah, E.; Wong, J.P.; Yin, H. Integrating GEDI, Sentinel-2, and Sentinel-1 Imagery for Tree Crops Mapping. *Remote Sens. Environ.* **2025**, *319*, 114644. [CrossRef]
98. Doyog, N.D.; Lin, C. Generating Wall-to-Wall Canopy Height Information from Discrete Data Provided by Spaceborne LiDAR System. *Forests* **2024**, *15*, 482. [CrossRef]
99. Wang, C.; Elmore, A.J.; Numata, I.; Cochrane, M.A.; Lei, S.; Hakkenberg, C.R.; Li, Y.; Zhao, Y.; Tian, Y. A Framework for Improving Wall-to-Wall Canopy Height Mapping by Integrating GEDI LiDAR. *Remote Sens.* **2022**, *14*, 3618. [CrossRef]

remote sensing

MDPI

Article

Extraction of Mangrove Community of *Kandelia obovata* in China Based on Google Earth Engine and Dense Sentinel-1/2 Time Series Data

Chen Lin [1,†], Jiali Zheng [1,†], Luojia Hu [2] and Luzhen Chen [1,*]

[1] State Key Laboratory of Marine Environmental Science, Key Laboratory of the Ministry of Education for Coastal and Wetland Ecosystems, College of the Environment and Ecology, Xiamen University, Xiamen 361102, China; chenlin1201@xmu.edu.cn (C.L.); jializheng@whu.edu.cn (J.Z.)

[2] Qian Xuesen Laboratory of Space Technology, China Academy of Space Technology, Beijing 100094, China; huluojia@spacechina.com

* Correspondence: luzhenchen@xmu.edu.cn

† These authors contributed equally to this work.

Abstract: Although significant progress has been made in the remote sensing extraction of mangroves, research at the species level remains relatively limited. *Kandelia obovata* is a dominant mangrove species and is frequently used in ecological restoration projects in China. However, owing to the fragmented distribution of *K. obovata* within mixed mangrove communities and the significant spectral and textural similarities among mangrove species, accurately extracting large-scale *K. obovata*-based remote sensing data remains a challenging task. In this study, we conducted extensive field surveys and developed a comprehensive sampling database covering *K. obovata* and other mangrove species across mangrove-distributing areas in China. We identified the optimal bands for extracting *K. obovata* by utilizing time-series remote sensing data from Sentinel-1 and Sentinel-2, along with the Google Earth Engine (GEE), and proposed a method for extracting *K. obovata* communities. The main conclusions are as follows: (1) The spectral-temporal variability characteristics of the blue and red-edge bands play a crucial role in the identification of *K. obovata* communities. The 90th percentile metric of the blue wavelength band ranks first in importance, while the 75th percentile metric of the blue wavelength band ranks second; (2) This method of remote sensing extraction using spectral-temporal variability metrics with time-series optical and radar remote sensing data offers significant advantages in identifying the *K. obovata* species, achieving a producer's accuracy of up to 94.6%; (3) In 2018, the total area of pure *K. obovata* communities in China was 4825.97 ha; (4) In the southern provinces of China, Guangdong Province has the largest *K. obovata* community area, while Macau has the smallest. This research contributes to the understanding of mangrove ecosystems and provides a methodological framework for monitoring *K. obovata* and other coastal vegetation using advanced remote sensing technologies.

Keywords: *Kandelia obovata*; Sentinel-1; Sentinel-2; Google Earth Engine; remote sensing; mangrove forests

1. Introduction

Mangrove forests are communities that flourish in the intertidal zones of tropical and subtropical coastlines [1]. *Kandelia obovata* Sheue, Liu & Yong, as one of the mangrove species recently separated from *Kandelia candel* (L.) Druce, is geographically isolated from *Kandelia candel* by the South China Sea [2]. In China, *K. obovata* stands out as the most cold-resistant mangrove species, naturally occurring at the northernmost fringe of mangrove distribution in Fuding, Fujian Province [3]. It is also the sole species that has been successfully introduced and continues to thrive at the highest latitude of mangrove growth, which is in Yueqing, Zhejiang Province [3]. Due to its environmental adaptability, *K. obovata* is the primary optional species in mangrove ecological restoration in China [4] and has been planted in a lot of restoration projects in Zhejiang, Fujian, Guangdong, and Guangxi, making the most significant contribution to the enhancement of blue carbon sequestration in mangrove ecosystems [5,6]. Thus, the rapid and efficient extraction of the distribution data of this species is a crucial research focus, offering substantial benefits for ecosystem restoration, management, and assessment.

With continuous advancement in remote sensing technology, satellite data has emerged as a pivotal tool for identifying and monitoring mangrove forests [7–9]. In its early stages, the identification of mangrove forests was primarily conducted through traditional aerial photography, a valuable method but still constrained by its limited resolution and scope [10]. By the 21st century, the landscape focus shifted with the advent of medium-resolution satellite remote sensing data, which, when combined with land reflectance features and vegetation indices, became the standard for mangrove detection [11–13]. The past decade has witnessed a paradigmatic advancement in remote sensing, transitioning from a reliance on optical imagery to a multi-source data integration approach [14,15]. Current research priorities are focused on mapping mangrove distributions globally and at species-specific levels [16]. However, earlier studies were largely confined to local-scale species mapping, resulting in a fragmented understanding of species distribution at broader scales [17]. Despite *K. obovata*'s widespread presence in China, comprehensive national-scale species distribution mapping remains a significant gap in the literature.

In 2024, the spatial distribution of *K. obovata* across China was mapped for the first time, marking a significant breakthrough in large-scale mangrove species distribution mapping research, estimating to be 1199 ha totally [18]. However, according to Wang et al. [5], based on the National Mangrove Resource Survey Report, *K. obovata* and *Rhizophora stylosa* dominate the mangrove mid-intertidal area, covering a total of 9396 ha in China. This implies that the former may have underestimated the actual area of *K. obovata*. To enhance the accuracy of *K. obovata* distribution estimates in China, additional research is warranted.

The spectral signatures of various mangrove species often overlap, and the intricate interplay between water, plant communities, and soil in mangrove ecosystems results in complex spectral reflectance patterns. This complexity poses a challenge in accurately discriminating *K. obovata* from other species when relying solely on optical data [19,20]. The synergy of optical imagery with Synthetic Aperture Radar (SAR) data provides a promising strategy to overcome these limitations. Sentinel-2, launched in 2015, boasts high spatial resolution (10 m) and a rich spectral profile, adeptly capturing vegetation dynamics across the visible, near-infrared, and shortwave infrared bands. In contrast, Sentinel-1, with its SAR sensor, is highly responsive to surface texture, compensating for the constraints of optical imagery in intricate environments. The integration of optical and SAR data not only provides a more holistic dataset but has also been shown to significantly enhance the precision and robustness of classification [21,22]. Consequently, this multi-modal data integration is instrumental in improving the accuracy and reliability of *K. obovata* detection.

This study aims to map the spatial distribution of *K. obovata* in China, improving the accuracy of *K. obovata* distribution estimates. Furthermore, compared to the current studies, it evaluates classification accuracy. Here, we established a comprehensive field sampling database for *K. obovata* and other mangrove species across China, leveraging extensive field surveys and fixed sample plots. By harnessing dense time-series remote sensing data from Sentinel-1 and Sentinel-2 and employing the Google Earth Engine (GEE), we identified the optimal spectral bands for the precise extraction of *K. obovata*. Our approach not only refines the method for large-scale, high-precision extraction of these communities but also enhances the accuracy and reliability of *K. obovata* mapping. The findings of this research delineate the spatial distribution of *K. obovata* across China, offering crucial data for mangrove ecosystem restoration, blue carbon assessments, ecosystem service evaluations, and the development of sustainable management strategies.

2. Research Area and Data Sources

2.1. Research Area

The study area is the distribution region of mangroves along the Chinese coast in 2018, sourced from Zhang et al. [23]. It was mapped using six satellites with a spatial resolution of 2 m, including Ziyuan-3-01, Ziyuan-3-02, Gaofen-1A, Gaofen-1B, Gaofen-1C, and Gaofen-1D, covering a total area of 25,683.88 ha (Figure 1). The geographic range spans from 18° to 31°N and 108° to 123°E, including Zhejiang, Fujian, Guangdong, Guangxi, Hainan, Hong Kong, Macau, and Taiwan. The mangroves in China are predominantly composed of cold-tolerant species, including *Avicennia marina*, *Kandelia obovata*, and *Aegiceras corniculatum*, along with other species such as *Bruguiera gymnorrhiza*, *Bruguiera sexangula*, *Rhizophora stylosa*, and *Sonneratia caseolaris* [5].

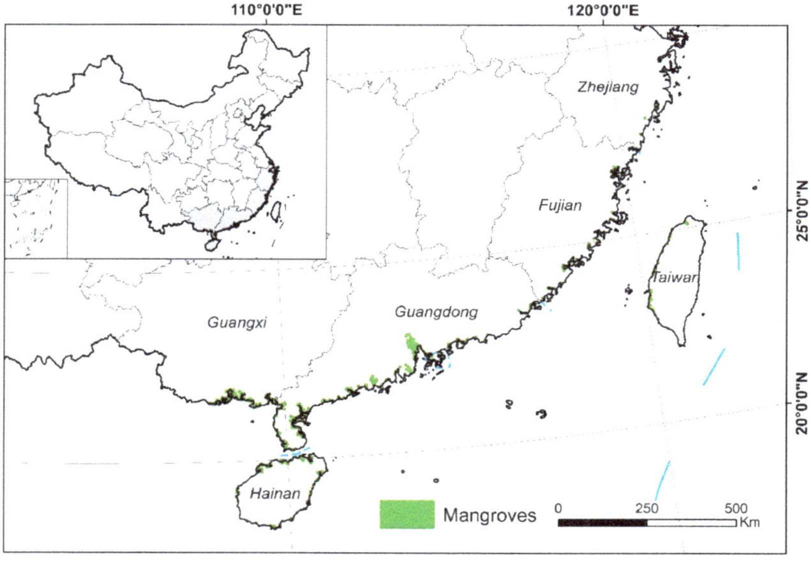

Figure 1. Location of the study area.

2.2. Data Source and Processing

We acquired dense time series remote sensing imagery from Sentinel-1 for 2018, which provided high-resolution radar data capable of penetrating cloud cover, and from Sentinel-2 for 2018, which offered multispectral optical data with high spatial resolution. This combination allowed us to effectively capture the spectral-temporal variability characteristics essential for identifying *K. obovata*. We integrated DEM data to account for elevation and topographic variations, which can influence mangrove distribution patterns. By incorporating this information, we enhanced the accuracy of our model in identifying *K. obovata*.

3. Methods

We established a mangrove species sample database across China through field surveys and visual interpretation, selecting a total of 5382 sample points for classification. A random forest classifier was utilized, executed four times, and the importance feature rankings from each run were exported. Subsequently, a band selection process was conducted to compare and analyze the reliability of the bands.

The overall workflow of the experiment is illustrated in Figure 2.

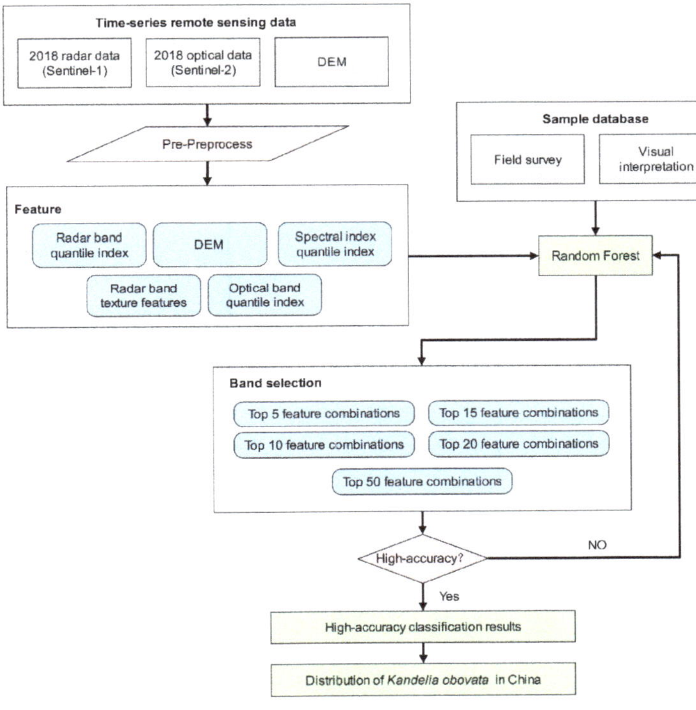

Figure 2. Experimental Workflow Diagram.

3.1. China's Mangroves Species Sample Database

We established 278 10 m × 10 m plots in various mangrove experimental areas across China and surveyed each mangrove species within these plots. If a plot contained only one species, such as pure *K. obovata* or another pure species, we recorded the latitude and longitude of the plot's center. Finally, a total of 115 pure species plots were selected from the 2018 community survey data. The details of these plots are presented in Table 1, Figure 3, and Table 2.

Table 1. Community survey sample data.

Species	Number of Plots
Kandelia obovata	36
Aegiceras corniculatum	35
Avicennia marina	33
Sonneratia apetala	10
Laguncularia racemosa	1
Total	115

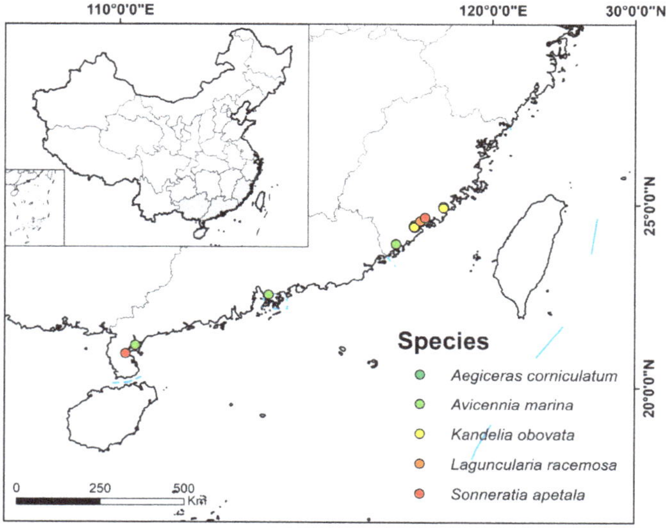

Figure 3. Sample plot survey purebred location.

Table 2. The survey photo and remote sensing images of sample points.

Location of Samples	Photos	Images	Characteristics
Aojiang, Wenzhou, Zhejiang		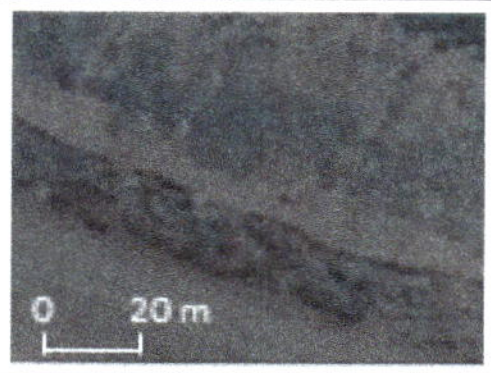	*Kandelia obovata*, artificially planted, young trees.
Luoyuan, Fuzhou, Fujian		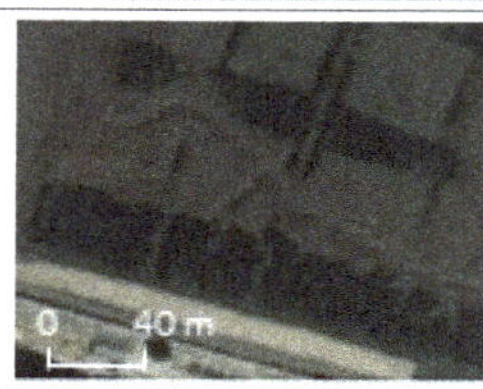	*Kandelia obovata*, artificially planted, young tree, with a height of about 1 m.
Eel Beach, Fujian, Fuzhou		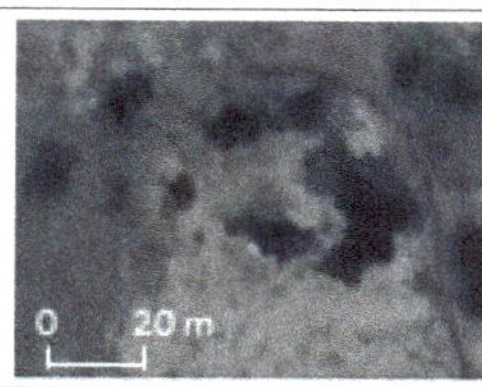	*Kandelia obovata*, natural, with a height of about 4–5 m.
Quanzhou Bay, Quanzhou, Fujian		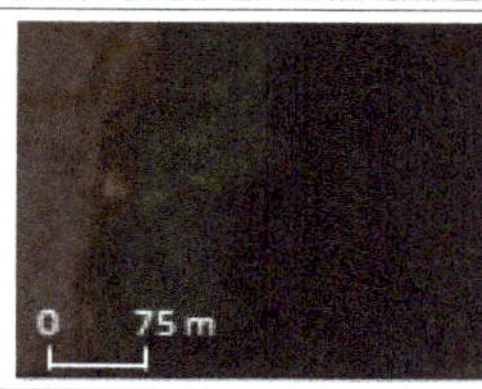	*Kandelia obovata*, artificially planted.

Table 2. *Cont.*

Location of Samples	Photos	Images	Characteristics
Fenglin, Xiamen, Fujian		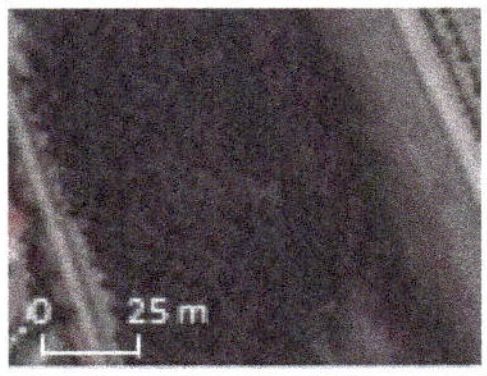	*Kandelia obovata*, artificially planted.
Jiulong River North Port, Zhangzhou, Fujian		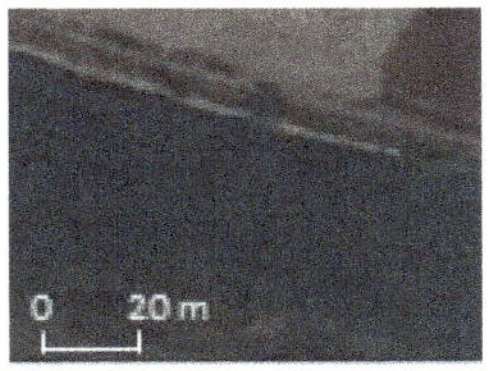	*Kandelia obovata*, artificially planted.
Sanjiang, Haikou, Hainan		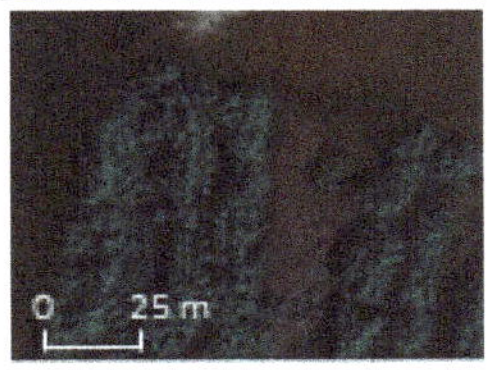	*Sonneratia apetala*, Artificially planted with a height of about 8 m.
Techeng, Zhanjiang, Guangdong		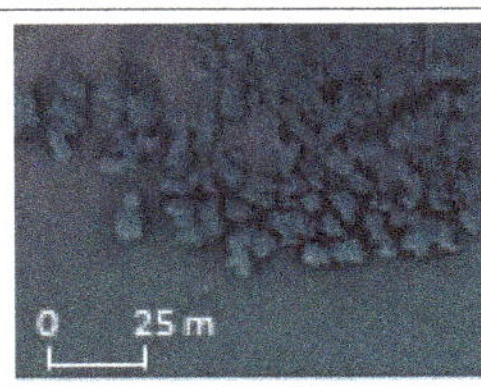	*Avicennia marina*, Natural growth of with a height of about 4 m.
Sanjiang, Hainan		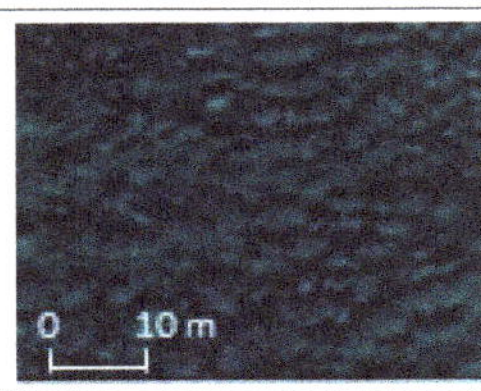	*Bruguiera sexangula*, with a height of about 5–7 m.
Tashi, Hainan		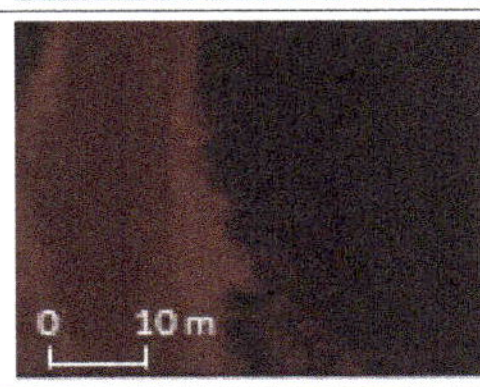	*Rhizophora stylosa*, with a height of about 7 m.
Muan, Guangxi		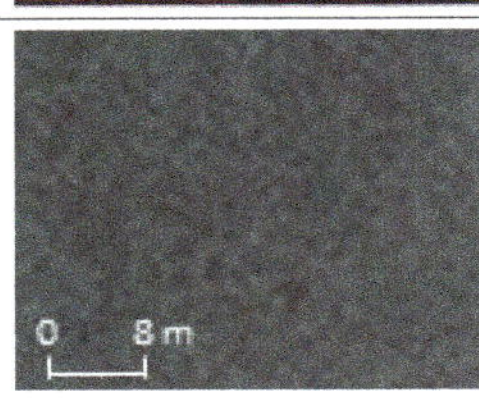	*Aegiceras corniculatum*, with a height of about 3 m.

Table 2 presents photos and remote sensing images of sampling points from different provinces. From the remote sensing imagery, it can be seen that in the early growth stages, the trees are small and sparse. In the mid-growth stages, the trees grow larger and denser, with higher vegetation indices, making the reflectance features more distinct in the remote sensing images. Due to differences in climate, soil types, and environmental conditions across provinces, the growth characteristics of *K. obovata* also vary. In tropical regions such as Hainan and Guangdong, *K. obovata* grows well, with larger canopy size and denser trees. As one moves north, temperatures drop, and *K. obovata* becomes shorter, especially in Zhejiang, where the species is mainly artificially introduced, resulting in smaller and sparser plants.

3.2. Sample Point Expansion Based on Field Survey and Google Earth Data

Visual interpretation is a method of remote-sensing image analysis that involves the direct observation of remote-sensing images to identify and interpret features and land cover classes based on experience and training. We utilized the similarity of the visual interpretation of *K. obovata* communities in Google Earth imagery from 2018 to identify *K. obovata* a and other mangrove species surrounding the plots, thereby expanding the sample points. The remote sensing identification markers and features are presented in Table 3.

We collected the pure species sample set of *K. obovata* and pure non-*K. obovata* around the existing plots based on experience and field surveys. A total of 6705 sample points were selected for classification, with 1671 training samples for *K. obovata* and 5034 training samples for other mangrove species (Figure 4). For validation sample points, we used ArcGIS to randomly select 350 *K. obovata* samples and 1011 other mangrove samples based on provincial area distribution. The points were then imported into Google Earth for visual interpretation, with field verification for uncertain points.

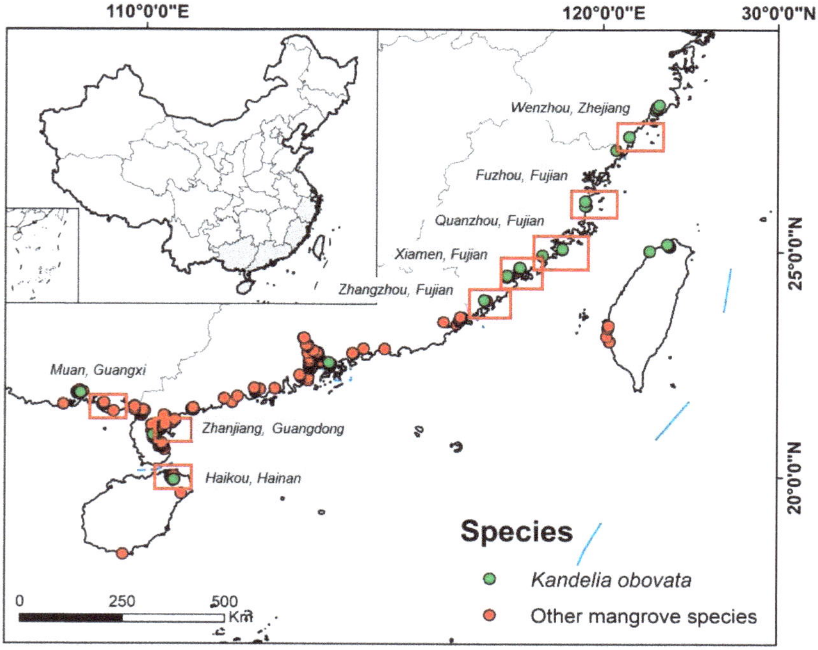

Figure 4. Sample points distribution map. The red boxes in the figure correspond to the plots in Table 2.

Table 3. Remote sensing identifiers and features.

Species	Images	Features
Kandelia obovata		The color is green, with a relatively uniform tone and a coarse texture, characterized by a flaky appearance.
Aegiceras corniculatum		The tone is complex, with a relatively smooth texture and a clustered appearance, accompanied by shaded edges.
Avicennia marina	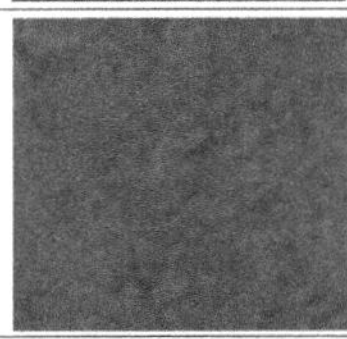	The color characteristics are prominent, displaying a light green hue with a relatively uniform tone and a coarse texture.
Sonneratia apetala	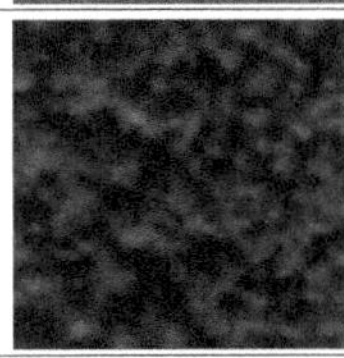	The color is dark green, with a complex tone and pronounced texture characteristics. It has a rough appearance and is distributed in a banded pattern, featuring shadows.
Rhizophora stylosa	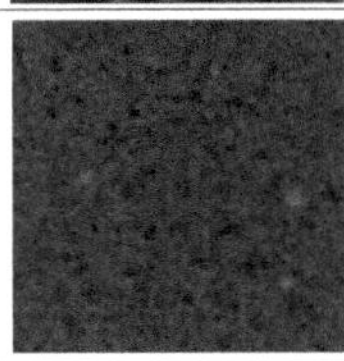	The color is gray-green, with a relatively uniform tone and a smooth, delicate texture.

3.3. Selection of Remote Sensing Features for K. obovata

3.3.1. Extraction of Temporal Variation Characteristics of Spectra

Vegetation indices enhance the characteristics of specific species through mathematical calculations involving two or more spectral bands, enabling distinction from other species. In this study, we utilized three classic indices: the Normalized Difference Vegetation Index (NDVI), the Normalized Difference Water Index (NDWI), and the Enhanced Vegetation Index (EVI). Additionally, based on existing research on mangrove species classification, we selected three commonly used indices: the Green Normalized Difference Vegetation Index (GNDVI), the Ratio Vegetation Index (RVI), and the Atmospherically Resistant Vegetation Index (ARVI). The calculation formulas and references are presented in Table 4. We further selected five spectral-temporal variability metrics—10%; 25%; 50%; 75%; and 90% percentiles—by arranging the spectral parameters of images captured throughout the year in ascending order. These percentiles reflect the unique temporal variations of *K. obovata* and other mangrove species (Figure 5). The indices reveal that *K. obovata* exhibits the widest reflectance range across all six vegetation indices, whereas *A. corniculatum* and *A. marina* display the narrowest, highlighting significant differences. Notably, in NDWI, the maximum reflectance values of *K. obovata* and *S. apetala* differ significantly, offering a clear criterion for distinguishing between the two species. The percentile reflectance means of *K. obovata* are distinctly higher than those of *A. corniculatum* across the six graphs, indicating that the two species are easily separable. Additionally, the ARVI reflectance percentiles show clear separability between *K. obovata* and both *A. corniculatum* and *S. apetala*. *K. obovata* consistently exhibits the highest maximum values in B7, B8, and B8A, significantly differing from *S. apetala*, facilitating differentiation. Conversely, *S. apetala* shows the highest maximum values in B2, B3, B4, B11, and B12, surpassing those of *K. obovata*, indicating a clear distinction between the two species.

Table 4. Calculation formula for spectral index.

Index	Calculation Formula	References
NDVI	(NIR − RED)/(NIR + RED)	[24]
NDWI	(GREEN − NIR)/(GREEN + NIR)	[25]
EVI	2.5 × (NIR − RED)/(NIR + 6.0 × RED − 7.5 × BLUE + 1)	[24]
GNDVI	(NIR − GREEN)/(NIR + GREEN)	[26]
RVI	NIR/RED	[22]
VARI	(GREEN − RED)/(GREEN + RED − BLUE)	[27]

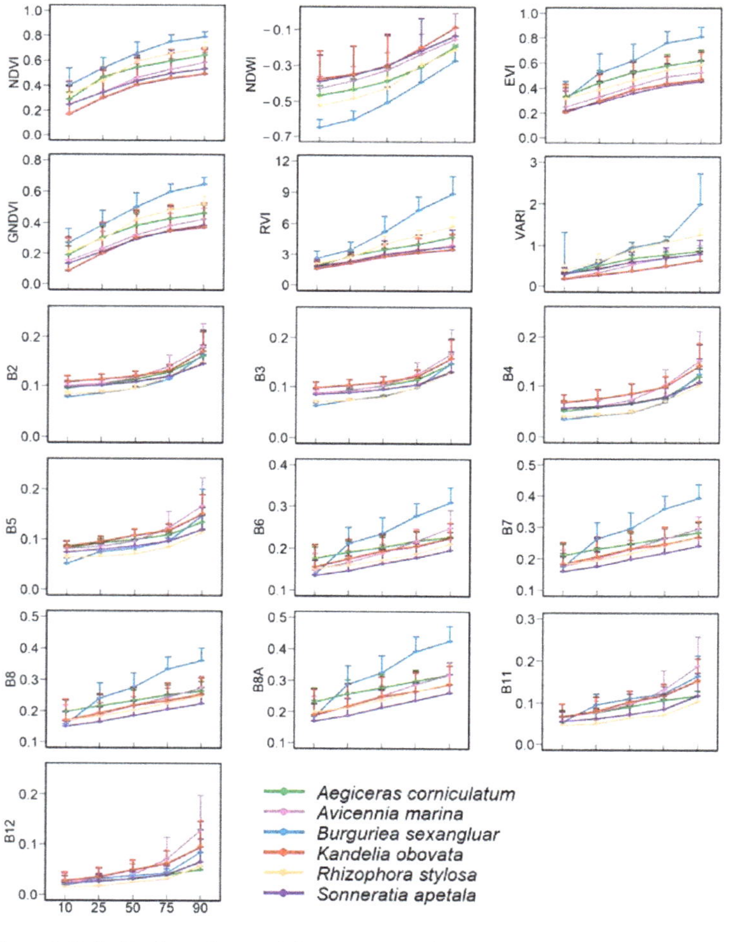

Figure 5. Quintile plots of spectral characteristics of *K. obovata* community with other species. The error bars in the figure represent the standard deviation of each species at different percentiles.

3.3.2. Feature Extraction Based on Texture Information

In remote sensing image classification, texture information is one of the essential features that aids in distinguishing between different land covers or land cover categories. The texture information in radar data encompasses spatial distribution characteristics, grayscale distribution, frequency features, and directional attributes. By extracting and analyzing texture information, remote sensing image classification algorithms can more effectively differentiate between various land cover types. A commonly used method for texture feature extraction is the Gray Level Co-occurrence Matrix (GLCM). GLCM reflects the spatial distribution patterns of gray levels in an image, providing insights into the texture structure. In this study, the statistical features used to extract texture information include contrast, entropy, correlation, standard deviation, and energy. Their calculation formulas and specific meanings are provided in Table 5.

Table 5. Texture feature information for grayscale covariance matrices.

Feature	Calculation Formula	Connotation
Contrast	$\sum_{i,j=1}^{Ng} (i-j)^2 GLCM(i,j)$	Whether the visual effect of the image is clear or not.
Entropy	$\sum_{i,j=1}^{Ng} GLCM(i,j) \times (-\ln(CLCM(i,j)))$	It is a measure of the information content contained within an image.
Correlation	$\sum_{i,j=1}^{Ng} \frac{\{i \times j\} \times CLCM(i,j) - \{\mu_x \times \mu_y\}}{\sigma_x \times \sigma_y}$	It reflects the local grayscale correlation—the larger the value; the stronger the correlation.
Standard deviation	$\sqrt{\sum_{i,j=1}^{Ng} (GLCM(i,j) - \mu)}$	It describes the dispersion of GLCM values around the mean.
Energies	$\sum_{i,j=1}^{Ng} (GLCM(i,j))^2$	It describes the uniformity of pixel intensity distribution within the image.

3.3.3. Pre-Experiment of Band Removal Based on Random Forest

In this study, the initial input consisted of 98 classification features. These included the percentile-based features of conventional optical image bands (B2, B3, B4, B5, B6, B7, B8, B8A, B11, B12), totaling 50 feature values. Additionally, six vegetation indices (NDVI, NDWI, EVI, GNDVI, RVI, AVRI) were included, with their corresponding percentile features contributing 30 feature values. For radar data, the percentiles of bands VV and VH, along with their contrast, entropy, correlation, standard deviation, and energy, provided a total of 15 feature values. Moreover, slope, aspect, and elevation derived from DEM data contributed 3 feature values, resulting in a total of 98 input features. After conducting four preliminary experiments, the feature importance rankings were aggregated. The importance proportion for each feature was calculated from high to low, and only those contributing 90% or more of the total importance were retained. The remaining features were pairwise matched to identify and eliminate those with minimal importance. Ultimately, 78 features were retained.

3.4. Extraction of *K. obovata*

Following the feature value screening in Section 3.3.3, 20 relatively unimportant bands were eliminated, leaving 78 feature values to be input into the random forest classifier for classification. Due to the inherent randomness of the classification algorithm, we ran the model 10 times with altered random seeds to obtain ten rankings of feature importance. The average of these rankings provided the final importance ordering for the 78 feature values. Then, select the top 50, 20, 15, 10, and 5 features for classification in sequence to determine the optimal number of bands for classification.

3.5. Accuracy Assessment

We employed ArcGIS to select random sample points within *K. obovata* patches using a stratified sampling method. Specifically, random sample points were chosen based on the area of *K. obovata* in each province. These sample points were then imported into Google Earth for visual interpretation. To assess the progress of the classification, a confusion matrix was utilized.

4. Results

4.1. Accuracy Evaluation Results

Following the method in Section 3.4, different numbers of bands were selected and input into the random forest classifier for classification, obtaining the confusion matrix and classification accuracy for each iteration. The results are shown in Table 6. The top 50 feature bands achieved the highest producer's accuracy for *K. obovata* at 94.6%, followed by the top 20 bands with an accuracy of 92.3%. The top 15 bands achieved 81.7% accuracy, while the top 10 and top 5 bands each yielded producer's accuracy below 80%. This indicates that the top 50 feature bands provide a producer's accuracy above 94% and contain critical information, making them suitable as feature bands for identifying *K. obovata*.

Table 6. Comparison of confusion matrices after classification of five, ten, fifteen, twenty, and fifty features.

	Attribute Data	Validation Data		User's Accuracy
		K. obovata	Non-*K. obovata*	
Top five	*K. obovata*	278	39	87.7
	Non-*K. obovata*	72	972	93.1
	Producer's accuracy	79.4	96.1	
Top ten	*K. obovata*	271	19	93.4
	Non-*K. obovata*	79	992	92.6
	Producer's accuracy	77.4	98.1	
Top fifteen	*K. obovata*	286	17	94.4
	Non-*K. obovata*	64	994	93.9
	Producer's accuracy	81.7	98.3	
Top twenty	*K. obovata*	323	8	97.6
	Non-*K. obovata*	27	1003	97.4
	Producer's accuracy	92.3	99.2	
Top fifty	*K. obovata*	331	8	97.6
	Non-*K. obovata*	19	1003	98.1
	Producer's accuracy	94.6	99.2	

4.2. Spatial Distribution of *K. obovata* in China

Figure 6 illustrates the distribution of *K. obovata* across China. The total area of *K. obovata* mangroves in China is 4825.97 ha. Specifically, the area of *K. obovata* communities is as follows: 47.34 ha in Zhejiang, 849.23 ha in Fujian, 1912.43 ha in Guangdong, 1367.58 ha in Guangxi, 501.12 ha in Hainan, 123.23 ha in Taiwan, 0.20 ha in Macau, and 24.85 ha in Hong Kong. Table 7 presents the proportion of *K. obovata* within the total mangrove area for each administrative region. It is evident that *K. obovata* constitutes the largest proportion of mangroves in Zhejiang, followed by Fujian, with the lowest proportion observed in Macau. The contrast between the proportion of *K. obovata* and the total mangrove area in Fujian underscores its status as the predominant mangrove species in the province.

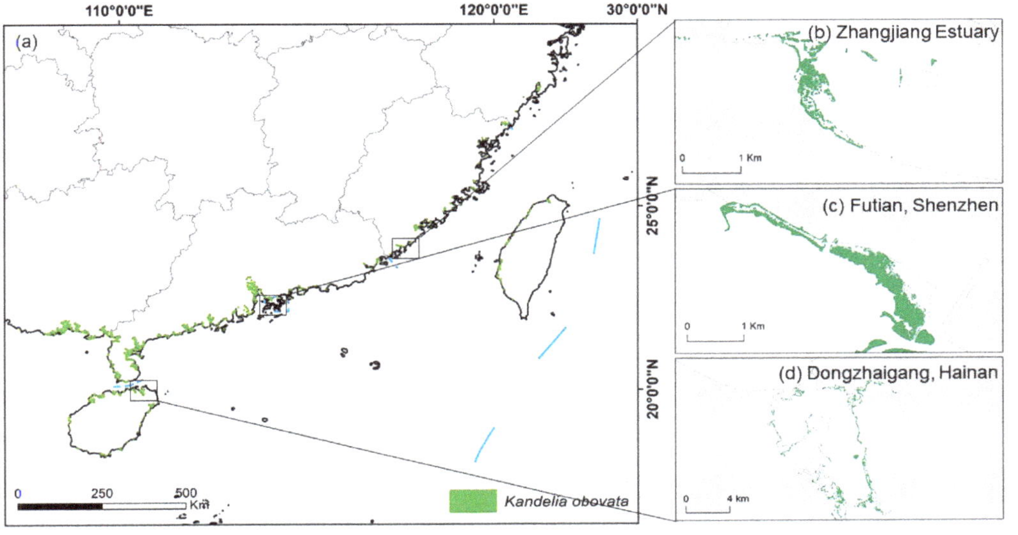

Figure 6. Distribution of *K. obovata* in China, 2018. (**a**) shows the distribution of *K. obovata* in China. (**b**) presents the extraction results in the Zhangjiang Estuary, Fujian province. (**c**) displays the extraction results in Futian, Shenzhen. (**d**) shows the extraction results in Dongzhaigang, Hainan province. In subplots (**b–d**), the yellow lines represent the distribution range of *K. obovata*.

Table 7. Distribution area of *K. obovata* community by province in China.

Province	Mangrove Area (ha) [22]	Proportion of Mangrove Forest Area (%)	Area of *K. obovata* (ha)	Number of Patches	Proportion of *K. obovata* Area (%)
Zhejiang	48.67	0.19	47.34	144	97.27
Fujian	1019.44	3.97	849.23	1573	83.30
Guangdong	10,323.51	40.19	1912.43	20,578	18.52
Guangxi	8451.47	32.91	1367.58	10,636	16.18
Hainan	4676.70	18.21	501.11	7025	10.72
Taiwan	605.06	2.36	123.23	1756	20.37
Macau	20.00	0.08	0.20	2	1.00
Hong Kong	539.03	2.10	24.85	601	4.61
Total	25,683.88	100	4825.97	42,315	

4.3. Comparison of the Research Results with Previous Studies

The latest research by Zhao et al. [18] first identified *K. obovata* across China. Figure 7 collectively demonstrates the comparison between Zhao et al.'s study and our work in mapping mangrove distributions across Jiulongjiang Estuary Reserve, Quanzhou Bay, and Wenzhou. In Jiulongjiang Estuary Reserve (Figure 7(a-1)~(a-5)), our method provided more accurate identification of *K. obovata*-dominated mangroves along riverbanks compared to Zhao et al.'s underestimation. Similarly, in Quanzhou Bay (Figure 7(b-1)~(b-5)), we detected smaller, early-growth-stage *K. obovata* more effectively. In Wenzhou (Figure 7(c-1)~(c-5)), our results better captured the distribution of artificially planted *K. obovata*, which Zhao et al. significantly underestimated. These findings demonstrate the superior accuracy of our method in mapping mangrove distributions across varied environments.

Figure 8 compares our results with previous studies by Li et al. [21] in Zhangjiangkou and Wang et al. [24] in Dongzhai Port. Li's classification (Figure 8(a-1)) achieved 84% accuracy for

K. obovata (dark green), while our results (Figure 8(a-2)) showed strong consistency, accurately identifying mixed communities of *K. obovata* with *A. corniculatum* and *A. marina*. Wang's classification (Figure 8(b-1)) achieved 78.6% accuracy but overestimated *K. obovata* (purple) coverage. Our conservative approach (Figure 8(b-2)) resulted in slightly lower coverage but demonstrated the effectiveness of our spectral band selection for *K. obovata* identification.

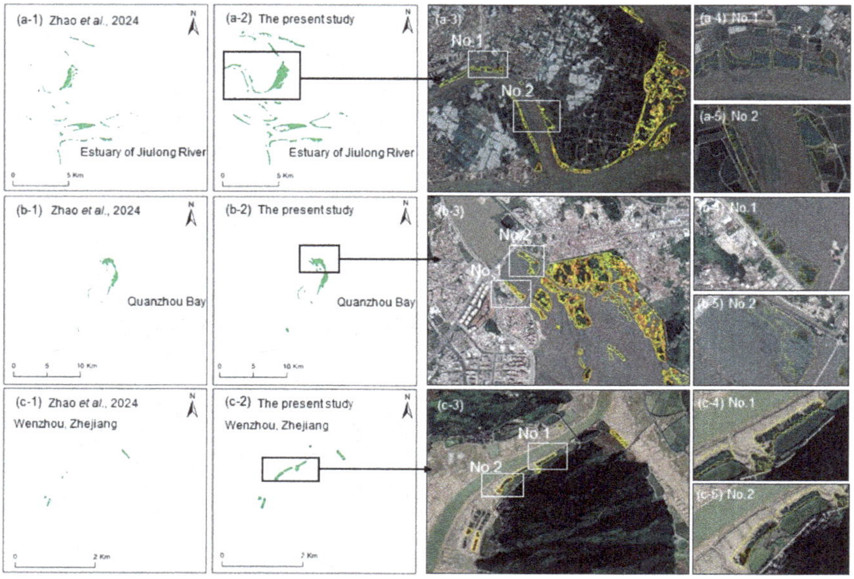

Figure 7. Comparison of *K. obovata* results. (**a-1**), (**b-1**), and (**c-1**) show the extraction results by Zhao et al. [18] in Estuary of Jiulong River, Quanzhou Bay, and Wenzhou, Zhejiang, respectively. (**a-2**), (**b-2**), and (**c-2**) show the extraction results of this study. (**a-3**)~(**a-5**), (**b-3**)~(**b-5**), and (**c-3**)~(**c-5**) represent the actual remote sensing features of the corresponding regions. The yellow lines indicate the *K. obovata* boundaries extracted by us, while the red lines represent the boundaries extracted by Zhao et al.

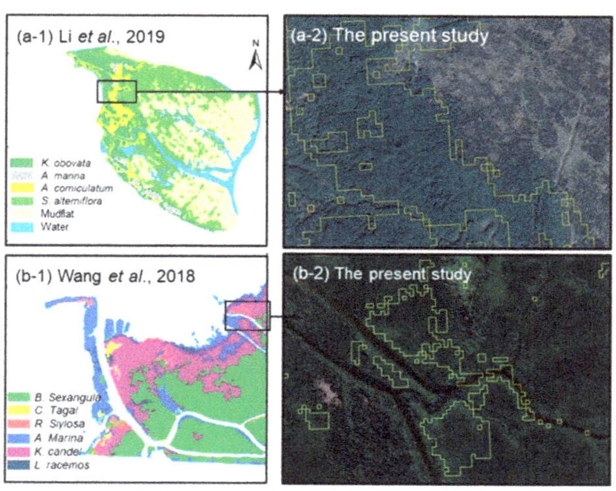

Figure 8. The performance of research results at a small scale. (**a-1**) shows the extraction results by Li et al. [21] at Zhangjiangkou, and (**a-2**) shows the present results. (**b-1**) shows the results by Wang et al. [24] at Sanjiang, Hainan, and (**b-2**) shows results of the present study.

5. Discussion

5.1. Comparison of Feature Band Selection Results

In Section 4.1, we performed random forest classification using combinations of different numbers of bands. The results show that the combination of 50 feature bands achieved the highest classification accuracy. To investigate which optimal bands among these 50 have a greater influence on the extraction of *K. obovata*, we ranked their importance, as shown in Table 8. The classification importance of B2_p90 (the 90th percentile of the blue wavelength band) ranks first. Among the top five features, three are blue wavelength band percentile metrics, while the top ten features are all spectral band percentile composite metrics, highlighting the clear advantage of spectral-temporal variability features in classification. Within the top twenty bands, six features include red-edge band percentile composites, and four include blue band percentile composites, indicating that blue and red-edge band percentiles are key feature combinations for achieving high-precision extraction of *K. obovata*. The blue wavelength band, with a range of 450–520 nm, effectively captures boundary details at intersections of ground objects, making it highly valuable in remote sensing extraction. Based on plant reflectance, the 350–500 nm band strongly correlates with the absorption spectra of carotenoids and chlorophyll, and spectral reflectance in this range is closely related to plant characteristics. The 430–450 nm band is commonly used for mapping water bodies and forests, as well as distinguishing soil from evergreen and deciduous vegetation. This study confirms that percentile metrics in the blue band offer an advantage for identifying fragmented and scattered patches of *K. obovata*.

Table 8. Importance ranking of the top 50 features.

Rank	Band	Values	Rank	Values	Values
1	B2_p90	5.9695	26	VV_ent	3.9414
2	B2_p75	5.9060	27	B8_p10	3.9279
3	B3_p25	5.5084	28	GNDVI_p10	3.9089
4	B12_p10	5.0019	29	B6_p10	3.8983
5	B2_p10	4.9815	30	B2_p25	3.8824
6	B3_p75	4.9332	31	ARVI_p25	3.8113
7	B4_p90	4.9205	32	B11_p75	3.7914
8	B3_p10	4.8683	33	slope	3.7810
9	B3_p90	4.6251	34	B7_p50	3.6994
10	B5_p75	4.5327	35	B4_p75	3.6917
11	B4_p10	4.5062	36	B8A_p90	3.6595
12	B5_p10	4.4497	37	B12_p75	3.6229
13	B5_p50	4.4473	38	B3_p50	3.6090
14	B2_p50	4.4364	39	ARVI_p50	3.6071
15	B12_p50	4.3657	40	GNDVI_p75	3.5584
16	B5_p90	4.2893	41	VV_p75	3.5025
17	B5_p25	4.2835	42	B4_p50	3.4975
18	B6_p25	4.2221	43	B6_p75	3.4935
19	VV_contrast	4.2136	44	EVI_p50	3.4899
20	B11_p10	4.2078	45	B12_p90	3.4687
21	B11_p50	4.1351	46	NDVI_p10	3.3940
22	B6_p50	4.1141	47	B7_p90	3.3925
23	B4_p25	4.0872	48	B7_p10	3.3481
24	ARVI_p75	4.0274	49	B6_p90	3.3168
25	VH_contrast	3.9914	50	RVI_p10	3.2959

In comparison, B7 and B8A bands are less effective for *K. obovata* extraction, while spectral-temporal variability metrics in the red-edge bands show greater sensitivity than the red-edge bands alone. Research by Misra et al. [28] demonstrates that Sentinel-2's red-edge bands enhance wetland vegetation mapping accuracy, as the optimal bandwidth and superior spatial resolution of Sentinel-2 bands 5 and 6 provide better red-edge estimates than other sensors. Unlike indices such as NDVI and EVI, which estimate chlorophyll, red-edge bands are more resistant to delays caused by canopy structure effects [29]. Additionally, Pinheiro et al. [30] found that red-edge bands are particularly important for distinguishing dense grasslands from broad-leaved shrubs during the rainy season. Beyond optical data, this study finds that the top fifty feature bands, compared to the top twenty, include additional radar backscatter percentile metrics and radar texture information. Integrating radar and optical data further enhances classification accuracy.

5.2. The Strengths and Limitations of This Study

Compared to previous studies, this research extracts time-variability indicators from spectral and backscatter data, leveraging the unique temporal profiles of mangroves and their species. A key advantage of this approach is its global applicability without the need for location-specific adjustments. In terms of data usage, the study combines optical and radar data. In cases where cloud cover obscures spectral imagery and information is lacking, Sentinel-1 images, which can penetrate clouds, provide a valuable supplement. This allows for the acquisition of high-resolution surface details, as well as broader surface coverage and more comprehensive information, ensuring the integrity of experimental patterns. Furthermore, the research establishes mangrove sample plots distributed evenly across coastal provinces. Pure mangrove species plots, obtained through community surveys, serve as accurate data sources. By visually interpreting surrounding mangrove species in remote sensing images, a reliable mangrove species database is constructed.

In the context of remote sensing vegetation extraction, seasonality is widely recognized as a critical factor influencing spectral characteristics due to the physiological changes plants undergo across different seasons [31–33]. However, *K. obovata* exhibits exceptional cold tolerance in China, maintaining its green foliage throughout the year without a distinct leaf-shedding period, resulting in relatively stable spectral characteristics [34]. Consequently, this study did not incorporate seasonal variation as a factor in the analysis. Future research applying these methods to other species should carefully evaluate the necessity of including seasonality based on the specific phenological characteristics of the target species.

Currently, the extent of co-existence between *K. obovata* and other species in China remains poorly understood. Our field surveys revealed distinct spatial patterns: *K. obovata* communities north of the Jiulong River are predominantly pure stands, while those south of the Jiulong River demonstrate varying degrees of symbiosis with *A. corniculatum*, *A. marina*, and *Acanthus ilicifolius* L. These symbiotic relationships manifest differently across spatial scales—primarily horizontal for *A. corniculatum* and *A. marina*; and vertical for *Acanthus ilicifolius*. Given *K. obovata*'s extensive canopy coverage and evergreen nature, the vertical-scale symbiosis with *Acanthus ilicifolius* has minimal impact on extraction accuracy due to year-round canopy coverage. However, at the horizontal scale, extraction accuracy is influenced by both the spatial resolution of remote sensing data and the structural complexity of mangrove communities.

The 10 m spatial resolution of Sentinel satellite imagery used in this study presents challenges in areas with mixed species composition, potentially leading to mixed pixels. The classification accuracy of these mixed pixels depends on the proportional composition of species within each pixel. When other species constitute a minor proportion within a 10 m × 10 m pixel, their spectral contribution is typically insufficient to alter the dominant species classification. Conversely, when the proportion of other species exceeds a certain threshold, the pixel may be classified as the alternative species, leading to a potential underestimation of the target species' extent.

It is crucial to note that the training samples in this study were derived from pure stands, meaning the resulting *K. obovata* distribution represents pure communities rather than mixed stands. This methodological approach likely results in a slightly conservative estimation of the actual *K. obovata* distribution. Future research directions should focus on incorporating forest structure parameters into classification algorithms, calibrating results using field-based mixed forest proportions, and exploring higher-resolution satellite imagery to enhance classification accuracy and better capture the complexity of mangrove community composition.

6. Conclusions

This study, based on extensive field surveys and the accumulation of fixed sample plots of mangroves, established a field sampling database covering the communities of *K. obovata* and other mangrove species across various provinces in China. Building on this, the study utilized dense time series remote sensing data from Sentinel-1 and Sentinel-2, along with the Google Earth Engine, to identify the optimal spectral bands for extracting *K. obovata*. The main conclusions are as follows: (1) The spectral-temporal variability characteristics of the blue and red-edge bands play a crucial role in the identification of *K. obovata*. The 90th percentile metric of the blue wavelength band ranks first in importance, while the 75th percentile metric of the blue wavelength band ranks second. (2) This method of remote sensing extraction using spectral-temporal variability metrics with time-series optical and radar remote sensing data offers significant advantages in identifying the *K. obovata* species, achieving a producer's accuracy of up to 94.6% and enhancing classification effectiveness. (3) In 2018, the total area of *K. obovata* in China's mangroves was 4825.97 ha. The area of *K. obovata* in various provinces/regions ranges from 0.2 to 1912.43 ha. The area of *K. obovata* is the largest in Guangdong Province, while Macau has the smallest area. Compared to previous studies, our findings are more aligned with field survey results. The classification method employed in this study is not only applicable for the extraction of *K. obovata* but also suitable for extracting various species, provided that sufficient ground samples are available. These results provide valuable data support for mangrove ecosystem restoration and blue carbon assessments.

Author Contributions: Conceptualization, L.C. and L.H.; methodology, L.H.; validation, C.L.; formal analysis, J.Z.; investigation, J.Z.; resources, L.C.; writing—original draft preparation, J.Z.; writing—review and editing, C.L., L.H. and L.C.; visualization, C.L.; project administration, L.C.; funding acquisition, L.C. All authors have read and agreed to the published version of the manuscript.

Funding: This research was supported by the National Key Research and Development (R&D) Program of China (2022YFC3105401), the National Natural Science Foundation of China (U22A20584, 42076176), the Monitoring and Assessment of Mudflat Elevation and Blue Carbon in Futian Mangrove International Important Wetland, and the Visiting Scientist Program of the Key Laboratory for Coastal and Wetland Ecosystems, Ministry of Education.

Data Availability Statement: Dataset available on request from the authors.

Acknowledgments: We are grateful to Xingzhu Lin and Shanghong Bao for providing Taiwan sample point data, and thanks to Jialin Zhang, ChangXin Liu, Liangxia Feng, Mei Li, and Xiaoxuan Gu for their assistance in collecting sample points. We are grateful to Deling Tang for his help in data processing.

Conflicts of Interest: The authors declare no conflicts of interest.

References

1. Field, C.D. Rehabilitation of mangrove ecosystems: An overview. *Mar. Pollut. Bull.* **1999**, *37*, 383–392. [CrossRef]
2. Sheue, C.-R.; Liu, H.-Y.; Yong, J.W.H. Kandelia obovata (Rhizophoraceae), a new mangrove species from Eastern Asia. *Taxon* **2003**, *52*, 287–294. [CrossRef]
3. Lin, P. *Chinese Mangrove Ecosystem*; Science Press: Beijing, China, 1999.
4. Chen, L.; Wang, W.; Zhang, Y.; Lin, G. Recent progresses in mangrove conservation, restoration and research in China. *J. Plant Ecol.* **2009**, *2*, 45–54. [CrossRef]
5. Wang, W.; Fu, H.; Lee, S.Y.; Fan, H.; Wang, M. Can strict protection stop the decline of mangrove ecosystems in China? From rapid destruction to rampant degradation. *Forests* **2020**, *11*, 55. [CrossRef]
6. Chen, J.; Zhai, G.; Chen, G.; Wu, J.; Liu, J.; Xue, Z.; Ye, Y. Differences in ecosystem organic carbon stocks due to species selection and site elevation of restored mangrove forests. *Catena* **2023**, *226*, 107089. [CrossRef]
7. Chen, B.; Xiao, X.; Li, X.; Pan, L.; Doughty, R.; Ma, J.; Dong, J.; Qin, Y.; Zhao, B.; Wu, Z.; et al. A mangrove forest map of China in 2015: Analysis of time series Landsat 7/8 and Sentinel-1A imagery in Google Earth Engine cloud computing platform. *ISPRS J. Photogramm. Remote Sens.* **2017**, *131*, 104–120. [CrossRef]
8. Jia, M.; Wang, Z.; Mao, D.; Ren, C.; Song, K.; Zhao, C.; Wang, C.; Xiao, X.; Wang, Y. Mapping global distribution of mangrove forests at 10-m resolution. *Sci. Bull.* **2023**, *68*, 1306–1316. [CrossRef]
9. Lu, Y.; Wang, L. How to automate timely large-scale mangrove mapping with remote sensing. *Remote Sens. Environ.* **2021**, *264*, 112584. [CrossRef]
10. Mitchell, A.; Lucas, R.M. Integration of aerial photography, hyperspectral and SAR data for mangrove characterization. In Proceedings of the IEEE 2001 International Geoscience and Remote Sensing Symposium, Sydney, Australia, 9–13 July 2001; Volume 5, pp. 2193–2195. [CrossRef]
11. Huang, X.; Zhang, L.; Wang, L. Evaluation of morphological texture features for mangrove forest mapping and species discrimination using multispectral IKONOS imagery. *IEEE Geosci. Remote Sens. Lett.* **2009**, *6*, 393–397. [CrossRef]
12. Jia, M.; Zhang, Y.; Wang, Z.; Song, K.; Ren, C. Mapping the distribution of mangrove species in the core zone of Mai Po sarshes nature reserve, Hong Kong, using hyperspectral data and high-resolution data. *Int. J. Appl. Earth Obs. Geoinf.* **2014**, *33*, 226–231. [CrossRef]
13. Viennois, G.; Proisy, C.; Féret, J.-B.; Prosperi, J.; Sidik, F.; Suhardjono; Rahmania, R.; Longépé, N.; Germain, O.; Gaspar, P. Multitemporal analysis of high-spatial-resolution optical satellite imagery for mangrove species mapping in Bali, Indonesia. *IEEE J. Sel. Top. Appl. Earth Obs. Remote Sens.* **2016**, *9*, 3680–3686. [CrossRef]
14. Li, H.; Han, Y.; Chen, J. Combination of Google Earth imagery and Sentinel-2 data for mangrove species mapping. *J. Appl. Remote Sens.* **2020**, *14*, 010501. [CrossRef]
15. Ghorbanian, A.; Ahmadi, S.A.; Amani, M.; Mohammadzadeh, A.; Jamali, S. Application of artificial neural networks for mangrove mapping using multi-temporal and multi source remote sensing imagery. *Water* **2022**, *14*, 244. [CrossRef]
16. Zimudzi, E.; Sanders, I.; Rollings, N.; Omlin, C.W. Remote sensing of mangroves using unmanned aerial vehicles: Current state and future directions. *J. Spat. Sci.* **2021**, *66*, 195–212. [CrossRef]
17. Lu, Y.; Wang, L. The current status, potential and challenges of remote sensing for large-scale mangrove studies. *Int. J. Remote Sens.* **2022**, *43*, 6824–6855. [CrossRef]
18. Zhao, C.; Jia, M.; Zhang, R.; Wang, Z.; Mao, D.; Zhong, C.; Guo, X. Distribution of mangrove species Kandelia obovata in China using time-series Sentinel-2 imagery for sustainable mangrove management. *J. Remote Sens.* **2024**, *4*, 0143. [CrossRef]
19. Heenkenda, M.K.; Joyce, K.E.; Maier, S.W.; Bartolo, R. Mangrove species identification: Comparing WorldView-2 with aerial photographs. *Remote Sens.* **2014**, *6*, 6064. [CrossRef]
20. Held, A.; Ticehurst, C.; Lymburner, L.; Williams, N. High resolution mapping of tropical mangrove ecosystems using hyperspectral and radar remote sensing. *Int. J. Remote Sens.* **2003**, *24*, 2739–2759. [CrossRef]
21. Li, H.; Jia, M.; Zhang, R.; Ren, Y.; Wen, X. Incorporating the plant phenological trajectory into mangrove species mapping with dense time series sentinel-2 imagery and the Google Earth Engine platform. *Remote Sens.* **2019**, *11*, 2479. [CrossRef]
22. Fu, B.; He, X.; Yao, H.; Liang, Y.; Deng, T.; He, H.; Fan, D.; Lan, G.; He, W. Comparison of RFE-DL and stacking ensemble learning algorithms for classifying mangrove species on UAV multispectral images. *Int. J. Appl. Earth Obs. Geoinf.* **2022**, *112*, 102890. [CrossRef]
23. Zhang, T.; Hu, S.; He, Y.; You, S.; Yang, X.; Gan, Y.; Liu, A. A Fine-Scale Mangrove Map of China Derived from 2-Meter Resolution Satellite Observations and Field Data. *ISPRS Int. J. Geo-Inf.* **2021**, *10*, 92. [CrossRef]
24. Wang, D.; Wan, B.; Qiu, P.; Su, Y.; Guo, Q.; Wang, R.; Sun, F.; Wu, X. Evaluating the performance of Sentinel-2, Landsat 8 and Pléiades-1 in mapping mangrove extent and species. *Remote Sens.* **2018**, *10*, 1468. [CrossRef]
25. Miraki, M.; Sohrabi, H.; Immitzer, M. Tree species mapping in mangrove ecosystems using UAV-RGB imagery and object-based image classification. *J. Indian Soc. Remote Sens.* **2023**, *51*, 2095–2103. [CrossRef]

26. Peng, L.; Liu, K.; Cao, J.; Zhu, Y.; Li, F.; Liu, L. Combining GF-2 and RapidEye satellite data for mapping mangrove species using ensemble machine-learning methods. *Int. J. Remote Sens.* **2020**, *41*, 813–838. [CrossRef]
27. Su, Y.; Geng, J.; Ma, X.; Wang, X.; Wang, X. Mangrove species classification based on multiple vegetation index extraction and joint sparse representation. *Mar. Environ. Sci.* **2017**, *36*, 114–120. [CrossRef]
28. Misra, G.; Cawkwell, F.; Wingler, A. Status of phenological research using Sentinel-2 data: A Review. *Remote Sens.* **2020**, *12*, 2760. [CrossRef]
29. Peng, Y.; Nguy-Robertson, A.; Arkebauer, T.; Gitelson, A.A. Assessment of canopy chlorophyll content retrieval in maize and soybean: Implications of hysteresis on the development of generic algorithms. *Remote Sens.* **2017**, *9*, 226. [CrossRef]
30. Pinheiro, H.S.K.; Barbosa, T.P.R.; Antunes, M.A.H.; Carvalho, D.C.d.; Nummer, A.R.; Carvalho Junior, W.d.; Chagas, C.d.S.; Fernandes-Filho, E.I.; Pereira, M.G. Assessment of phytoecological variability by Red-Edge spectral indices and soil-landscape relationships. *Remote Sens.* **2019**, *11*, 2448. [CrossRef]
31. Caparros-Santiago, J.A.; Rodriguez-Galiano, V.; Dash, J. Land surface phenology as an indicator of global terrestrial ecosystem dynamics: A systematic review. *ISPRS J. Photogramm. Remote Sens.* **2021**, *171*, 330–347. [CrossRef]
32. Joiner, J.; Yoshida, Y.; Vasilkov, A.P.; Schaefer, K.; Jung, M.; Guanter, L.; Zhang, Y.; Garrity, S.; Middleton, E.M.; Huemmrich, K.F.; et al. The seasonal cycle of satellite chlorophyll fluorescence observations and its relationship to vegetation phenology and ecosystem-atmosphere carbon exchange. *Remote Sens. Environ.* **2014**, *152*, 375–391. [CrossRef]
33. Zhu, X.; Liu, D. Improving forest aboveground biomass estimation using seasonal Landsat NDVI time-series. *ISPRS J. Photogramm. Remote Sens.* **2015**, *102*, 222–231. [CrossRef]
34. Chen, L.; Wang, W.; Li, Q.Q.; Zhang, Y.; Yang, S.; Osland, M.J.; Huang, J.; Peng, C. Mangrove species' responses to winter air temperature extremes in China. *Ecosphere* **2017**, *8*, e01865. [CrossRef]

remote sensing

MDPI

Article

The Change Detection of Mangrove Forests Using Deep Learning with Medium-Resolution Satellite Imagery: A Case Study of Wunbaik Mangrove Forest in Myanmar

Kyaw Soe Win [1,2,*] and Jun Sasaki [1]

1 Department of Socio-Cultural Environmental Studies, Graduate School of Frontier Sciences, The University of Tokyo, Kashiwa 277-8561, Japan; jsasaki@k.u-tokyo.ac.jp
2 Environmental Conservation Department, Ministry of Natural Resources and Environmental Conservation, Naypyitaw 15011, Myanmar
* Correspondence: kyawsoewin.ecd@gmail.com or 1282270769@edu.k.u-tokyo.ac.jp

Abstract: This paper presents the development of a U-Net model using four basic optical bands and SRTM data to analyze changes in mangrove forests from 1990 to 2024, with an emphasis on the impact of restoration programs. The model, which employed supervised learning for binary classification by fusing multi-temporal Landsat 8 and Sentinel-2 imagery, achieved a superior accuracy of 99.73% for the 2020 image classification. It was applied to predict the long-term mangrove maps in Wunbaik Mangrove Forest (WMF) and to detect the changes at five-year intervals. The change detection results revealed significant changes in the mangrove forests, with 29.3% deforestation, 5.75% reforestation, and −224.52 ha/yr of annual rate of changes over 34 years. The large areas of mangrove forests have increased since 2010, primarily due to naturally recovered and artificially planted mangroves. Approximately 30% of the increased mangroves from 2015 to 2024 were attributed to mangrove plantations implemented by the government. This study contributes to developing a deep learning model with multi-temporal and multi-source imagery for long-term mangrove monitoring by providing accurate performance and valuable information for effective conservation strategies and restoration programs.

Keywords: mangrove; Landsat; Sentinel-2; CNN; U-Net; change detection; restoration

1. Introduction

Mangroves are intertidal and salt-tolerant evergreen forests that grow in the tropical and subtropical regions of the world [1,2]. They are critical in providing valuable ecosystem services, blue carbon conservation, and nature-based solutions for climate change adaptation and mitigation [1–5]. Despite their invaluable roles, approximately 3.4% of global mangroves have disappeared over the past 24 years [6] due to both natural drivers, such as climate change-induced sea level rise and extreme weather events, and anthropogenic factors, such as increasing human populations, industrialization, aquaculture expansion, natural retraction, and the expansion of paddy fields [1,7–9]. Therefore, accurate monitoring of mangrove ecosystems is imperative for understanding their changing extent and sustainable management and conservation practices.

Mangrove forests are in the global attention to conserve and manage them continuously. Despite their invaluable matters, it is difficult to access the remote and wide forest with tidal and mudflat existence, which causes difficulties in collecting reliable data, needs many workers, and is costly to monitor regularly [7,10–12]. Currently, leveraging the advancements of machine learning algorithms and the accessibility of satellite imagery facilitates the comprehensive mapping of mangrove ecosystems, enabling effective long-term monitoring [10,13]. Among the freely available satellite imagery, Landsat imagery has been widely used to conduct the historical mapping of mangrove forests than Sentinel-2 [14], although Sentinel-2 provides a higher spatial resolution and accurate mapping of mangrove species and extent than Landsat imagery [15,16]. The global distribution of mangrove forests in 2020 was generated with Sentinel-2 at 10 m resolution to allow for better detection of mangrove forests [17]. Accordingly, applying multi-source data fusing Landsat and Sentinel-2 imagery is a current trend to provide consistent spatial resolution for historical and continuous monitoring in cloudy regions [18]. However, according to our knowledge, no study has focused on a classification task fusing Landsat and Sentinel-2 imagery for mangrove mapping and other forest classifications.

Satellite image classification has been attempted using the multispectral bands of optical images, the polarizations of radar images, vegetation indices, and elevation data. Many studies have documented the usefulness of the Normalized Difference Vegetation Index (NDVI) in mangrove studies because of its significance in detecting healthy mangroves [13,19–22]. Normalized Difference Water Index (NDWI) and Soil-adjusted Vegetation Index (SAVI) are widely used indices for the change detection of mangrove forests while some studies combined with the mangrove recognition indices [13]. Elevation data have been applied to improve the discrimination of mangrove forests stacked with the Canopy Height Model (CHM) and slope [21–23] or applied to mask the input images with the high elevation and coastal water areas before classification [24,25]. However, recent researchers have not utilized numerous input features when training deep learning models because of the ability of convolutional layers to extract the distinct features of the input images [26,27]. The model can be trained efficiently, as fewer input features need less computation power. Accordingly, this study aims to explore the development of a deep learning model with fewer input features, considering its applicability to multi-source satellite imagery.

Remote sensing has been utilized to produce mangrove maps combined with a series of machine learning classification techniques, including random forests [28–30], support vector machine (SVM) [31], decision trees [19], and iterative self-organizing data analysis (ISODATA) [32]. With the recent development of computational power, deep learning algorithms have been popular for researchers in remote sensing image analyses [33]. Deep learning models have proven to perform outstandingly compared to traditional models, with improved environmental monitoring with remote sensing data [34]. Relatively few studies have applied deep learning to mangrove extent mapping with different satellite imagery from high to medium resolution [26,35,36], focusing on attempts to classify mangroves with a short-temporal [23,37] and single-source satellite datasets with multiple input features [23,37]. Although the multi-source models fusing Sentinel-1 and Sentinel-2 were applied to mangrove mapping for short-term distribution [29,38], there remains a need for a multi-source deep learning model based on Landsat imagery to monitor the mangrove forests on a long-term scale.

Recently, Guo et al. (2021) [26] proposed a Capsules–Unet model for large-scale and long-term mangrove mapping from 1990 to 2015 using Landsat imagery to achieve the precise extraction of mangroves. They achieved a low accuracy of 85.7–88.7% when they utilized a large dataset available in 1990, 2000, 2010, and 2015 [26]. Their model had challenges with low accuracy, although it can be applied for large-scale and long-term mangrove mapping. Studies using Landsat 8 or Sentinel-2 imagery with a temporal dataset for small-scale areas achieved a higher accuracy of 97.64% and 97.48% when developing the deep learning models, namely MSNet and ME-Net [37,39]. According to the study by Ghorbanian et al. (2022), multi-source datasets of Sentinel-1 and Sentinel-2 improved the Artificial Neural Networks (ANN) model and provided accurate mangrove maps compared to single-source datasets of Sentinel-2 [38]. Combining Landsat 8 and Sentinel-2 could improve image quality by reducing the spatial resolution gaps between them and providing temporally short-term observations for environmental applications [40]. Therefore, the present study will explore the higher performance of a deep learning model using multi-temporal and multi-source datasets using Landsat and Sentinel-2 for long-term mangrove mapping.

Myanmar ranked as the top country with the highest annual rate of mangrove loss from 2000 to 2012 [41] despite decreasing the net loss of global mangrove areas [1]. The Wunbaik Mangrove Forest (WMF) in Myanmar is one of the largest remnant mangrove ecosystems endowed with ecologically important and endangered species, providing invaluable ecosystem services to local communities [42]. It experienced a large hectare of mangrove loss due to changing land uses to paddy fields and shrimp ponds in the past [43] and has been recorded as one of the hotspots of mangrove changes worldwide [6]. However, existing studies on long-term mangrove distribution in WMF need to be updated [42,43], and recent information is a gap. They highlight the need for continuous monitoring using accurate techniques to understand the historical changes and current conditions of WMF, especially following extensive restoration efforts by the government.

The study by Rahman et al. (2024) observed an improving trend of mangroves in Southeast Asia, including Myanmar, from 2015 to 2020 when they analyzed the dataset of Global Mangrove Watch (GMW), although most of the existing studies discussed a high rate of deforestation in Myanmar [20,21,32,43–45]. A recent study by Maung and Sasaki (2021) observed a slight decline in WMF due to the results of change detection from 2015 to 2020, while they detected mangrove gains in the plantation sites and approximately 50% of the naturally recovered mangroves in the abandoned sites [23]. Meanwhile, the government of Myanmar has implemented restoration programs, such as the Myanmar Rehabilitation and Restoration Programme (MRRP), across the country to reach the national climate change and biodiversity targets [46,47]. Therefore, there is an urgent need to study the contributions of restoration programs to the changes in mangrove forests. This study will investigate the long-term distribution of mangrove forests, emphasizing the drivers of mangrove losses and gains in WMF.

Considering the advantages of multi-source data, this study aimed to develop a multi-temporal and multi-source deep learning model fusing medium-resolution satellite imagery of Landsat 8 and Sentinel-2 for predicting the long-term distribution of mangrove forests and to investigate the patterns of change distribution in WMF by discussing the drivers behind mangrove losses and role of mangrove plantations on increasing mangrove coverage. The proposed deep learning model provided accurate and reliable performance in predicting long-term mangrove mapping in WMF. This approach helped to better understand the changing patterns of mangrove forests, filling a critical gap of outdated information on historical mangrove distribution in WMF and identifying the losses and gains of mangrove forests due to anthropogenic factors in long-term periods.

2. Materials and Methods

2.1. Study Site

The study area is WMF, located between 19°07′02″N–19°23′30″N and 93°51′00″E–94°02′30″E in Yambye township, Rakhine State, Myanmar. WMF consists of Wunbaik Reserved Mangrove Forest (WRMF), Mingyaung Public Mangrove Forest, and Extended Mingyaung Mangrove Forest (see Figure 1), protected by the Forest Department of Myanmar, and it plays a vital role in providing numerous ecosystem benefits to residents living in and near the forest and enriching important and endangered mangrove species [48]. The topography of the study area is extensively flat and connects to the hill in the western part. A road 20 miles long across WRMF was constructed and occupied 0.4% of the total mangrove area in WMF [42].

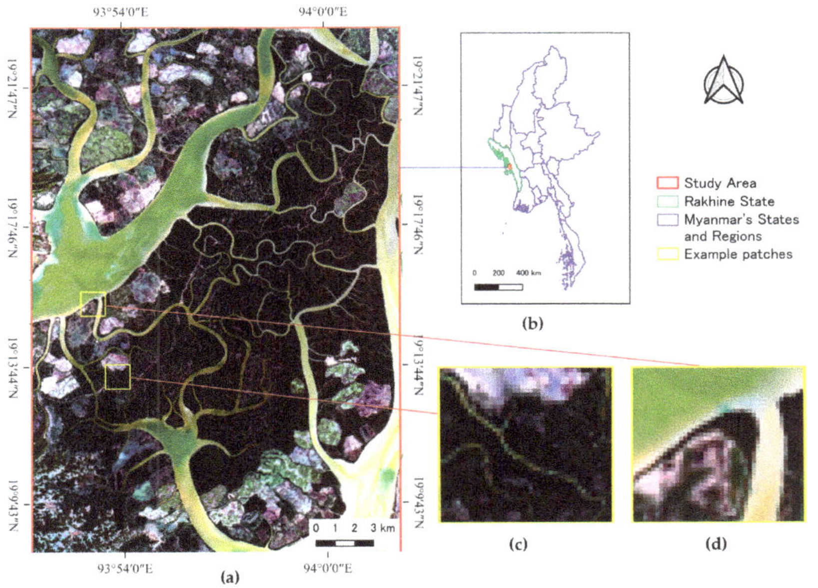

Figure 1. The location of the study area: (**a**) Wunbaik Mangrove Forest (Landsat 8 composite image of 4 bands); (**b**) Myanmar's States and Regions (Myanmar Information Management Unit—MIMU); and (**c**,**d**) Example patches used in U-Net model training (Landsat 8 composite image).

2.2. Ground Truth Image

This study utilized a ground truth image or label image (Figure 2), created by Maung and Sasaki (2021) [23], who manually drew it based on the ground truth points collected from September to October 2019. As the original ground truth image was not available from the researchers, the current image was created by copying it from the original manuscript [49], upscaling it in OpenCV, georeferencing it in QGIS v3.32.0, validating it in high-resolution Google Earth images, setting the spatial resolution to 10 m and 30 m, and reclassifying it as 255 for mangroves and 0 for non-mangroves.

Figure 2. Ground truth image used for model training.

2.3. Earth Observation Data

Considering the historical changes, Landsat series imagery was applied to investigate the important input features to develop deep learning models and to predict long-term mangrove maps despite its medium resolution of 30 m spatial resolution. Additionally, Sentinel-2 imagery with a higher resolution of 10 m was utilized to build a multi-source model with greater precision, and the Harmonized Landsat and Sentinel-2 imagery (HLS) were employed to compare the area differences provided by different sources.

Multiple Landsat series images, including Landsat 5 for 1990, 1995, 2000, 2005, and 2010; Landsat 8 for 2015, 2019, and 2020; and Landsat 9 for 2024 and Sentinel-2 images for each year from 2015 to 2024, were downloaded, considering level 2 with the least cloud coverage (Appendix A). We conducted atmospheric correction using the Remotior Sensus [50] in Google Collab; Landsat 7 images were not collected because of scan line errors in the study area. HLS images of 2017, 2020, 2022, and 2024 [51] were available from the EARTHDATA (https://search.earthdata.nasa.gov/, accessed on 28 June 2024). Resampled Landsat images with 10 m resolution were used to compare deep learning models, build multi-temporal and multi-source models, and predict long-term mangrove mappings from 1990 to 2024.

For topographic and canopy height information, the Shuttle Radar Topography Mission (SRTM) 1 arc-second global Digital Elevation Model (DEM) [52] was downloaded from the United States Geological Survey in Earth Resources Observation and Science (USGS) EarthExplorer website (https://earthexplorer.usgs.gov/, accessed on 2 September 2023). The Multi-Error-Removed Improved-Terrain DEM (MERIT) [53] was available through the MERIT DEM website (https://hydro.iis.u-tokyo.ac.jp/~yamadai/MERIT_DEM/, accessed on 2 September 2023). Preprocessing processes such as band combination, image clipping, resampling, and calculation of vegetation indices and canopy heights were conducted in QGIS v3.32.0.

2.4. Input Feature Selections

Properly selecting important features is critical to building an accurate model in machine learning, as irrelevant features can reduce the model's performance [54]. For model training based on 2020, the important input features were selected among the seven multispectral bands of Landsat 8 images; four vegetation indices, including NDVI [55], NDWI [56], SAVI [57], Combined Mangrove Recognition Index (CMRI) [58], and topographic height information, including SRTM, MERIT, Canopy Height Model (CHM), and Slope by conducting multiple experiments combined with different input features using the ANN; and Convolutional Neural Networks (CNN) models. Red, Green, Blue, Near-infrared (NIR), and Short-wave infrared (SWIR) bands were mainly applied in the input feature selections because of the unavailability of all bands in all Landsat series imagery. CHM was obtained from the difference between SRTM and MERIT, where SRTM was utilized as the Digital Surface Model (DSM) and MERIT as the Digital Terrain Model (DTM) [23,59–61]. The vegetation indices and CHM were calculated using the following equations. The default value of 0.5 was used as the soil-adjusted value (L) in SAVI.

$$\mathrm{NDVI} = \frac{\mathrm{NIR} - \mathrm{RED}}{\mathrm{NIR} + \mathrm{RED}} \quad (1)$$

$$\mathrm{NDWI} = \frac{\mathrm{Green} - \mathrm{NIR}}{\mathrm{Green} + \mathrm{NIR}} \quad (2)$$

$$\mathrm{SAVI} = \left(\frac{\mathrm{NIR} - \mathrm{RED}}{\mathrm{NIR} + \mathrm{RED} + \mathrm{L}}\right) \times (1 + \mathrm{L}) \quad (3)$$

$$\mathrm{CMRI} = \mathrm{NDVI} - \mathrm{NDWI} \quad (4)$$

$$\text{CHM} = \text{SRTM DEM} - \text{MERIT DEM} \quad (5)$$

2.5. CNN

CNN, a specialized form of ANN, is uniquely designed for tasks involving grid-structured data, such as images. Its ability to learn spatial information, including textures, edges, and distinct patterns, makes it particularly useful for remote sensing image analysis, such as scene-based classification and pixel-wise segmentation [33,62]. The basic architecture of CNN consists of three types of layers: convolutional layer, pooling layer, and fully connected layer. The convolutional layer, operating with a three-dimensional structure with height, width, and depth or channels, includes numerous optimizable filters that transform the input features into the important features that describe the targets. Pooling layers perform to reduce the number of parameters by down-sampling while preserving discriminant information. The fully connected layer interprets the features extracted by the convolutional and pooling layers, facilitating final classification tasks.

Multiple CNN architectures have been published, mainly for scene-based image classifications [62], and they are not appropriate for pixel-wise satellite image classifications. Therefore, the architecture (Figure 3) for CNN was designed on three convolutional layers, one pooling layer, and one flattened layer, and parameters were selected through hyper-parameter tuning. The model was trained with the parameters of 96, 64, 64 filters in three convolutional layers, 96 nodes in dense layers, four dropout layers with 0.25, Rectified Linear Unit (ReLU) activation functions for the hidden layers, sigmoid activation function for the output layer, Adaptive Moment Estimation (Adam) optimizer, and binary cross-entropy for loss function. Then, the results were compared with the U-Net model to identify the strengths and weaknesses of each model. To train the CNN model, the input features were clipped into the shape of 7 × 7 using the Pyrsgis library.

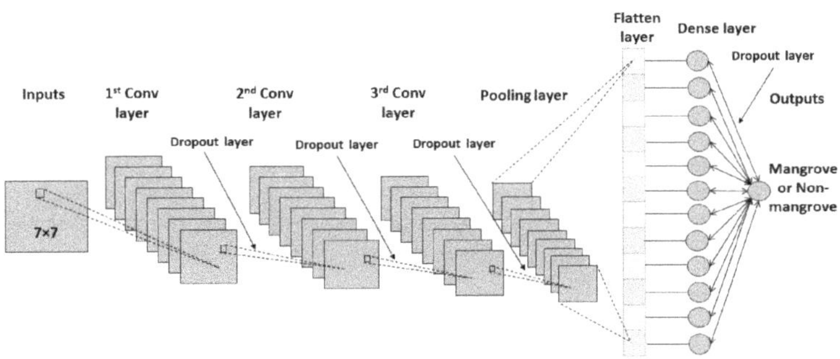

Figure 3. The architecture of the CNN model.

2.6. U-Net

The U-Net, a CNN designed specifically for medical image segmentation, stands out with its U-shaped architecture. This unique design includes a contracting path (encoder) for feature extraction and an expanding path (decoder) for precise localization facilitated by skip connections. Similar to a convolutional network, the contracting path comprises convolutional and pooling layers for down-sampling. The expanding path, on the other hand, up-samples the features learned in the contracting path using up-pooling operations, followed by convolutional layers with the activation functions [63].

In this study, the architecture of the U-Net model (Figure 4) was designed using the convolutional layers, max-pooling layers, convolutional transpose layers, concatenate layers, and dropout layers. For training the model, the images were clipped into the shape of 128 × 128 using the Geotile library. On the encoder side, the model was set with the parameters of 32 filters in the first two convolutional layers, 64 in the second two convolutional layers, and 128 in the third two convolutional layers, where three dropout layers were added between two convolutional layers, followed by max-pooling layers and 256 filters in the fourth two convolutional layers with a dropout layer. On the decoder side, 256 filters were applied in the first two convolutional layers, 128 in the second two convolutional layers, and 64 in the third two convolutional layers, where three dropout layers were added between two convolutional layers, followed by a concatenate layer and 32 filters in the fourth two convolutional layers with a dropout layer. The loss function, optimizer, and learning rate were utilized in the same way as the parameters of the CNN model: binary cross entropy for the loss function and Adam optimizer with 0.0001 learning rate.

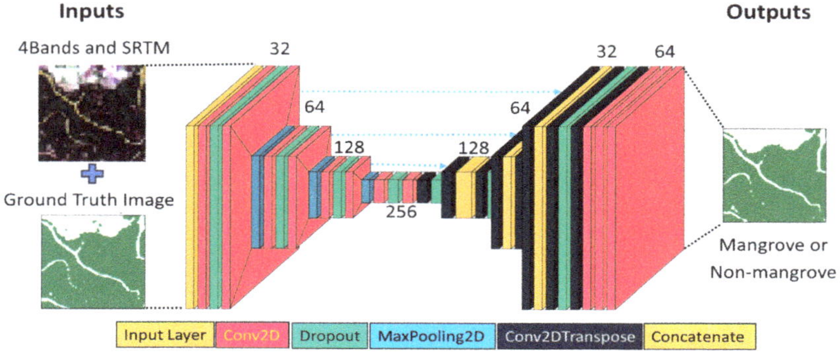

Figure 4. The architecture of the U-Net model (visualized using visualkeras).

2.7. Model Training and Evaluation

The input dataset in each model was split into three parts: training data, representing 60%; testing data, 20%; and validation data, 20%. The training data and validation data were utilized in the training process, and the model performance was evaluated using the testing data after training. Model training and evaluation were conducted using TensorFlow 2.4.1 with 1 GPU in the Wisteria/BDEC-01 Supercomputer System of the University of Tokyo. The results of each model were evaluated using several metrics calculated in the following equations: overall accuracy, precision, recall, F1-score, and intersection over union (IoU), and the predicted results were summarized in the confusion matrix describing the true positives (TP), true negatives (TN), false positives (FP), and false negatives (FN).

$$\text{Accuracy} = \frac{\text{Number of Correct Predictions}}{\text{Number of Total Predictions}} \tag{6}$$

$$\text{Precision} = \frac{\text{TP}}{\text{TP} + \text{FP}} \tag{7}$$

$$\text{Recall} = \frac{\text{TP}}{\text{TP} + \text{FN}} \tag{8}$$

$$\text{F1} - \text{Score} = 2 \times \frac{\text{Recall} \times \text{Precision}}{\text{Recall} + \text{Precision}} \tag{9}$$

$$\text{Intersection over Union (IoU)} = \frac{\text{TP}}{\text{TP} + \text{FP} + \text{FN}} \tag{10}$$

2.8. Change Detection

Change detection is useful for identifying the differences in an image by observing it on different dates and is widely applied to several applications based on remote sensing [64,65]. The present study utilized the image differencing method among the multiple post-classification change detection techniques. The results of change detection were demonstrated in mangrove gains, mangrove losses, and unchanged areas and were visually checked using high-resolution Google Earth images. The number of pixels showing the gains, losses, and unchanged were extracted and used to calculate the long-term changes in mangrove forests and the annual rate of changes.

3. Results

3.1. Important Input Features

Important input features were first identified using feature importance scores derived from the Random Forest Classifier and correlation values between input features and targets. However, these techniques were not useful in providing a proper combination of input features. Subsequently, the proper combinations were investigated by executing multiple experiments using ANN and CNN models.

The feature importance scores of the Random Forest Classifier demonstrated that NDVI, CMRI, NDWI, and SAVI were the most important ones, followed by MERIT, SRTM, Slope, and CHM. According to experiments based on the ANN model, the highest accuracy of 93.80% was achieved with the combination of input features consisting of NDVI, NDWI, SAVI, CMRI, SRTM, and CHM. On the other hand, the CNN model provided the highest accuracy of 95.99% with the combination of four bands of Landsat 8 (Blue, Green, Red, and NIR) and SRTM (Appendix B). Therefore, four bands and SRTM were applied to compare the deep learning models of CNN and U-Net and to develop a multi-temporal and multi-source model using U-Net architecture.

3.2. Multi-Temporal and Multi-Source Model

3.2.1. Comparison of CNN and U-Net Using Landsat 8 and Sentinel-2

The CNN and U-Net models were compared using Landsat 8 and Sentinel-2 imagery to observe the strengths and weaknesses of each model and to select an optimal model for the long-term distribution of mangrove forests.

The comparison results (Table 1) revealed that the U-Net model using Sentinel-2 outperformed the CNN model, while similar accuracy was produced by both CNN and U-Net trained with Landsat 8. The U-Net model using Sentinel-2 produced an accuracy of 98.25%, while the CNN result was 97.24%. The most significant difference was training time; about 250–260 s was needed for the U-Net model, while the CNN model took about 10,000–12,000 s. Although the architecture of the CNN model was simple with 152,193 training parameters, it needed a huge amount of training time compared to the U-Net model, which utilized many convolutional layers with many filters, providing 3,133,921 parameters. This comparison has important implications, suggesting that the U-Net model is more efficient and effective than the CNN model for this application.

Table 1. The comparison of CNN and U-Net models using Landsat 8 and Sentinel-2.

Satellites	Models	Accuracy	IoU	Precision	Recall	F1-score	Training Time (s)	Inputs	Epochs
Landsat 8 (10 m)	CNN	96.65%	0.93	0.96	0.97	0.96	10,988.99	4 images	200
	U-Net	96.64%	0.93	0.96	0.97	0.96	260.59	4 images	200
Sentinel-2 (10 m)	CNN	97.34%	0.94	0.96	0.98	0.98	11,231.63	4 images	200
	U-Net	98.25%	0.96	0.98	0.98	0.98	257.36	4 images	200

Four images consist of two pairs of identical images, with each pair captured on different dates.

3.2.2. Performance of Multi-Temporal and Multi-Source Model

A multi-temporal and multi-source deep learning model was built on the U-Net architecture using the multi-temporal images available in November 2019, December 2019, and January 2020 of Landsat 8 and Sentinel-2. The model was trained for 500 epochs using twelve images by doubling the six temporal images of Landsat 8 and Sentinel-2. As a result, it achieved the highest accuracy of 99.73%, IoU 0.99, precision 1, recall 1, and F1-score

1 (Table 2). Then, it was applied to produce the mangrove maps with Landsat 8 and Sentinel-2 of 2020 to draw a confusion matrix (Table 3) describing TP, TN, FP, and FN and visualize them (Figure 5). The confusion matrix described a few false positives and false negatives given by the model, where more false positives and false negatives were found in Landsat 8 than in the results of Sentinel-2. After model evaluation, mangrove distribution maps from 1990 to 2024 were predicted using the model. The resulting maps (Figure 6) were visually checked using Google Earth images. The results were promising, and the changes in mangrove forests were significantly identified due to the visual assessment with high-resolution Google Earth images.

Table 2. The classification results of the multi-temporal and multi-source U-Net model.

Accuracy	IoU (MG)	IoU (nMG)	Precision (MG)	Precision (nMG)	Recall (MG)	Recall (nMG)	F1-Score (MG)	F1-Score (nMG)
99.73%	0.99	1	1	1	1	1	1	1

MG = Mangrove, nMG = Non-Mangrove.

Table 3. Confusion matrix of multi-temporal and multi-source U-Net model on Landsat 8 and Sentinel-2 images.

Landsat 8 (21 January 2020)				Total Pixels	Sentinel-2 (21 January 2020)			
		Predicted nMG	MG				Predicted nMG	MG
Actual	nMG	2,951,560	4304	2,955,864	Actual	nMG	2,952,392	2157
	MG	7620	2,422,512	2,430,132		MG	6788	2,424,659

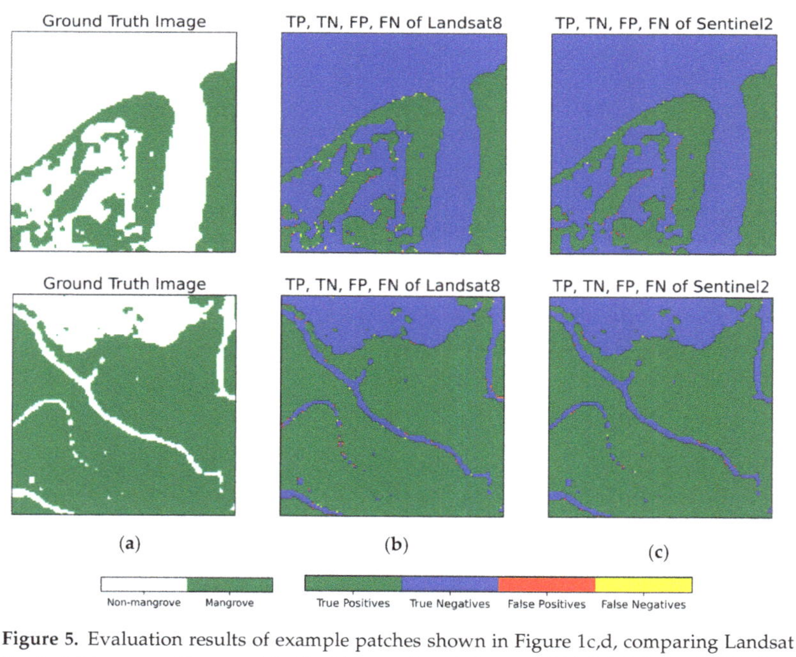

Figure 5. Evaluation results of example patches shown in Figure 1c,d, comparing Landsat 8 and Sentinel-2 images with ground truth: (**a**) ground truth images; (**b**) TP, TN, FP, and FN of Landsat 8 images; and (**c**) TP, TN, FP, and FN of Sentinel-2 images.

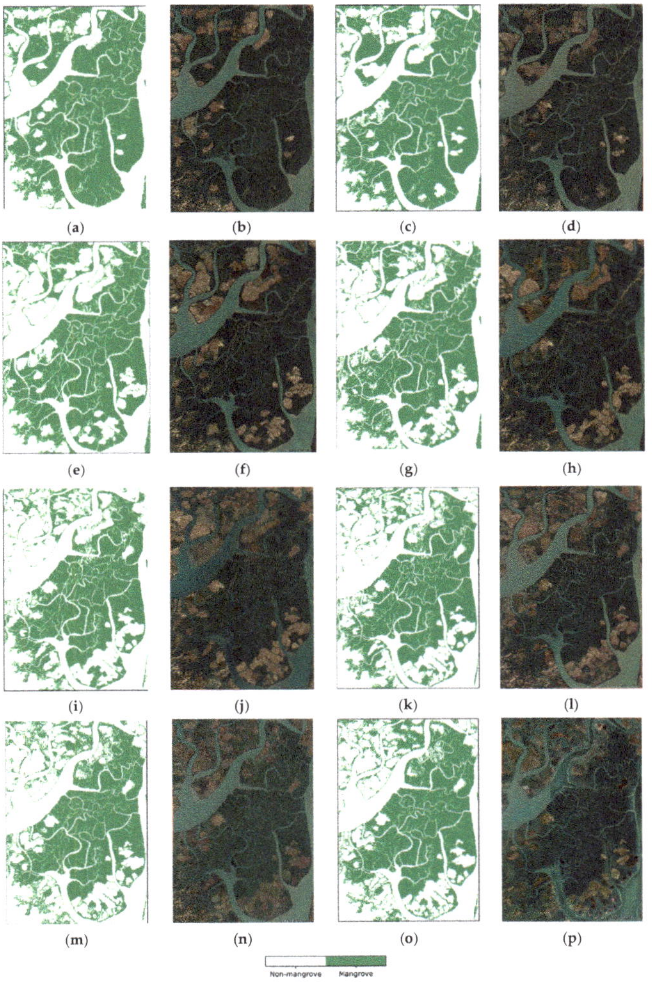

Figure 6. Mangrove maps predicted by the multi-temporal and multi-source model and Google Earth images from 1990 to 2024: (**a**) the mangrove map of 1990; (**b**) the Google Earth image of 1990; (**c**) the mangrove map of 1995; (**d**) the Google Earth image of 1995; (**e**) the mangrove map of 2000; (**f**) the Google Earth image of 2000; (**g**) the mangrove map of 2005; (**h**) the Google Earth image of 2005; (**i**) the mangrove map of 2010; (**j**) the Google Earth image of 2010; (**k**) the mangrove map of 2015; (**l**) the Google Earth image of 2015; (**m**) the mangrove map of 2020; (**n**) the Google Earth image of 2020; (**o**) the mangrove map of 2024; and (**p**) the Google Earth image of 2024. (The dark green is mangroves, light and dark brown are agricultural and aquacultural areas, and light green is water bodies).

3.3. *Long-Term Mangrove Changes in WMF*

Monitoring the distribution and changes in mangrove forests over time is crucial for understanding the health of their ecosystems and the intervention of humans in these ecosystems, and informing conservation efforts. The present study utilized the image differencing between two different temporal maps provided by the model to detect the mangrove changes in WMF. Then, the mangrove and non-mangrove areas were extracted from mangrove maps for each year, and the changes in mangrove areas for each year were visualized for 34 years. The result of mangrove changes demonstrated the abnormal fluctuation of mangrove distribution in WMF (Figure 7). Mangrove areas in WMF were 32,402.40 ha in 1990 and had steadily decreased to 23,561.64 ha by 2010. Afterward, they started increasing slowly in 2015 and have still increased to 24,768.77 ha in 2024.

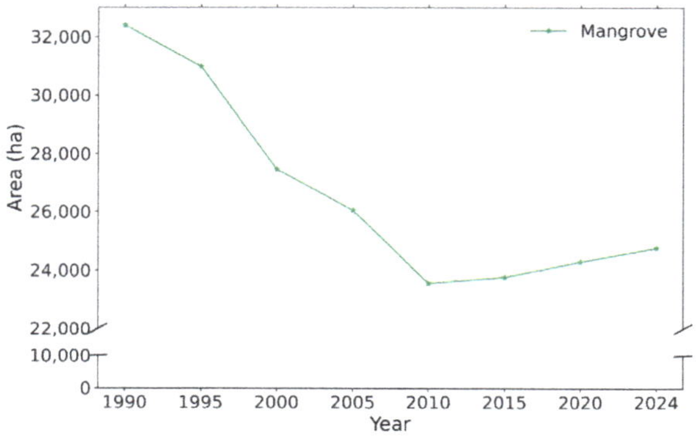

Figure 7. The annual changes of mangrove forests in WMF from 1990 to 2024.

Change detection results highlighted that a significant area of mangroves has been changed in WMF over a period of 34 years (Figure 8). A substantial portion of the mangrove forests have been lost since 1990 and a minor recovery by 2024. Specifically, 29.3% of the mangrove extent in WMF has been lost, while only 5.75% has been gained. The most significant deforestation occurred during the period of 2005–2010, while the most substantial reforestation was observed during the period of 2015–2020. Notably, the expansion of mangrove forests has been consistent since 2010. The annual rate of change is −224.52 ha/yr, equivalent to a decrease of 0.69% per year over the 34 years. The highest negative rate of change was recorded at −703.4 ha/yr, representing a severe 2.27% per year decrease over 5 years between 1995 and 2000, with a positive rate of change observed from 2010, peaking between 2020 and 2024 (Figure 9).

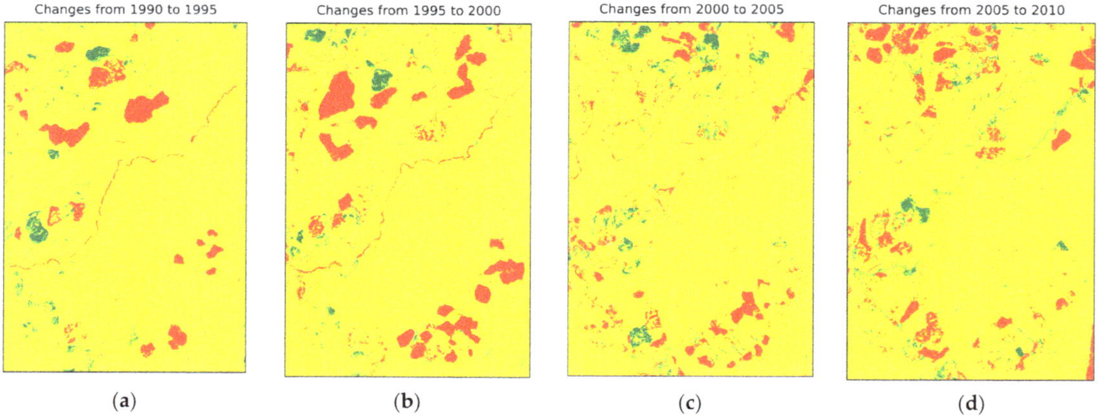

(a) (b) (c) (d)

Figure 8. *Cont.*

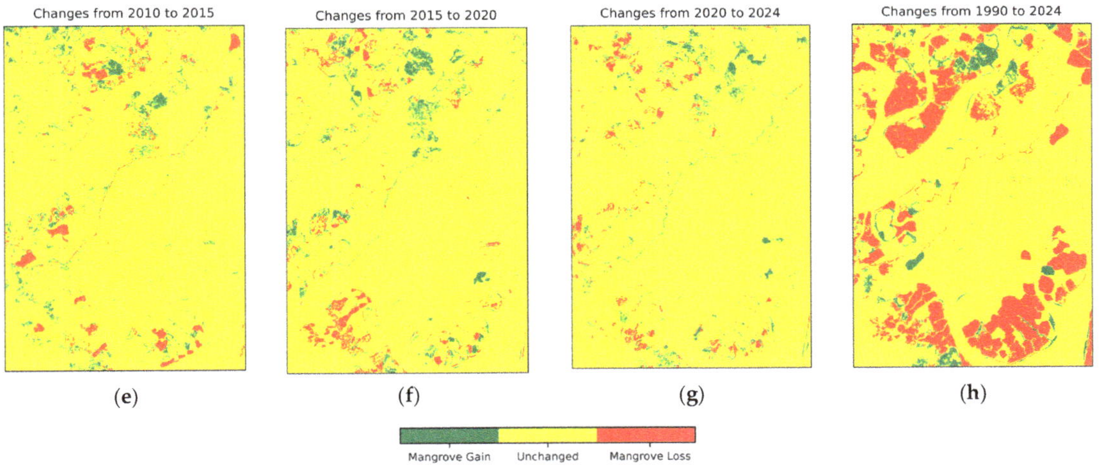

Figure 8. Change detection maps for each period: (**a**) 1990–1995; (**b**) 1995–2000; (**c**) 2000–2005; (**d**) 2005–2010; (**e**) 2010–2015; (**f**) 2015–2020; (**g**) 2020–2024; and (**h**) 1990–2024.

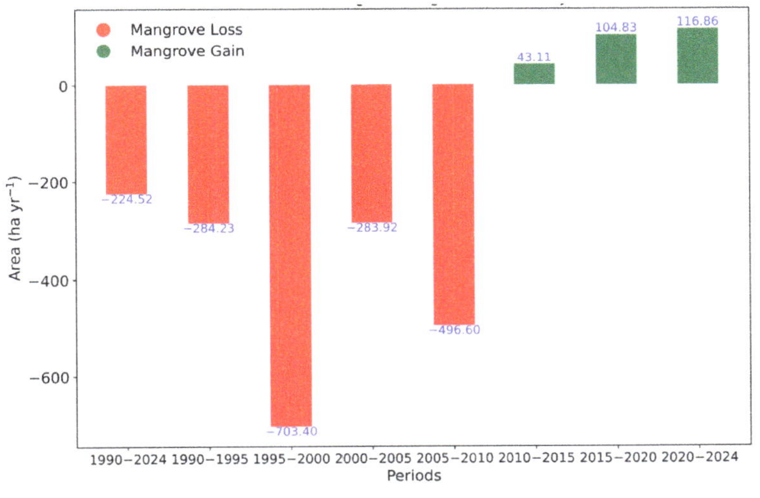

Figure 9. Annual rate of mangrove changes in WMF.

4. Discussion

4.1. Selection of Key Input Features

Input feature selection is important in preprocessing data before training the model. Properly selecting important features influences the model's performance by reducing the data complexity and irrelevant features among multiple input features [54]. To select the optimal combination of input features for model training, the present study utilized the Random Forest Classifier and conducted manual experiments using ANN and CNN models. Although the Random Forest Classifier demonstrated NDVI, CMRI, NDWI, and SAVI as the most important ones, followed by MERIT, SRTM, Slope, and CHM, it did not identify the optimal combination for model training. Therefore, multiple experiments were conducted using the ANN model for 30 epochs, which was designed with two hidden layers with 50 and 40 nodes for each layer. These results provided the highest accuracy of

93.80%, with the combination of input features consisting of NDVI, NDWI, SAVI, CMRI, SRTM, and CHM.

These findings suggested that the application of four vegetation indices, such as NDVI, NDWI, SAVI, and CMRI, was sufficient for mangrove classification with the ANN model, while SRTM and CHM contributed to higher accuracy in discriminating the terrestrial forests in high-elevation areas. However, these findings differed from those of Maung and Sasaki (2021), who discussed the significance of MERIT and CHM for model improvement by integrating ten bands of Sentinel-2, NDVI, and NDWI [23]. In the present study, SRTM emerged as significantly more important in almost every combination of input features than MERIT (Appendix B). Although CHM was observed as an important feature of the ANN model based on 2020 data, it did not represent 2020 because it was estimated using SRTM and MERIT, which were created on elevation data in 2000. For historical change detection, it does not represent different years and could produce a bias in the historical map prediction. Moreover, the above input combination demonstrated overfitting when used to train the CNN model with 50 epochs (Figure 10). Therefore, experiments for feature selection were conducted using the CNN model, considering fewer input features. The results of these experiments provided the highest accuracy of 95.99%, with the combination of four bands and SRTM when training for 100 epochs.

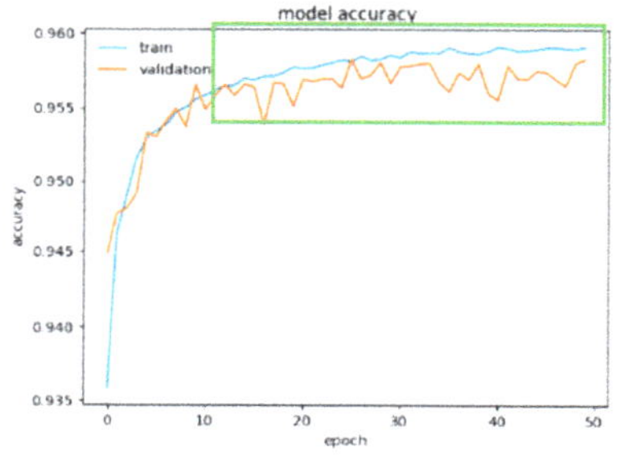

(a)

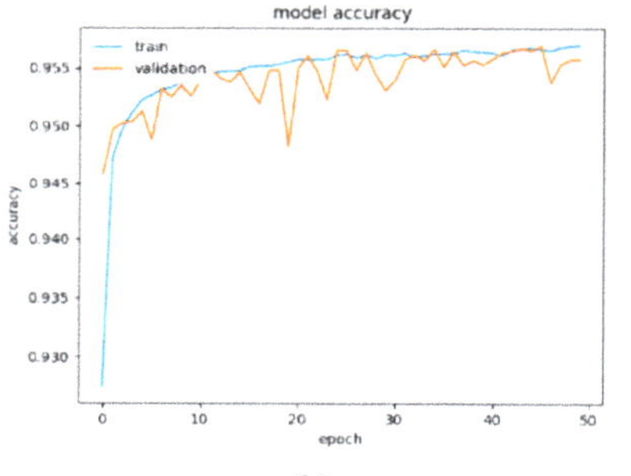

(b)

Figure 10. The training flow of the CNN model with different input features: (**a**) accuracy flow trained with NDVI, NDWI, SAVI, CMRI, SRTM, and CHM; (**b**) accuracy flow trained with 4 bands and SRTM.

Existing studies proved the success of deep learning-based mangrove classification using three bands, including Red, NIR, and SWIR; and four bands, including Blue, Green, Red, and NIR combined with VH and VV of Sentinel-1; seven bands; and multiple bands combined with vegetation indices and elevation data [23,26,37–39]. The present study observed that four bands with SRTM were useful for mangrove classification, where SRTM effectively removed the high-elevation areas, such as terrestrial forests. Moreover, the present study found that three bands, including Green, Red, and NIR with SRTM, could be utilized in long-term mangrove monitoring except for 1995 images of Landsat 5, in which open forests with less mangrove cover were misclassified as non-mangroves (Figure 11). We did not identify the causes of misclassification in 1995 images, possibly due to the low image quality available in 1995 and their original image preprocessing. Therefore, this model training was conducted using the input features of four multispectral bands and SRTM.

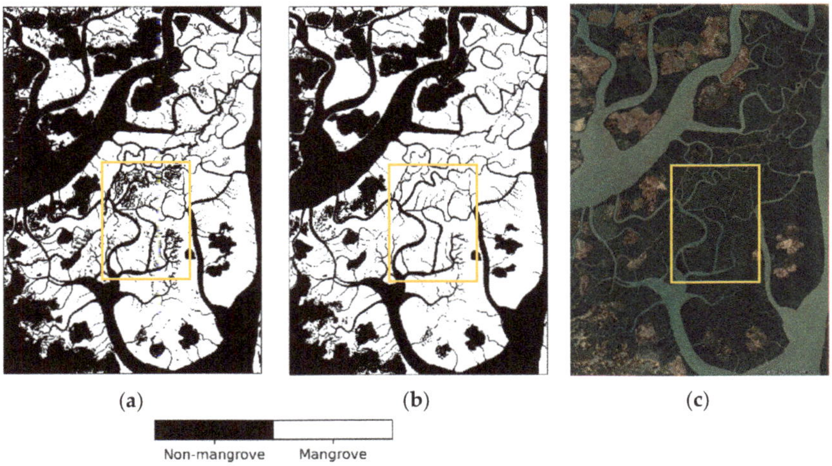

Figure 11. Mangrove maps predicted by CNN model trained with three bands and SRTM for 1995: (**a**) mangrove map predicted using 3 bands and SRTM; (**b**) mangrove map predicted using 4 bands and SRTM; and (**c**) Google Earth image for 1995. (The dark green is mangroves, light brown is agricultural and aquacultural areas, and light green is water bodies).

4.2. Application of Multi-Temporal and Multi-Source Imagery to U-Net

The application of multi-temporal and multi-source imagery is a pivotal aspect of mangrove mapping. Numerous studies have demonstrated the benefits of multi-temporal images in developing machine learning and deep learning models [19,26,38], and some studies have utilized multi-source imagery of radar and optical satellites to enhance the model's accuracy by reducing the effects of weather conditions [21,29,38]. While some studies have explored multi-source imagery, our study is unique in its focus on fusing Landsat and Sentinel-2 imagery.

Given the availability of a single ground truth image for 2020, we initially developed the model using a temporal dataset available in January 2020. However, the model's performance was not consistent across different temporal images, leading to misclassified results when predicting the images in December 2019, when the conditions of paddy fields were different (Figure 12). The farmers in WMF grow the paddy fields from June to August and harvest them from December to March [43]. The paddy fields in November and December appeared green in the true color images, which were before harvesting, while white colors appeared in January after harvesting.

The fluctuation of NDVI values in different months confirmed the significant differences in paddy fields. The NDVI values of paddy fields were slightly different, mainly in January. These values of paddy fields were lower than those of mangroves in January and became higher in December than in January. These values were nearly identical to those of mangroves in November. The images from February to May showed conditions similar to those in January, and those before November contained a high percentage of clouds in the satellite images because of the monsoon rainy season. Therefore, three temporal images available in November, December, and January were utilized to train a multi-temporal and multi-source model.

The comparison of CNN and U-Net highlighted the strengths and weaknesses of each model when using Landsat 8 and Sentinel-2 images. The strength of the CNN model was that it could be trained using a single satellite image, but it took longer than U-Net. The CNN model is best suited if a study emphasizes a small area using medium-resolution satellite images. On the other hand, the U-Net model achieved higher accuracy when training for many epochs and needed less training time than the CNN model. However, the weakness of the U-Net model is its requirement for large training samples. Data augmentation techniques, such as doubling the same images, were employed during the training processes to address it. Moreover, we encountered overfitting when conducting multiple experiments to obtain an optimal model and solved it by changing the number of filters in each layer and adding or reducing the number of layers in the U-Net architecture.

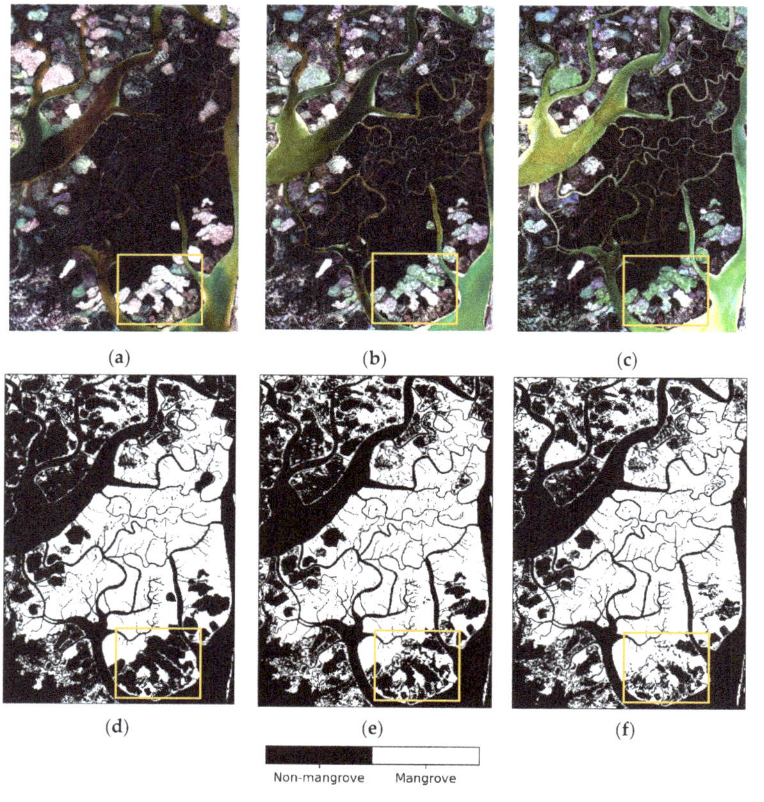

Figure 12. Different temporal conditions of paddy fields: (**a**) Landsat composite image of January 2020; (**b**) Landsat composite image of December 2019; (**c**) Landsat composite image of November 2019; (**d**) CNN result for January 2020; (**e**) CNN result for December 2019; and (**f**) CNN result for November 2019. (The dark brown is mangroves, light green and white are agricultural and aquacultural areas, and yellow-green is water bodies).

The results of comparing different satellite imagery demonstrated that Sentinel-2 outperformed Landsat 8. The U-Net model with Sentinel-2 achieved higher accuracy than that of Landsat 8. Sentinel-2 imagery could be the best choice for mangrove mapping when focusing on data since 2015. However, this study needed to utilize the Landsat imagery for historical changes in mangrove forests. Therefore, the fusion of resampled Landsat 8 and Sentinel-2 images was applied to train the U-Net model using the multi-temporal datasets.

4.3. Model Performance and Limitations

The U-Net model trained with multi-temporal and multi-source imagery achieved the highest accuracy of 99.73%, which outperformed the ANN model trained on the same ground truth image [23] with Sentinel-2 in the same site, and all existing models, such as Capsules-Unet [26], MSNet [39], and ANN [38], studied in different locations with different satellite imagery. Moreover, it can be applied to multi-temporal datasets from medium-resolution optical satellite imagery, including Landsat, Sentinel-2, and HLS. Its higher accuracy was achieved when training the model for 500 epochs.

The U-Net model provided less accuracy when training with single-source datasets for 200 epochs; it provided an accuracy of 98.25% with Sentinel-2 and 96.64% with Landsat 8 (Table 1). The result could be about 98% if the present study focused on training the model with Landsat imagery for 500 epochs. It highlighted the usefulness of fusing Landsat 8 and Sentinel-2 imagery for mangrove mapping. The combination of Landsat and Sentinel-2 optical imagery enhanced the accurate performance by reducing the spatial resolution gap between them. This approach allows the model to leverage the higher spatial resolution of Sentinel-2 while benefiting from the broader spectral bands of Landsat 8.

However, the model has limitations in generalization. When the model was applied to different locations in Myanmar, the results were evaluated using the mangrove maps of the High-Resolution Global Mangrove Forest (HGMF), and it performed well in all mangrove areas located near Rakhine State, Myanmar. However, it failed to accurately classify mangrove forests located in the Ayeyarwady Delta and the Tanintharyi Region, Myanmar. To address these issues, we conducted multiple experiments using different input features and larger training samples from the above three regions. The testing model trained using four bands, NDVI and NDWI, classified mangrove forests in different regions, though it had a weakness in distinguishing mangroves from terrestrial forests. The model trained using four bands and SRTM with larger training datasets available from three regions performed well for all regions and accurately distinguished between mangroves and terrestrial forests. Therefore, more ground truth images representing different locations should be included to train a generalized model for nationwide maps. Additionally, it should be tested with data from different countries to ensure broader generalization and better reliability.

The present model trained on multi-source imagery provided a slight discrepancy in mangrove extents when predicting their specific images and with different spatial resolutions. The resampled Landsat 8 images of 2020 provided an estimated extent of 2430.13 ha, while Sentinel-2 images estimated 2431.45 ha. A significant difference was identified when predicting the same Landsat 8 images without resampling, which covered an extent of 2405.74 ha (Figure 13). Its resulting map provided a coarser distribution of mangrove forests without showing distinct features of the river network (Figure 14).

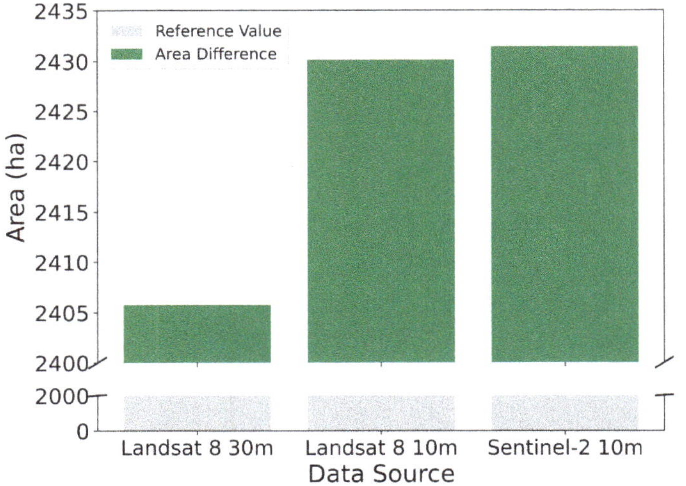

Figure 13. Area differences between Landsat 8 (30 m), Landsat 8 (10 m), and Sentinel-2 (10 m).

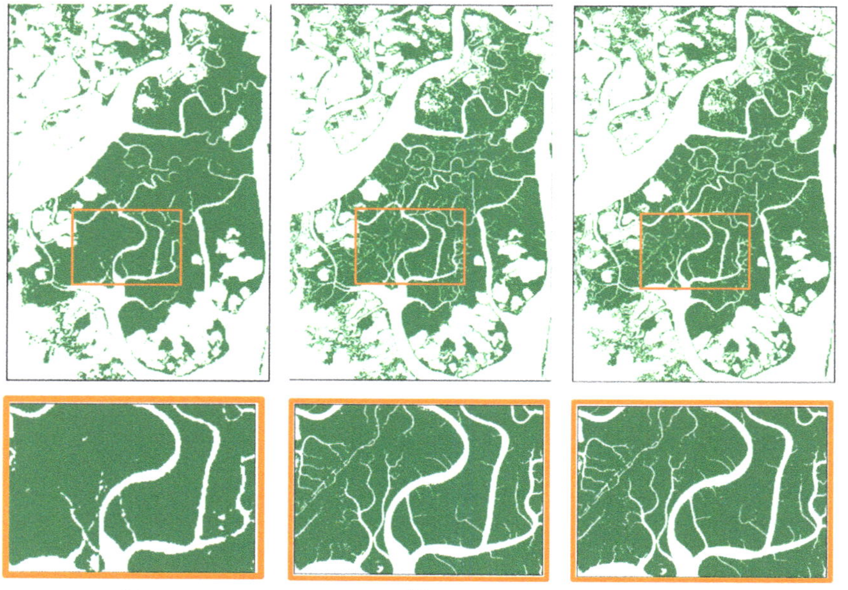

(**a**) (**b**) (**c**)

Figure 14. Model results on satellite images with different spatial resolutions: (**a**) U-Net result on Landsat 8 (30 m); (**b**) U-Net result on Landsat 8 (10 m); and (**c**) U-Net result on Sentinel-2 (10 m).

Similar discrepancies were observed when interpreting the mangrove extents of Landsat 8 at 30 m resolution in 2024, resampled Landsat 8 (10 m resolution), and Sentinel-2 (10 m resolution). The largest extent was observed using resampled Landsat 8, with about a 30-hectare difference compared to the results of Sentinel-2, followed by the least extent of Landsat 8 without resampling. Therefore, the U-Net model of this study should not be used to predict the Landsat images without resampling because it can provide a less accurate map. Instead, it is crucial to resample the Landsat images into 10 m to match the spatial resolution of Sentinel-2 images, ensuring consistent and reliable predictions.

4.4. Extent Changes in WMF

The change detection analysis revealed that 29.3% of the mangrove extent in the study area was deforested from 1990 to 2024, including a steady increase in mangroves since 2010. However, Saw and Kanzaki (2015) reported that 40% of the mangrove extent in WRMF was lost from 1990 to 2011 [43], and Maung and Sasaki (2021) observed a slight decrease in WMF from 2015 to 2020 [23]. The difference between the study of Saw and Kanzaki (2015) and the present study depends on the extent of the study area. The research of Saw and Kanzaki (2015) focused on the extent of WRMF, which covers 22,919 ha, while the total area of this study is 53,859.95 ha. When the change detection was analyzed for the period from 1990 to 2010, it was found that 30.75% of the mangrove forests had been deforested.

Our findings from 2015 to 2020 differ from those of Maung and Sasaki (2021), in which they observed a slight decline from 254.30 to 249.83 km2 in WMF, while our study observed a slight increase. To validate the results of the present study, we downloaded available mangrove maps of the Global Mangrove Watch (GMW) and analyzed their long-term distribution. GMW data showed a similar pattern of mangrove changes as the present study, illustrating deforestation from 1996 to 2015 and reforestation from 2015 to 2020 (Figure 15). The declined results of Maung and Sasaki (2021) could be due to the usage of different preprocessing levels of Sentinel-2: level 1 C images for 2015 and level 2 A for

2020 and the application of transfer learning in a small dataset, representing mainly the agricultural areas.

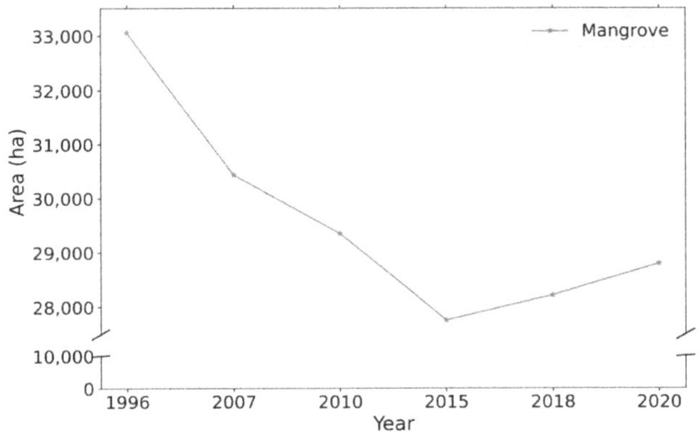

Figure 15. Annual area changes of WMF given by GMW data.

4.5. Extent Comparison with Different Datasets

Evaluating the resulting mangrove map with the available data is a crucial step to demonstrate the reliability of the proposed model. We compared the mangrove map predicted by the U-Net model with the existing global datasets for 2020, such as the Global Mangrove Watch (GMW) [6] and the High-resolution Global Mangrove Forest (HGMF) [17], by referencing the ground truth image used for model training. The comparison highlighted huge discrepancies: the area difference between the ground truth image (24,268.16 ha) and the GMW data (28,785.78 ha) was substantial. In contrast, the gap between the ground truth image and the HGMF data was much smaller, around 1000 ha (Table 4). The GMW data overestimated the distribution of mangrove forests in the study area by approximately 4000 ha. The mangrove map produced by the U-Net model is closely aligned with the ground truth image, indicating the reliability of the deep learning approach.

Table 4. The extent comparison with global datasets in 2020 for the study area (ha).

Ground Truth Image	U-Net Result	GMW Data	HGMF Data
24,268.16	24,301.32	28,785.78	25,882.29

The long-term changes in mangrove forests were examined using Sentinel-2 and HLS images to compare the results of Landsat images with different medium-resolution satellite images. The mangrove distribution maps of Sentinel-2 predicted by the U-Net model indicated a relatively stable trend with slight fluctuations (Figure 16). The mangrove extent of the study area in 2015 was 24,541.09 ha, reaching a peak around 2017, followed by a slight decrease in 2020 to 24,314.47 ha and gradually increasing to 24,469.76 ha in 2024. The above data provided by Sentinel-2 showed a slight overall decrease in WMF from 2015 to 2020, with a marginal increase in 2024.

To verify the extent of changes in WMF, we applied the model to both HLS L30 and HLS S30. HLS L30 was created using Landsat surface reflectance by adjusting the Sentinel-2 tiling system. At the same time, HLS S30 images were created based on the Sentinel-2 imagery by resampling to 30 m and adjusting to Landsat 8 and 9 spectral functions. The mangrove distributions of HLS L30 demonstrated an abnormal increase from 2015 to 2024, with the least coverage in 2017 (Figure 17a). Meanwhile, HLS S30 images illustrate a marginal decrease from 2015 to 2020, followed by a slight increase in 2024 (Figure 17b). Their results are also different from those of Sentinel-2; however, they confirmed an increase in mangrove forests by 2024.

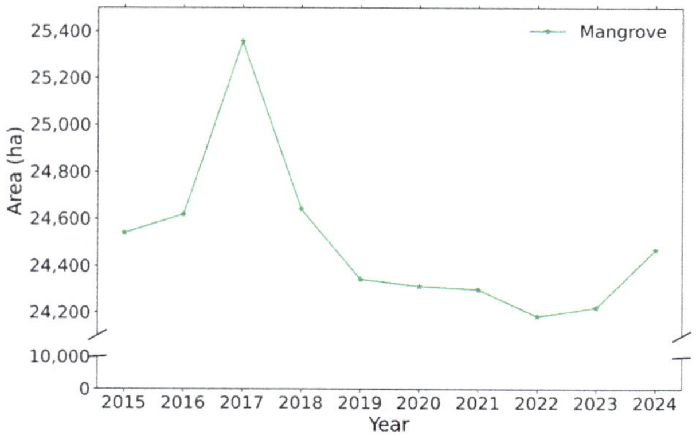

Figure 16. Annual area changes of WMF given by Sentinel-2 data.

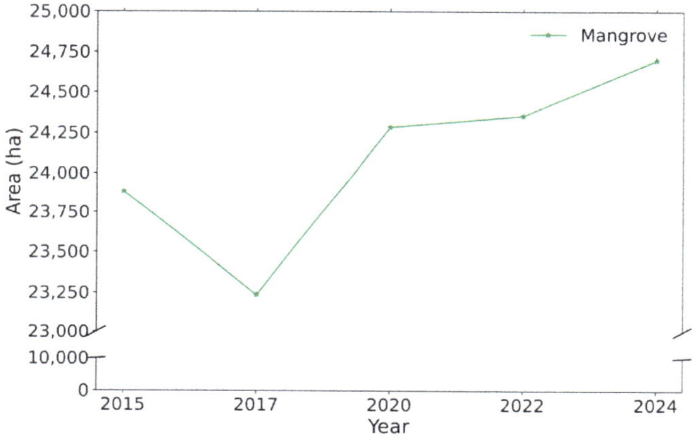

(**a**)

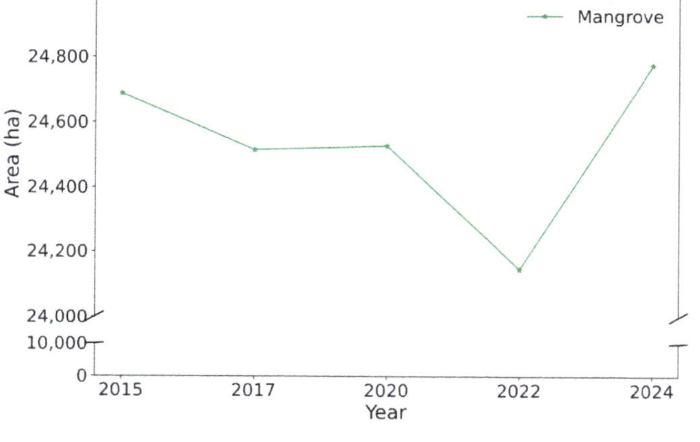

(**b**)

Figure 17. Annual area changes of WMF given by HLS data: (**a**) HLS L30 data, (**b**) HLS S30 data.

Different satellite images provided different results, possibly due to spatial resolutions, sensors, and preprocessing techniques, although they were available from the same source using the Remotior Sensus library. This study utilized the Landsat series images due to the availability of historical data. Moreover, our results align similarly to those from the global GMW dataset. Therefore, the results produced by Landsat images are considered reliable.

4.6. Drivers of Mangrove Losses and Gains

The existing studies reported that the drivers of mangrove changes in WMF were anthropogenic factors, such as the expansion of shrimp ponds and paddy fields [42,43]. Although the present study did not directly detect the drivers of mangrove changes, it identified the drivers of mangrove losses and gains through high-resolution Google Earth images (Figure 18). Due to change detection results, mangrove forests had been steadily lost from 1990 to 2010. These losses were caused by constructing a crossing road through the reserved forest in 1994, the significantly increased number of farmers and shrimp-pond operators from 1990 to 2000, the expanding paddy fields between 1994 and 2003, and illegal logging with weak regulations [43].

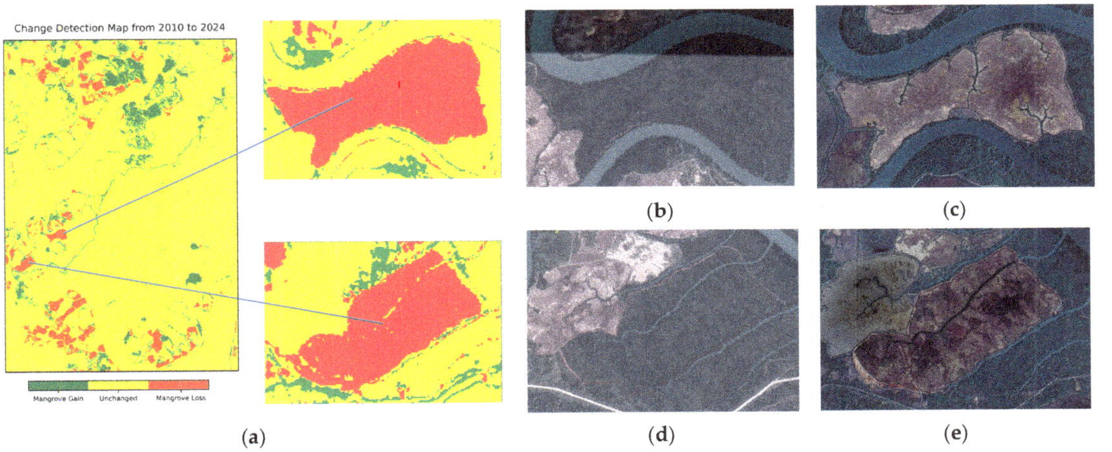

Figure 18. Mangrove losses: (**a**) changes from 2010 to 2024; (**b**) land use in 2009; (**c**) land use in 2024; (**d**) land use in 2009; and (**e**) land use in 2024. The drivers of losses from (**b**) to (**c**) are shrimp ponds and from (**d**) to (**e**) are paddy fields due to Google Earth images and the study by Maung et al. (2024) [48].

The mangrove forests have increased forward since 2010. The drivers of mangrove gains were identified as artificially planted and naturally recovered mangroves in abandoned sites, in sedimentation areas, and along roadsides (Figure 19). The coverage of mangrove trees has increased along roadsides; the road crossing through the study area was significantly detected in the mangrove maps from 1995 to 2015, while it was detected in some parts of 2020 and 2024. The Forest Department of Myanmar implemented 855 acres (346 ha) of mangrove plantations from 2017 to 2023 within WRMF. Moreover, Maung and Sasaki (2021) confirmed that mangrove gains were due to mangrove plantations and the natural recovery ability of mangroves [23].

The present study identified that some areas with recently recovered mangroves had the risk of being lost again due to anthropogenic disturbances. Maung and Sasaki (2021) found that mangroves naturally recovered at approximately 50% of the three abandoned sites. However, mangroves in two of these sites were lost again by 2024 (Figure 20). Accordingly, the natural recovery rate of mangroves is insufficient to ensure the long-term stability of these ecosystems. Anthropogenic disturbances pose a significant threat to the sustainability of naturally recovered mangroves. Protecting naturally recovered mangroves can increase mangrove coverage at a low cost. Therefore, sustainable management and conservation strategies are essential to protecting these vital ecosystems, ensuring their resilience and ability to thrive in the face of environmental and human-induced challenges.

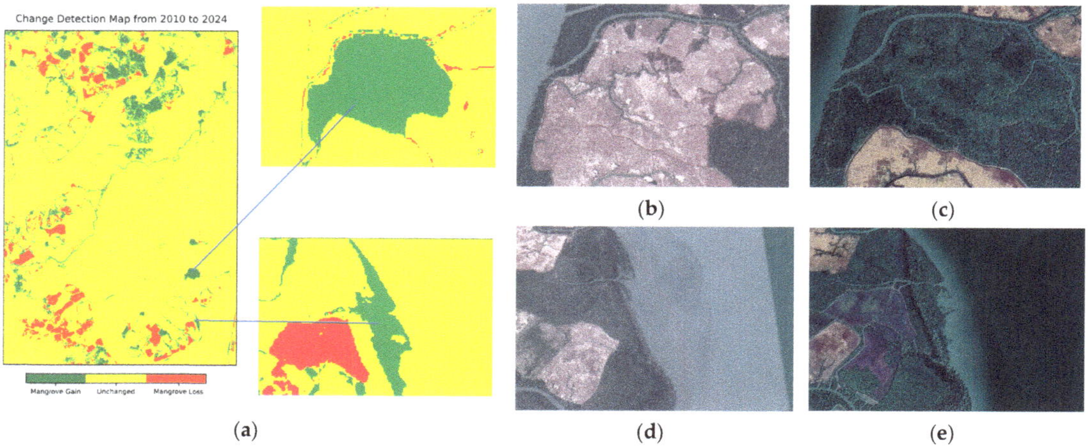

Figure 19. Mangrove gains: (**a**) changes from 2010 to 2024; (**b**) land use in 2009; (**c**) land use in 2023; (**d**) land use in 2009; and (**e**) land use in 2023. The drivers of gains from (**b**) to (**c**) are mangrove plantations and from (**d**) to (**e**) are natural mangroves due to Google Earth images and Maung and Sasaki (2021) [23].

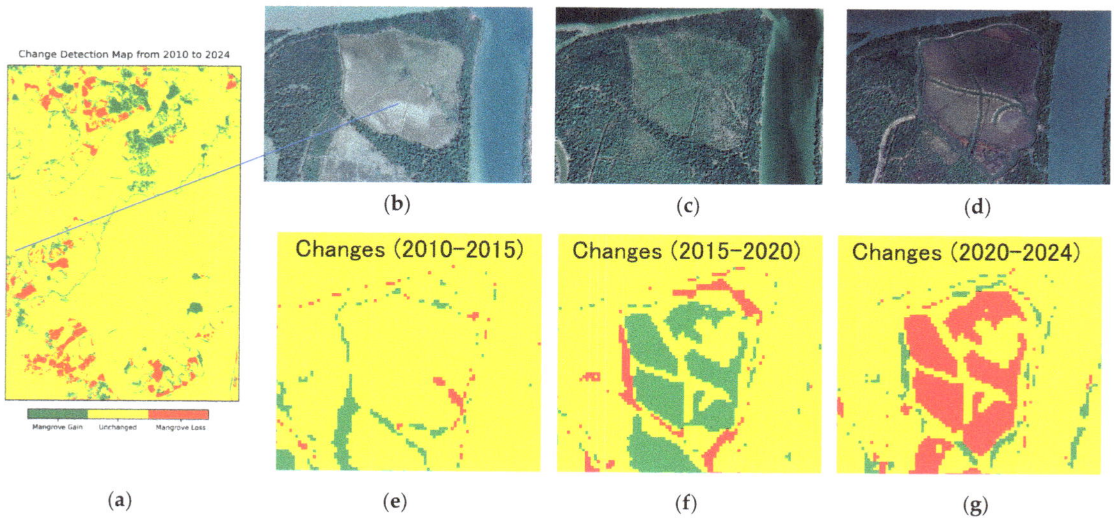

Figure 20. Mangrove losses after recovery: (**a**) changes from 2010 to 2024; (**b**) land use in 2015; (**c**) land use in 2018; (**d**) land use in 2024; (**e**) changes from 2010 to 2015; (**f**) changes from 2015 to 2020; and (**g**) changes from 2020 to 2024 (Google Earth images for historical land uses).

Monitoring the present conditions in mangrove forests is vital for detecting changes, pinpointing their locations, and identifying the drivers behind them. Saw and Kanzaki (2015) reported that illegal woodcutting occurred in the reserved forests to obtain charcoal and firewood [43]. The present study observed that mangrove losses have occurred since

2020 and onward, both within and outside the reserved area. Over half of the lost areas from 2020 to 2024 have been lost within a year since 2023. As shown in Figure 21, some areas covered with mangroves in 2020 disappeared by 2024; however, over half of those mangroves were still present in the Google Earth images from January 2023, indicating rapid and recent cutting.

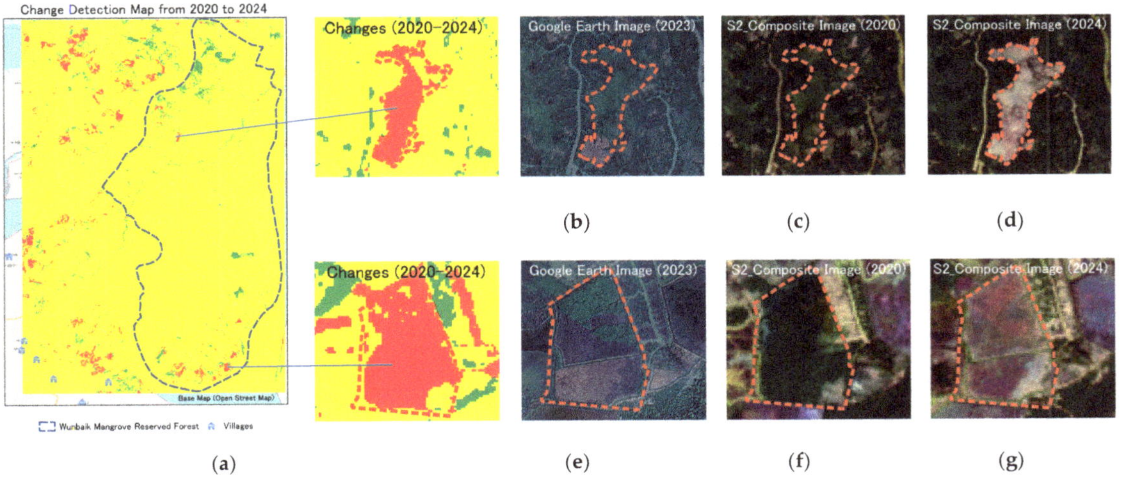

Figure 21. Mangrove losses after 2020: (**a**) changes from 2020 to 2024; (**b**) Google Earth image of 2023; (**c**) Sentinel-2 composite image of 2020; (**d**) Sentinel-2 composite image of 2024; (**e**) Google Earth image of 2023; (**f**) Sentinel-2 composite image of 2020; and (**g**) Sentinel-2 composite image of 2024. (The red line represents the areas of change with dark green for mangroves, light and dark brown indicating newly cut areas for aquaculture).

According to change detection from 2020 to 2024, most losses were outside the reserved area. Approximately half of those losses were detected close to the local villages. The proximity of those losses near the villages suggests that local communities could contribute to deforestation. The impact of local villages on mangrove loss underscores the need for community-based mangrove conservation. Therefore, the deep learning model effectively identified all changes and can be applied to real-time mangrove mapping.

4.7. Mangrove Gains After Restoration

Mangrove restoration is a nature-based solution (NbS) that supports biodiversity and climate change mitigation. Mangrove reforestation also benefits blue carbon storage. Therefore, mangrove restoration should be set as a priority when designing NbS [4]. Despite the numerous benefits of restoration, the systematic assessment and documentation of its achievements are still limited, especially in Myanmar [66].

Mangroves have been increasing in the plantation sites within WMF, initiated by the Forest Department [23]. The present study also identified the numerous patches of mangrove plantations in WMF, referencing the Google Earth images. Additionally, change detection summarized the increased areas of mangrove forests in WMF since 2010. Therefore, the increased areas were compared with the data on the implementation of mangrove plantations by the Forest Department.

The Forest Department planted 855 acres (346 ha) of mangrove trees in WRMF from 2017 to 2023. The model predicted 991.59 ha of mangrove gains from 2015 to 2024. Mangrove plantations of 346 ha accounted for 34.89% of the mangrove gains observed from 2015 to 2024. These percentages of planted mangroves did not fully represent all the increased mangroves identified in the change detection results because the model failed to predict newly planted mangroves accurately, classifying them as non-mangroves due to high soil reflection and low vegetation reflection. Therefore, the restoration efforts could represent approximately 30% of the total increased areas of mangrove forests from 2015 to 2024, with the remaining areas likely consisting of naturally recovered mangroves.

A lot of restoration programs have been implemented by the Forest Department, not only in WMF but also across the country. The present model detected the increased extent of mangrove forests introduced by restoration programs and noticed relatively small areas for 10 m spatial resolution despite having limitations for newly cultivated mangroves. Therefore, our research could be useful for designing mangrove restoration strategies for conservationists and policymakers by utilizing the present model for the methodology to develop an advanced model and using the results of the change detection analysis.

One of the possibilities for increasing mangrove areas could be the development of a forest restoration program. The Myanmar Rehabilitation and Restoration Programme (MRRP), developed by the Ministry of Natural Resources and Environmental Conservation (MONREC), is a 10-year restoration program (2017/2018 to 2026/2027) to increase the forest land up to 30% of the total country area by 2030 and to reduce the deforestation around the country. This program aimed to achieve the targets of Nationally Determined Contributions (NDC) with the guidance of the Myanmar Forest Policy (1995) [46,47]. Therefore, the beginning year of MRRP was observed in the same year, 2017, when the Forest Department initiated mangrove plantations in WMF. This highlights the restoration policy's contribution to increasing mangrove areas in WMF.

5. Conclusions

This study developed a U-Net model using the multi-temporal and multi-source imagery of Landsat 8 and Sentinel-2 to predict mangrove maps for long-term periods from 1990 to 2024. The model utilized four optical bands and SRTM as input features and achieved an accuracy of 99.73%, outperforming the existing models for mangrove classification. The development of a deep learning model fusing Landsat 8 and Sentinel-2 imagery for mangrove mapping could be documented as the first attempt due to our knowledge, although some studies existed with deep learning models fusing Sentinel-1 and Sentinel-2 for mangrove mapping. The present model still needed additional data to improve generalization and to support national-scale mangrove mapping. While fusing Landsat 8 and Sentinel-2 images slightly enhanced the model by reducing the need for huge datasets, discrepancies in the estimated mangrove extents persisted across different satellite imagery, such as Landsat series, Sentinel-2, HLS L30 and HLS S30, and different spatial resolutions, such as Landsat 8 30 m and Landsat 8 10 m with resampling. These significant variations highlight the need for further investigation to address these challenges in future studies.

Change detection highlighted significant losses of mangrove forests from 1990 to 2010 and steady gains from 2010 to 2024. In WMF, 29.3% of the mangrove extent has been deforested, while only 5.75% has been reforested with −224.52 ha/yr of annual rate of changes over 34 years. Anthropogenic activities, such as shrimp ponds and paddy fields with weak regulations, were drivers of mangrove losses. By contrast, mangrove gains can be classified as planted mangroves, naturally recovered mangroves grown in abandoned sites, sedimentation areas, and along roadsides. Mangrove losses have continued in WMF despite gains where 30% were attributed to mangrove plantations and the remaining percentage to natural regeneration. The naturally recovered mangroves in the abandoned sites near the villages and outside the reserved area can potentially be lost again by human encroachment. Therefore, mangrove forests should be monitored continuously using the improved model, which helps reduce deforestation, monitors the healthy conditions of planted mangroves, provides accurate information, and supports blue carbon conservation strategies. Further research should develop a model to classify the natural and planted mangroves to better understand the contribution of restoration programs to mangrove gains.

Author Contributions: Conceptualization, methodology, modeling, analysis, and visualization, K.S.W.; writing—original draft preparation, K.S.W.; writing—review and editing, J.S.; supervision and conceptualization, J.S. All authors have read and agreed to the published version of the manuscript.

Funding: This study received no external funding.

Data Availability Statement: The data presented in the study are available at the request of the first author.

Acknowledgments: The first author is indebted to the Asian Development Bank for financial assistance during the study in Japan. The authors would like to express sincere thanks to Win Sithu Maung for permission to use his ground truth image, the Forest Department of Myanmar for plantation data, and the anonymous reviewers for their careful reading of the manuscript and insightful comments. The computation was carried out using the computer resource offered under the category of General Projects by the Research Institute for Information Technology, Kyushu University, and the FUJITSU Server PRIMERGY GX2570 (Wisteria/BDEC-01) at the Information Technology Center, the University of Tokyo.

Conflicts of Interest: The authors declare no conflicts of interest.

Appendix A

Table A1. Specifications of the satellite imagery used in the study.

Satellite Name	Acquired Date	Data Type/Product Type	Cloud Coverage
Landsat 5	2 January 1990	TM_L2SP	0
Landsat 5	5 March 1995	TM_L2SP	0
Landsat 5	14 January 2000	TM_L2SP	0
Landsat 5	27 January 2005	TM_L2SP	0
Landsat 5	10 February 2010	TM_L2SP	0
Landsat 8	23 January 2015	OLI_TIRS_L2SP	2
Landsat 8 *	18 November 2019	OLI_TIRS_L2SP	0
Landsat 8 *	4 December 2019	OLI_TIRS_L2SP	2
Landsat 8 *	21 January 2020	OLI_TIRS_L2SP	0
Landsat 9	8 January 2024	OLI_TIRS_L2SP	0
Sentinel-2	23 November 2015	S2MSI2A	5
Sentinel-2	27 December 2016	S2MSI2A	0
Sentinel-2	2 December 2017	S2MSI2A	0
Sentinel-2	31 January 2018	S2MSI2A	0
Sentinel-2 *	22 November 2019	S2MSI2A	0
Sentinel-2 *	22 December 2019	S2MSI2A	1
Sentinel-2 *	21 January 2020	S2MSI2A	0
Sentinel-2	25 January 2021	S2MSI2A	0
Sentinel-2	9 February 2022	S2MSI2A	0
Sentinel-2	4 February 2023	S2MSI2A	0
Sentinel-2	20 January 2024	S2MSI2A	0
HLS L30	25 December 2015	HLS Landsat OLI	0
HLS L30	1 March 2017	HLS Landsat OLI	0
HLS L30	21 January 2020	HLS Landsat OLI	0
HLS L30	10 January 2022	HLS Landsat OLI	0
HLS L30	8 January 2024	HLS Landsat OLI	0
HLS S30	23 December 2015	HLS Sentinel-2 MSI	0
HLS S30	16 January 2017	HLS Sentinel-2 MSI	0
HLS S30	5 February 2020	HLS Sentinel-2 MSI	0
HLS S30	25 January 2022	HLS Sentinel-2 MSI	0
HLS S30	20 January 2024	HLS Sentinel-2 MSI	0

Satellite imagery with * was used for training the multi-source model.

Appendix B

Table A2. Experimental results of input feature selection using ANN model with 30 epochs.

No.	Groups of Input Features	Accuracy (%)
1	7 Bands (B1, B2, B3, B4, B5, B6, B7)	57.06
2	5 Bands (B2, B3, B4, B5, B6) (BGRNIR, SWIR)	54.77
3	4 Bands (B2, B3, B4, B5)	54.77
4	NDVI, NDWI, SAVI, CMRI	92.04
5	NDVI, NDWI, SAVI, CMRI, MERIT	93.41
6	NDVI, NDWI, SAVI, CMRI, SRTM	93.67
7	NDVI, SAVI, CMRI, NDWI, Slope	92.85
8	NDVI, NDWI, SAVI, CMRI, CHM	92.42
9	NDVI, NDWI, SAVI, CMRI, SRTM, CHM	93.80
10	NDVI, NDWI, SAVI, CMRI, MERIT, CHM	93.37
11	NDVI, NDWI, SAVI, CMRI, SRTM, MERIT, CHM	93.63
12	NDVI, NDWI, SAVI, CMRI, SRTM, MERIT, CHM, Slope	93.34
13	NDVI, NDWI, SAVI, SRTM	93.43
14	NDVI, SAVI, CMRI, SRTM	93.71
15	NDVI, NDWI, CMRI, SRTM	93.65
16	4 Bands, NDVI, NDWI, SAVI, CMRI, SRTM	93.53
17	5 Bands, NDVI, NDWI, SAVI, CMRI, SRTM	93.72
18	4 Bands, NDVI, NDWI, SAVI, CMRI, MERIT	93.47
19	5 Bands, NDVI, NDWI, SAVI, CMRI, MERIT	93.59
20	4 Bands, NDVI, NDWI, SAVI, CMRI, SRTM, CHM	93.74
21	5 Bands, NDVI, NDWI, SAVI, CMRI, SRTM, CHM	93.58
22	4 Bands, NDVI, SAVI, CMRI, SRTM	93.64
23	5 Bands, NDVI, SAVI, CMRI, SRTM	93.59
24	4 Bands, NDVI, NDWI, CMRI, SRTM	93.25
25	5 Bands, NDVI, NDWI, CMRI, SRTM	93.44
26	NDVI, SAVI, CMRI, MERIT	93.43
27	NDVI, SAVI, CMRI, CHM	92.81
28	4 Bands, SRTM	69.79
29	3 Bands (B3, B4, B5), SRTM	68.90
30	7 Bands, NDVI, NDWI, SAVI, CMRI, SRTM, CHM	93.47

Table A3. Experimental results of input feature selection using CNN model with 100 epochs.

No.	Groups of Input Features	Accuracy (%)
1	NDVI, NDWI, SAVI, CMRI, SRTM, CHM	95.93
2	4 Bands	54.91
3	4 Bands, SRTM	95.99
4	3 Bands, SRTM	95.85

References

1. FAO. *The World's Mangroves 2000–2020*; FAO: Rome, Italy, 2023; ISBN 978-92-5-138004-8.
2. Mitra, A. Mangroves: A Unique Gift of Nature. In *Sensitivity of Mangrove Ecosystem to Changing Climate*; Mitra, A., Ed.; Springer: New Delhi, India, 2013; pp. 33–105. ISBN 978-81-322-1509-7.
3. Spalding, M.; Kainuma, M.; Collins, L. *World Atlas of Mangroves*; Routledge: London, UK, 2011; ISBN 978-1-84977-660-8.
4. Song, S.; Ding, Y.; Li, W.; Meng, Y.; Zhou, J.; Gou, R.; Zhang, C.; Ye, S.; Saintilan, N.; Krauss, K.W.; et al. Mangrove Reforestation Provides Greater Blue Carbon Benefit than Afforestation for Mitigating Global Climate Change. *Nat. Commun.* **2023**, *14*, 756. [CrossRef] [PubMed]
5. van Hespen, R.; Hu, Z.; Borsje, B.; De Dominicis, M.; Friess, D.A.; Jevrejeva, S.; Kleinhans, M.G.; Maza, M.; van Bijsterveldt, C.E.J.; Van der Stocken, T.; et al. Mangrove Forests as a Nature-Based Solution for Coastal Flood Protection: Biophysical and Ecological Considerations. *Water Sci. Eng.* **2023**, *16*, 1–13. [CrossRef]
6. Bunting, P.; Rosenqvist, A.; Hilarides, L.; Lucas, R.M.; Thomas, N.; Tadono, T.; Worthington, T.A.; Spalding, M.; Murray, N.J.; Rebelo, L.-M. Global Mangrove Extent Change 1996–2020: Global Mangrove Watch Version 3.0. *Remote Sens.* **2022**, *14*, 3657. [CrossRef]

7. Kuenzer, C.; Bluemel, A.; Gebhardt, S.; Quoc, T.V.; Dech, S. Remote Sensing of Mangrove Ecosystems: A Review. *Remote Sens.* **2011**, *3*, 878–928. [CrossRef]
8. Veettil, B.K.; Pereira, S.F.R.; Quang, N.X. Rapidly Diminishing Mangrove Forests in Myanmar (Burma): A Review. *Hydrobiologia* **2018**, *822*, 19–35. [CrossRef]
9. Giri, C.; Ochieng, E.; Tieszen, L.L.; Zhu, Z.; Singh, A.; Loveland, T.; Masek, J.; Duke, N. Status and Distribution of Mangrove Forests of the World Using Earth Observation Satellite Data. *Glob. Ecol. Biogeogr.* **2011**, *20*, 154–159. [CrossRef]
10. Maurya, K.; Mahajan, S.; Chaube, N. Remote Sensing Techniques: Mapping and Monitoring of Mangrove Ecosystem—A Review. *Complex Intell. Syst.* **2021**, *7*, 2797–2818. [CrossRef]
11. Lu, Y.; Wang, L. The Current Status, Potential and Challenges of Remote Sensing for Large-Scale Mangrove Studies. *Int. J. Remote Sens.* **2022**, *43*, 6824–6855. [CrossRef]
12. Wang, L.; Jia, M.; Yin, D.; Tian, J. A Review of Remote Sensing for Mangrove Forests: 1956–2018. *Remote Sens. Environ.* **2019**, *231*, 111223. [CrossRef]
13. Vasquez, J.; Acevedo-Barrios, R.; Miranda-Castro, W.; Guerrero, M.; Meneses-Ospina, L. Determining Changes in Mangrove Cover Using Remote Sensing with Landsat Images: A Review. *Water. Air. Soil Pollut.* **2023**, *235*, 18. [CrossRef]
14. Wang, D.; Wan, B.; Qiu, P.; Su, Y.; Guo, Q.; Wang, R.; Sun, F.; Wu, X. Evaluating the Performance of Sentinel-2, Landsat 8 and Pléiades-1 in Mapping Mangrove Extent and Species. *Remote Sens.* **2018**, *10*, 1468. [CrossRef]
15. Nasiri, V.; Deljouei, A.; Moradi, F.; Sadeghi, S.M.M.; Borz, S.A. Land Use and Land Cover Mapping Using Sentinel-2, Landsat-8 Satellite Images, and Google Earth Engine: A Comparison of Two Composition Methods. *Remote Sens.* **2022**, *14*, 1977. [CrossRef]
16. Bouslihim, Y.; Kharrou, M.H.; Miftah, A.; Attou, T.; Bouchaou, L.; Chehbouni, A. Comparing Pan-Sharpened Landsat-9 and Sentinel-2 for Land-Use Classification Using Machine Learning Classifiers. *J. Geovisualization Spat. Anal.* **2022**, *6*, 35. [CrossRef]
17. Jia, M.; Wang, Z.; Mao, D.; Ren, C.; Song, K.; Zhao, C.; Wang, C.; Xiao, X.; Wang, Y. Mapping Global Distribution of Mangrove Forests at 10-m Resolution. *Sci. Bull.* **2023**, *68*, 1306–1316. [CrossRef]
18. Chaves, M.E.D.; Picoli, M.C.A.; Sanches, I.D. Recent Applications of Landsat 8/OLI and Sentinel-2/MSI for Land Use and Land Cover Mapping: A Systematic Review. *Remote Sens.* **2020**, *12*, 3062. [CrossRef]
19. Ma, C.; Ai, B.; Zhao, J.; Xu, X.; Huang, W. Change Detection of Mangrove Forests in Coastal Guangdong during the Past Three Decades Based on Remote Sensing Data. *Remote Sens.* **2019**, *11*, 921. [CrossRef]
20. Baltezar, P.; Murillo-Sandoval, P.J.; Cavanaugh, K.C.; Doughty, C.; Lagomasino, D.; Tieng, T.; Simard, M.; Fatoyinbo, T. A Regional Map of Mangrove Extent for Myanmar, Thailand, and Cambodia Shows Losses of 44% by 1996. *Front. Mar. Sci.* **2023**, *10*, 1127720. [CrossRef]
21. De Alban, J.D.T.; Jamaludin, J.; Wong De Wen, D.; Than, M.M.; Webb, E.L. Improved Estimates of Mangrove Cover and Change Reveal Catastrophic Deforestation in Myanmar. *Environ. Res. Lett.* **2020**, *15*, 034034. [CrossRef]
22. Tran, T.V.; Reef, R.; Zhu, X. A Review of Spectral Indices for Mangrove Remote Sensing. *Remote Sens.* **2022**, *14*, 4868. [CrossRef]
23. Maung, W.S.; Sasaki, J. Assessing the Natural Recovery of Mangroves after Human Disturbance Using Neural Network Classification and Sentinel-2 Imagery in Wunbaik Mangrove Forest, Myanmar. *Remote Sens.* **2021**, *13*, 52. [CrossRef]
24. Bunting, P.; Rosenqvist, A.; Lucas, R.M.; Rebelo, L.-M.; Hilarides, L.; Thomas, N.; Hardy, A.; Itoh, T.; Shimada, M.; Finlayson, C.M. The Global Mangrove Watch—A New 2010 Global Baseline of Mangrove Extent. *Remote Sens.* **2018**, *10*, 1669. [CrossRef]
25. Weber, S.; Keddell, L.; Kemal, M.S. *Myanmar Ecological Forecasting: Utilizing NASA Earth Observations to Monitor, Map, and Analyze Mangrove Forests in Myanmar for Enhanced Conservation*; NASA: Greenbelt, MD, USA, 2014.
26. Guo, Y.; Liao, J.; Shen, G. Mapping Large-Scale Mangroves along the Maritime Silk Road from 1990 to 2015 Using a Novel Deep Learning Model and Landsat Data. *Remote Sens.* **2021**, *13*, 245. [CrossRef]
27. Dong, Y.; Yu, K.; Hu, W. GC-UNet: An Improved UNet Model for Mangrove Segmentation Using Landsat8. In Proceedings of the 2021 3rd International Conference on Big Data Engineering, Cox's Bazar, Bangladesh, 23–25 September 2021; Association for Computing Machinery: New York, NY, USA, 2021; pp. 58–63.
28. Toosi, N.B.; Soffianian, A.R.; Fakheran, S.; Pourmanafi, S.; Ginzler, C.; Waser, L.T. Comparing Different Classification Algorithms for Monitoring Mangrove Cover Changes in Southern Iran. *Glob. Ecol. Conserv.* **2019**, *19*, e00662. [CrossRef]
29. Sharifi, A.; Felegari, S.; Tariq, A. Mangrove Forests Mapping Using Sentinel-1 and Sentinel-2 Satellite Images. *Arab. J. Geosci.* **2022**, *15*, 1593. [CrossRef]
30. Elmahdy, S.I.; Ali, T.A.; Mohamed, M.M.; Howari, F.M.; Abouleish, M.; Simonet, D. Spatiotemporal Mapping and Monitoring of Mangrove Forests Changes from 1990 to 2019 in the Northern Emirates, UAE Using Random Forest, Kernel Logistic Regression and Naive Bayes Tree Models. *Front. Environ. Sci.* **2020**, *8*, 102. [CrossRef]
31. Rosmasita; Siregar, V.P.; Agus, S.B.; Jhonnerie, R. An Object-Based Classification of Mangrove Land Cover Using Support Vector Machine Algorithm. *IOP Conf. Ser. Earth Environ. Sci.* **2019**, *284*, 012024. [CrossRef]
32. Estoque, R.C.; Myint, S.W.; Wang, C.; Ishtiaque, A.; Aung, T.T.; Emerton, L.; Ooba, M.; Hijioka, Y.; Mon, M.S.; Wang, Z.; et al. Assessing Environmental Impacts and Change in Myanmar's Mangrove Ecosystem Service Value Due to Deforestation (2000–2014). *Glob. Change Biol.* **2018**, *24*, 5391–5410. [CrossRef]
33. Kattenborn, T.; Leitloff, J.; Schiefer, F.; Hinz, S. Review on Convolutional Neural Networks (CNN) in Vegetation Remote Sensing. *ISPRS J. Photogramm. Remote Sens.* **2021**, *173*, 24–49. [CrossRef]
34. Reichstein, M.; Camps-Valls, G.; Stevens, B.; Jung, M.; Denzler, J.; Carvalhais, N. Prabhat Deep Learning and Process Understanding for Data-Driven Earth System Science. *Nature* **2019**, *566*, 195–204. [CrossRef] [PubMed]

35. Wei, Y.; Cheng, Y.; Yin, X.; Xu, Q.; Ke, J.; Li, X. Deep Learning-Based Classification of High-Resolution Satellite Images for Mangrove Mapping. *Appl. Sci.* **2023**, *13*, 8526. [CrossRef]
36. Liu, Y.; Zhang, Y.; Cheng, Q.; Feng, J.; Chun Chao, M.; Yeu Tsou, J. Mangrove Monitoring and Change Analysis with Landsat Images: A Case Study in Pearl River Estuary (China). *Ecol. Indic.* **2024**, *160*, 111763. [CrossRef]
37. Guo, M.; Yu, Z.; Xu, Y.; Huang, Y.; Li, C. ME-Net: A Deep Convolutional Neural Network for Extracting Mangrove Using Sentinel-2A Data. *Remote Sens.* **2021**, *13*, 1292. [CrossRef]
38. Ghorbanian, A.; Ahmadi, S.A.; Amani, M.; Mohammadzadeh, A.; Jamali, S. Application of Artificial Neural Networks for Mangrove Mapping Using Multi-Temporal and Multi-Source Remote Sensing Imagery. *Water* **2022**, *14*, 244. [CrossRef]
39. Xu, C.; Wang, J.; Sang, Y.; Li, K.; Liu, J.; Yang, G. An Effective Deep Learning Model for Monitoring Mangroves: A Case Study of the Indus Delta. *Remote Sens.* **2023**, *15*, 2220. [CrossRef]
40. Shao, Z.; Cai, J.; Fu, P.; Hu, L.; Liu, T. Deep Learning-Based Fusion of Landsat-8 and Sentinel-2 Images for a Harmonized Surface Reflectance Product. *Remote Sens. Environ.* **2019**, *235*, 111425. [CrossRef]
41. Friess, D.A.; Rogers, K.; Lovelock, C.E.; Krauss, K.W.; Hamilton, S.E.; Lee, S.Y.; Lucas, R.; Primavera, J.; Rajkaran, A.; Shi, S. The State of the World's Mangrove Forests: Past, Present, and Future. *Annu. Rev. Environ. Resour.* **2019**, *44*, 89–115. [CrossRef]
42. Stanley, O.; Broadhead, J.; Myint, A.A. *The Atlas and Guidelines for Mangrove Management in Wunbaik Reserved Forest: Sustainable Community Based Mangrove Management in Wunbaik Forest Reserve TCP/MYA/3204 (2009–2011)*; FAO-UN Myanmar: FAO Representation Office, Seed Division Compound: Yangon, Myanmar, 2011.
43. Saw, A.A.; Kanzaki, M. Local Livelihoods and Encroachment into a Mangrove Forest Reserve: A Case Study of the Wunbaik Reserved Mangrove Forest, Myanmar. *Procedia Environ. Sci.* **2015**, *28*, 483–492. [CrossRef]
44. Gandhi, S.; Jones, T.G. Identifying Mangrove Deforestation Hotspots in South Asia, Southeast Asia and Asia-Pacific. *Remote Sens.* **2019**, *11*, 728. [CrossRef]
45. Gaw, L.Y.F.; Linkie, M.; Friess, D.A. Mangrove Forest Dynamics in Tanintharyi, Myanmar from 1989–2014, and the Role of Future Economic and Political Developments. *Singap. J. Trop. Geogr.* **2018**, *39*, 224–243. [CrossRef]
46. Sann, B.; Brunner, J.; Brander, L. *Report on Cost-Benefit Analysis of Forest Restoration Interventions in Sagaing Region, Myanmar*; IUCN: Viet Nam Country Office: Hanoi, Vietnam, 2021; p. 23.
47. MONREC. *Myanmar Updated Nationally Determined Contributions-NDC; Ministry of Natural Resources and Environmental Conservation (MONREC), Naypyitaw, Myanmar.* 2021. Available online: https://unfccc.int/sites/default/files/NDC/2022-06/Myanmar%20Updated%20%20NDC%20July%202021.pdf (accessed on 8 September 2024).
48. Maung, W.S.; Tsuyuki, S.; Guo, Z. Improving Land Use and Land Cover Information of Wunbaik Mangrove Area in Myanmar Using U-Net Model with Multisource Remote Sensing Datasets. *Remote Sens.* **2024**, *16*, 76. [CrossRef]
49. Maung, W.S. *Assesment of Natural Recovery of Mangrove from Anthropogenic Disturbance Using Neural Network-Based Classification of Satellite Images*; The University of Tokyo: Tokyo, Japan, 2020.
50. Congedo, L.; Barsukov, I. Semiautomaticgit/Remotior_sensus: V0.3.5 2024. Zenedo. Available online: https://zenodo.org/records/10456752.
51. Claverie, M.; Ju, J.; Masek, J.G.; Dungan, J.L.; Vermote, E.F.; Roger, J.-C.; Skakun, S.V.; Justice, C. The Harmonized Landsat and Sentinel-2 Surface Reflectance Data Set. *Remote Sens. Environ.* **2018**, *219*, 145–161. [CrossRef]
52. Farr, T.G.; Kobrick, M. Shuttle Radar Topography Mission Produces a Wealth of Data. *Eos Trans. Am. Geophys. Union* **2000**, *81*, 583–585. [CrossRef]
53. Yamazaki, D.; Ikeshima, D.; Tawatari, R.; Yamaguchi, T.; O'Loughlin, F.; Neal, J.C.; Sampson, C.C.; Kanae, S.; Bates, P.D. A High-Accuracy Map of Global Terrain Elevations. *Geophys. Res. Lett.* **2017**, *44*, 5844–5853. [CrossRef]
54. Bharathi, N.; Rishiikeshwer, B.S.; Shriram, T.A.; Santhi, B.; Brindha, G.R. The Significance of Feature Selection Techniques in Machine Learning. In *Fundamentals and Methods of Machine and Deep Learning*; John Wiley & Sons, Ltd.: Hoboken, NJ, USA, 2022; pp. 121–134. ISBN 978-1-119-82190-8.
55. Rouse, J.W.; Haas, R.H.; Schell, J.A.; Deering, D.W. *Monitoring Vegetation Systems in the Great Plains with ERTS*; NASA: Greenbelt, MD, USA, 1974.
56. McFEETERS, S.K. The Use of the Normalized Difference Water Index (NDWI) in the Delineation of Open Water Features. *Int. J. Remote Sens.* **1996**, *17*, 1425–1432. [CrossRef]
57. Huete, A.R. A Soil-Adjusted Vegetation Index (SAVI). *Remote Sens. Environ.* **1988**, *25*, 295–309. [CrossRef]
58. Gupta, K.; Mukhopadhyay, A.; Giri, S.; Chanda, A.; Datta Majumdar, S.; Samanta, S.; Mitra, D.; Samal, R.N.; Pattnaik, A.K.; Hazra, S. An Index for Discrimination of Mangroves from Non-Mangroves Using LANDSAT 8 OLI Imagery. *MethodsX* **2018**, *5*, 1129–1139. [CrossRef]
59. Hawker, L.; Uhe, P.; Paulo, L.; Sosa, J.; Savage, J.; Sampson, C.; Neal, J. A 30 m Global Map of Elevation with Forests and Buildings Removed. *Environ. Res. Lett.* **2022**, *17*, 024016. [CrossRef]
60. Simard, M.; Rivera-Monroy, V.H.; Mancera-Pineda, J.E.; Castañeda-Moya, E.; Twilley, R.R. A Systematic Method for 3D Mapping of Mangrove Forests Based on Shuttle Radar Topography Mission Elevation Data, ICEsat/GLAS Waveforms and Field Data: Application to Ciénaga Grande de Santa Marta, Colombia. *Remote Sens. Environ.* **2008**, *112*, 2131–2144. [CrossRef]
61. Aslan, A.; Aljahdali, M.O. Characterizing Global Patterns of Mangrove Canopy Height and Aboveground Biomass Derived from SRTM Data. *Forests* **2022**, *13*, 1545. [CrossRef]

62. Chen, L.; Li, S.; Bai, Q.; Yang, J.; Jiang, S.; Miao, Y. Review of Image Classification Algorithms Based on Convolutional Neural Networks. *Remote Sens.* **2021**, *13*, 4712. [CrossRef]
63. Ronneberger, O.; Fischer, P.; Brox, T. U-Net: Convolutional Networks for Biomedical Image Segmentation. In Proceedings of the Medical Image Computing and Computer-Assisted Intervention–MICCAI 2015: 18th International Conference, Munich, Germany, 5–9 October 2015.
64. Singh, A. Review Article Digital Change Detection Techniques Using Remotely-Sensed Data. *Int. J. Remote Sens.* **1989**, *10*, 989–1003. [CrossRef]
65. Wu, C.; Du, B.; Cui, X.; Zhang, L. A Post-Classification Change Detection Method Based on Iterative Slow Feature Analysis and Bayesian Soft Fusion. *Remote Sens. Environ.* **2017**, *199*, 241–255. [CrossRef]
66. Gerona-Daga, M.E.B.; Salmo, S.G.I. A Systematic Review of Mangrove Restoration Studies in Southeast Asia: Challenges and Opportunities for the United Nation's Decade on Ecosystem Restoration. *Front. Mar. Sci.* **2022**, *9*, 987737. [CrossRef]

remote sensing MDPI

Article

Integrating SAR, Optical, and Machine Learning for Enhanced Coastal Mangrove Monitoring in Guyana

Kim Chan-Bagot [1,2,*], Kelsey E. Herndon [3,4], Andréa Puzzi Nicolau [2,5], Vanesa Martín-Arias [3,4], Christine Evans [3,4], Helen Parache [6], Kene Mosely [1], Zola Narine [1] and Brian Zutta [2,5]

[1] National Agricultural and Research Extension Institute (NAREI), Georgetown, Guyana; kmoseley@narei.gov.gy (K.M.); znarine@narei.gov.gy (Z.N.)
[2] SERVIR Amazonia, Cali 76001, Colombia; apnicolau@sig-gis.com (A.P.N.); bzutta@sig-gis.com (B.Z.)
[3] Earth System Science Center, University of Alabama in Huntsville, Huntsville, AL 35899, USA; keh0023@uah.edu (K.E.H.); vm0043@uah.edu (V.M.-A.); cae0004@uah.edu (C.E.)
[4] NASA SERVIR Science Coordination Office, NASA Marshall Space Flight Center, Huntsville, AL 35812, USA
[5] Spatial Informatics Group (SIG), San Francisco, CA 94566, USA
[6] NASA IMPACT, NASA Marshall Space Flight Center, Huntsville, AL 35812, USA; helen.blue.parache@nasa.gov
* Correspondence: kbagot19@gmail.com

Abstract: Mangrove forests are a biodiverse ecosystem known for a wide variety of crucial ecological services, including carbon sequestration, coastal erosion control, and prevention of saltwater intrusion. Given the ecological importance of mangrove forests, a comprehensive and up-to-date mangrove extent mapping at broad geographic scales is needed to define mangrove forest changes, assess their implications, and support restoration activities and decision making. The main objective of this study is to evaluate mangrove classifications derived from a combination of Landsat-8 OLI, Sentinel-2, and Sentinel-1 observations using a random forest (RF) machine learning (ML) algorithm to identify the best approach for monitoring Guyana's mangrove forests on an annual basis. Algorithm accuracy was tested using high-resolution planet imagery in Collect Earth Online. Results varied widely across the different combinations of input data (overall accuracy, 88–95%; producer's accuracy for mangroves, 50–87%; user's accuracy for mangroves, 13–69%). The combined optical–radar classification demonstrated the best performance with an overall accuracy of 95%. Area estimates of mangrove extent ranged from 908.4 to 3645.0 hectares. A ground-based validation exercise confirmed the extent of several large, previously undocumented areas of mangrove forest loss. The results establish that a data fusion approach combining optical and radar data performs marginally better than optical-only approaches to mangrove classification. This ML approach, which leverages free and open data and a cloud-based analytics platform, can be applied to mapping other areas of mangrove forests in Guyana. This approach can also support the operational monitoring of mangrove restoration areas managed by Guyana's National Agricultural and Research Extension Institute (NAREI).

Keywords: remote sensing; blue carbon; wetlands; sustainable development goals; Google Earth Engine; random forest

1. Introduction

Mangroves are a woody halophyte inhabiting the upper intertidal zones of saltwater areas, primarily in tropical and subtropical regions [1–3]. These forests grow along muddy beaches with low hydrodynamic intensity, particularly in areas with extensive river estuaries and deltas where sediments are continuously supplied by upland catchments [4]. Globally, mangroves are estimated to occupy approximately 152,000 km^2 across 124 countries [5]. Mangrove species can be found in almost 75% of the world's coastlines, between latitudes of 25° north and south [1,5,6]. Mangroves have a number of attributes that make them physically and ecologically distinct. These characteristics include aerial roots, high rates of canopy production, extremely efficient nutrient storage mechanisms, capacity to adapt to saline water, and rapid carbon sequestration, among other morphological characteristics [7].

Mangrove forests are among the most productive and important ecosystems in the world because they provide a wide range of ecological and socioeconomic benefits to human society [7,8]. They provide valuable ecosystem services, including the prevention of coastal erosion, saltwater intrusion, and flooding of coastal communities. Mangrove forests also help to stabilize the coastline and subsequently mitigate the devastating impacts of natural disasters, such as tsunamis, erosion caused by hurricanes, and flooding [9]. In addition, mangroves regulate nutrients and sequester carbon. In fact, mangroves have a potential carbon sequestration rate that is 50 times greater than that of many other types of tropical forests due to their ability to store high levels of carbon below ground and in mangrove sediment soils [10]. Mangroves also support biological diversity by providing vital breeding and nursing grounds for many fishes, invertebrate species, and coastal animals. By providing habitats for a diverse array of marine and terrestrial animals, mangrove forests have the potential to boost ecotourism and support the local economy [11,12].

The three primary species of mangroves in Guyana are black mangroves (Avicennia germinans), red mangroves (Rhizophora mangle), and white mangroves (Laguncularia racemosa). Significant portions of the coast are productive monoculture stands of black mangroves [13,14], with secondary establishments of red mangrove usually within riverine environments. White mangroves vary in density along the coast [14]. Conocarpus erectus, popularly known as buttonwood, is considered a mangrove associate and usually grows at higher elevations within the mangrove ecosystem [15].

In Guyana, mangrove ecosystems play a vital role in forming a natural soft defense for coastal protection [16]. Approximately 90% of Guyana's population and economic activities reside on the coastal plain, which lies between 0.5 to 1 m below high and spring tide levels from the Atlantic Ocean, making the coast very vulnerable to flooding, sea level rise, and erosion from the ocean [17]. The mapping and monitoring of mangroves are important steps in gathering information on geomorphologic and environmental conditions [18,19]. This paper seeks to address the challenges of mapping mangrove ecosystems by using free and open remote sensing datasets and leveraging a cloud computing platform. Mangroves have distinct spectral and spatial characteristics that are distinguishable in satellite images, making remote sensing an effective technique for mapping and monitoring this important ecosystem [9]. Remote sensing techniques can be used to estimate the mangrove forest cover that extends over large areas, especially in areas that are inaccessible or costly to access in field surveys [19].

Background

The National Agricultural Research and Extension Institute (NAREI) in Guyana has mapped mangroves manually from 2010 to 2013 as part of the Guyana Mangrove Restoration Project initiative. Subsequently, the mandate to restore and manage mangroves was permanently integrated into NAREI [20], with 2018 being the first year remote sensing was introduced into NAREI's monitoring operations. While fieldwork is an essential component of management efforts, the personnel, time, equipment, and logistical organization needed to conduct fieldwork make it a costly initiative. Remote sensing provides a cost-effective method by which to map mangroves, expanding the scope and scale of traditional fieldwork. The ability to use freely available data on cloud-based platforms can revolutionize the workflow of regularly monitoring environments like mangrove forests. Prior remote sensing work for mangroves in Guyana was carried out by Conservation International (CI), who developed a 2018 mangrove extent map based on a random forest (RF) classification of Sentinel-2 optical imagery [21]. The 2018 CI classification was run with multiple vegetation indices, and the resulting polygon of mangrove extent was manually edited using expert, in-region knowledge. This initial effort of mapping Guyana's mangrove extent using remote sensing noted the exclusive use of optical data as a limitation in the study and recommended incorporating lidar or radar data in mapping work, given the considerable cloud cover over Guyana's coastline.

In partnership with SERVIR, a joint program between the United States Agency for International Development (USAID) and NASA aimed at connecting decision makers across the world with earth observations, NAREI, in the context of SERVIR-Amazonia, is working to use remote sensing technology to develop a national mangrove monitoring system. This study supports these efforts by evaluating an optical, synthetic aperture radar (SAR), and fusion approach to generate annual mangrove extent maps as part of NAREI's monitoring and restoration efforts.

There have been extensive efforts to map global distributions of mangrove ecosystems [9,21–26]. Giri et al. (2005) [27] produced a global mangrove extent product for the year 2000 using Landsat Thematic Mapper (TM) imagery by applying an unsupervised ISODATA clustering algorithm combined with an iterative labeling process. Since then, numerous projects have attempted to improve upon these results and have expanded the types of remotely sensed data used to create mangrove classifications [21–23,26]. More recently, global products evaluating drivers of mangrove loss [24,28] have been developed to support decision making within conservation and restoration initiatives. The publication led by Bunting et al. (2018) [21] described the use of an extremely randomized tree classification to both SAR and optical data to produce the 2010 baseline for Global Mangrove Watch (GMW). Some studies employed a combination of optical and SAR data applied to a supervised machine learning (ML) approach [26,29]. Gumbricht et al. (2017) [25] exclusively used optical data in a supervised ML approach to map wetlands and peatlands throughout South America. Thomas et al. (2017) [24] conducted a manual visual interpretation of exclusively L-band SAR composites validated with optical imagery. Other studies extracted mangrove extent from several pre-existing databases, such as Terrestrial Ecosystems of the World, to create synthesized databases [22,23].

Techniques for mapping mangrove extent are clearly varied. In many cases, approaches exclusively using optical remote sensing [9,30] to map mangroves are insufficient due to persistent cloud cover in the tropics, the primary location of mangrove forests. Cloud-free optical imagery for mangrove mapping on an annual or subannual basis can be difficult to obtain [31]. Figure 1 illustrates the availability of Landsat-8 observations after cloud and cloud shadow are masked [32]. While Landsat-8 and Sentinel-2 data availability is relatively scarce here, the number of valid observations also varies greatly by year, demonstrating an opportunity to incorporate data unencumbered by persistent cloud cover to help fill these gaps. Specifically, as Figure 1 shows for the year 2020, there are many areas with fewer than approximately 10 cloud-free observations.

SAR data present multiple options and advantages for mapping mangrove forest ecosystems. Two critical characteristics of SAR data are different wavelengths, particularly X-band, C-band, and L-band listed from shortest to longest wavelength, and a variety of backscatter mechanisms [33]. The different wavelengths have different strengths and optimal applications, with L-band having the greatest capacity to penetrate vegetation canopies to reach the ground. Scattering mechanisms, namely, rough surface scattering, volume scattering, and double-bounce scattering, vary based on the wavelength and the surface characteristics of the terrain interacting with the radar signal [33]. Double-bounce scattering occurs when the signal bounces from a tall vertical structure to the ground, then back to the sensor. The double-bounce mechanism can be leveraged to detect understory floods or vegetated areas with high moisture content, such as mangrove forests [34].

Several other studies have demonstrated the suitability of SAR data across a variety of wavelengths for a wide range of applications, including mangrove mapping [29,35–39]. The study presented in this paper builds on these previous efforts by comparing a data fusion approach with optical-only and radar-only approaches. It also expands the evaluation of the algorithms to include an independent accuracy assessment and field validation exercise.

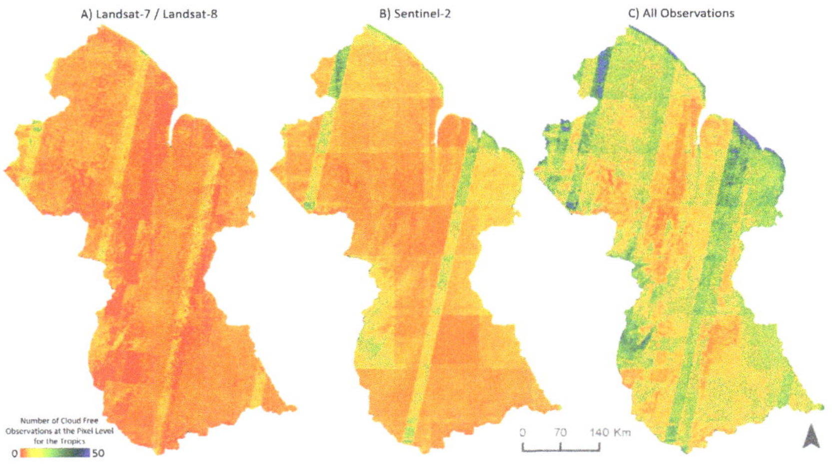

Figure 1. Cloud-free optical imagery (Landsat-7, Landsat-8, Sentinel-2) over Guyana for the year 2020.

The extent of Guyana's coastal mangroves varies greatly across global mangrove extent products, and guidelines for identifying which dataset is the most appropriate are limited (see the Sustainable Development Goals [40] for one perspective). In addition, these global datasets may not support local, national, or regional mangrove management and monitoring efforts as localized accuracy of the classification may not be optimized. The definitions and uncertainty of what is actually being mapped may also not meet local requirements (e.g., definition of mappable mangrove, appropriate time step, etc.). This emphasizes the need for more localized mangrove mapping products in some cases.

Since work on the restoration effort began in 2010, there has been a noticeable gap in Guyana's mangrove ecosystem systematic monitoring. For sustainable mangrove management, it is essential to comprehend the extent of the mangrove cover, the change (gains and losses), and the factors for change. Prior to this study, NAREI's mangrove monitoring program relied on field observations and Google Earth data to locate possible mangrove restoration locations and areas of natural deterioration and regeneration. Because losses were not identified until they were apparent from the coastline, reliance on ground assessments led to reactive rather than proactive management. This resulted in coastal flooding and the loss of arable land owing to saltwater intrusion in numerous cases. The study's methodology will enable yearly systematic monitoring of Guyana's coastal mangroves, providing NAREI information to support decision making regarding critical sites that need additional protection, eroding sites that need intervention, baseline data to support carbon storage evaluation, and areas that can support community livelihood projects.

This paper also provides the first comparison of mangrove extent mapping approaches in Guyana using remotely sensed data. This analysis is needed to identify the most appropriate technique to develop a mangrove monitoring service for Guyana in the context of NAREI and the SERVIR-Amazonia program. We evaluate the performance of an ML algorithm to classify three different data combinations to determine the best approach for mapping annual mangrove extent in a tropical, cloud-persistent, coastal ecosystem. The results from this analysis will inform an operational national mangrove monitoring system for Guyana.

2. Materials and Methods

2.1. Study Area

Guyana consists of 6 coastal regions: Barima-Waini (Region 1), Pomeroon-Supenaam (Region 2), Essequibo Islands-West Demerara (Region 3), Demerara-Mahaica (Region 4), Mahaica-Berbice (Region 5), and East Berbice-Corentyne (Region 6). Our study area in

Pomeroon-Supenaam (Region-2) is approximately 6195 km^2 and is bounded by the Atlantic Ocean to the north, the Pomeroon River Estuary to the west, the Essequibo River Estuary to the east, and extends all the way to the Supenaam River to the south (Figure 2) [41]. This region is of interest given that NAREI began mangrove restoration efforts in this region in 2013.

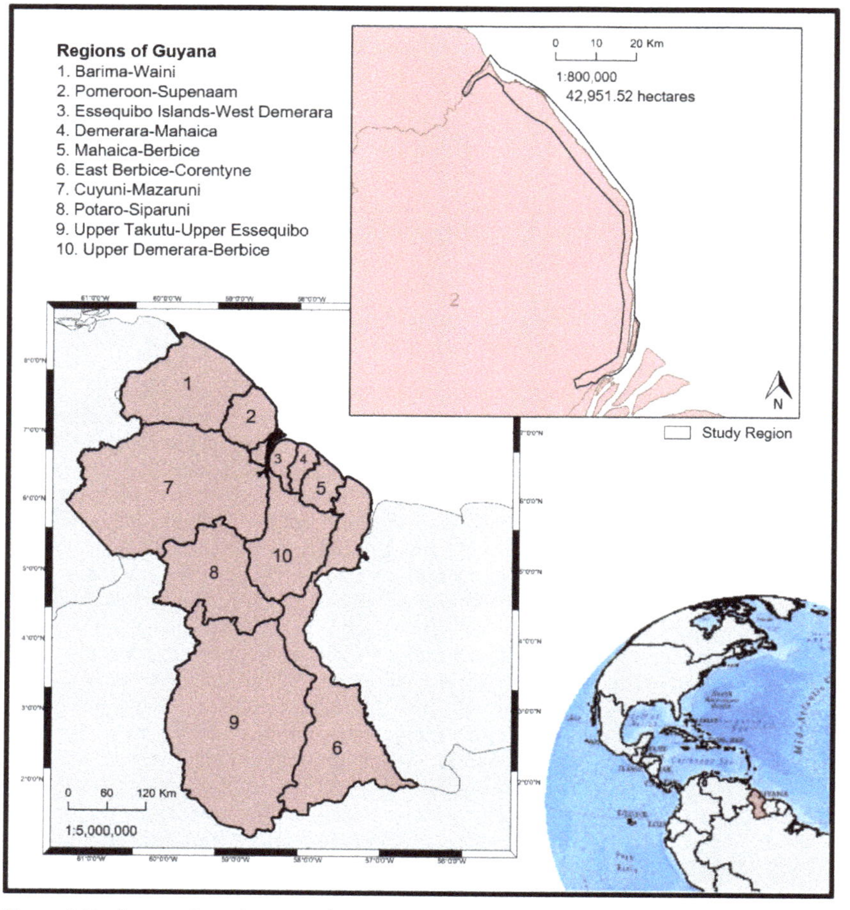

Figure 2. Study area along the coast of Pomeroon-Supenaam (Region 2), Guyana.

The coast of Pomeroon-Supenaam experiences a humid tropical climate, with an annual average rainfall that varies between 1778 and 2800 mm and an average daily temperature of 25 degrees Celsius. The warmest months are usually September and October, while the coldest are January and February [42]. The area has a relatively flat, low-lying elevation, with most of the population concentrated along the coast. The major land use and land cover (LULC) types include mangrove forests, wetlands, cropland, water, and human settlements. Rice cultivation is the primary economic activity in the region [41].

2.2. Data Acquisition and Preprocessing

The mapping workflow (Figure 3) encompasses the use of optical and radar datasets and an ML algorithm to map the extent of mangrove forests in the study area. The workflow includes dataset acquisition and preprocessing, ML model training and classification, model validation, and accuracy assessment.

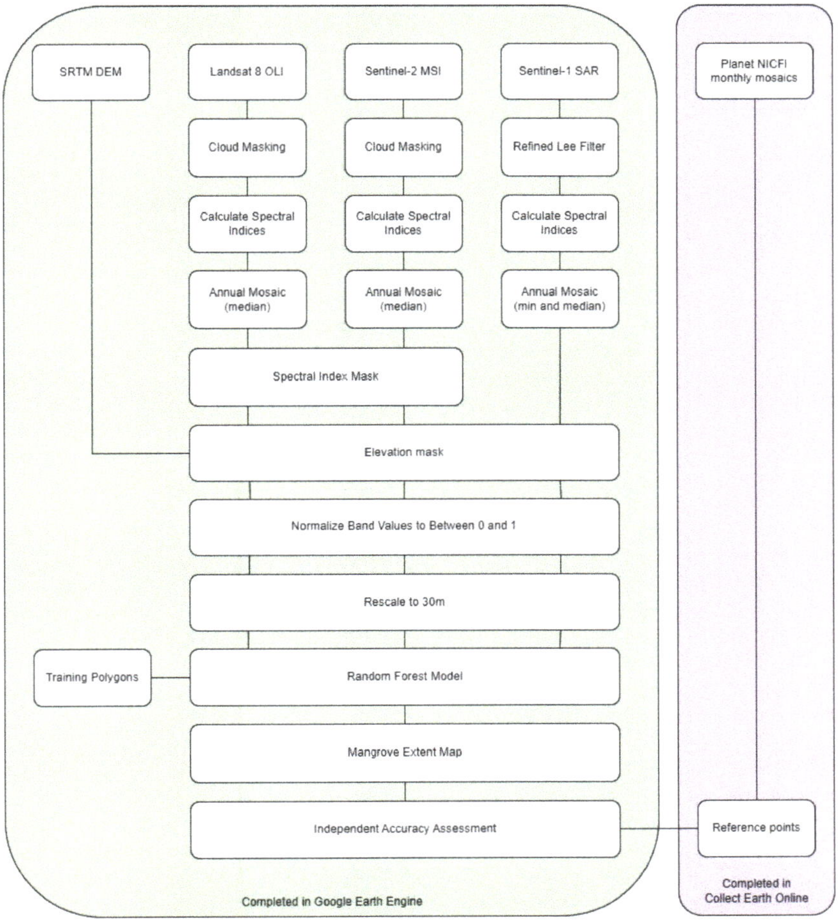

Figure 3. Mangrove extent mapping workflow.

Data from the Landsat-8 Operational Land Imager (OLI), Sentinel-2 MultiSpectral Instrument (MSI), and Sentinel-1 C-band SAR ground-range detected (GRD) product were acquired through Google Earth Engine (GEE) [43]. GEE is a cloud-based analysis and visualization platform with access to a large catalog of geospatial datasets. Given that satellite imagery, particularly SAR imagery, makes for very large files that can be slow to download and process on a local machine, a cloud-based platform was essential for this analysis and is required for NAREI's future mangrove monitoring system.

In GEE, the Landsat-8 Collection 2 Tier 1 dataset and the Sentinel-2 datasets were filtered to the year 2020. We applied a cloud mask to each scene using the quality assessment bands. We created annual composites using the median value of each pixel across the full year of imagery (Table 1).

Table 1. Number of scenes acquired from Landsat-8, Sentinel-2, and Sentinel-1 satellites, along with the sensors' repeat cycle and acquisition dates.

Dataset	Repeat Cycle	Number of Scenes	Acquisition Date
Landsat-8 (optical)	16 days	7	01-01-2020 to 31-12-2020
Sentinel-2 (optical)	10 days	216	01-01-2020 to 31-12-2020
Sentinel-1 (radar)	12 days	59	01-01-2020 to 31-12-2020

In GEE, the Sentinel-1 dataset contains GRD scenes that are preprocessed with the Sentinel-1 Toolbox [44] using the following steps: thermal noise removal, radiometric calibration, and terrain correction using Shuttle Radar Topography Mission (SRTM) 30. The final terrain-corrected values are converted to decibels via logarithmic scaling. We used the Sentinel-1 analysis-ready data (ARD) script made available by [45] to filter and process Sentinel-1 images for our area of study. The script was set to filter images acquired in 2020 with VV and VH polarizations and from both ascending and descending orbits, resulting in 59 total images. Furthermore, we applied the refined Lee speckle filter [46] using a 7 by 7 kernel to each image individually with the aim of despeckling the SAR data [47]. We did not apply terrain flattening as our area of study is generally flat and recent work suggests minimal impact on SAR data values [48]. The resulting images were exported as an GEE asset on the linear scale. We calculated the modified Radar Forest Degradation Index [49] for each image (reference Section 2.2.1) and created a final composite with six bands as follows: (1) VV ascending using a minimum reducer, (2) VH ascending using a minimum reducer, (3) VV descending using a minimum reducer, (4) VH descending using a minimum reducer, (5) mRFDI ascending using a median reducer, and (6) mRFDI descending using a median reducer. Each band was then normalized to values from zero to one per best practices [50]. The final composite was reprojected to have a 30 m spatial resolution to maintain consistency across datasets.

2.2.1. Spectral and Backscatter Indices

Multiple red–green–blue composites and standard spectral indices were created to employ in the ML classification as input variables (Table 2). The spectral indices include the Normalized Difference Vegetation Index (NDVI), which measures vegetation greenness as a proxy for health [51]; Normalized Difference Moisture Index (NDMI), which quantifies vegetation with high water content [52]; Modified Normalized Difference Water Index (MNDWI), which distinguishes open water features [53]; Green Chlorophyll Vegetation Index (GCVI), to quantify leaf chlorophyll content [54]; and modified Radar Forest Degradation Index (mRFDI), which can distinguish between different vegetation varieties using the difference between the VV and VH polarizations in SAR data [49] to potentially distinguish mangroves from other vegetation types. We also calculated additional indices, including the simple ratio (SR), to quantify vegetation density [55]; Band Ratio 54, which is helpful in distinguishing water bodies from vegetation [56]; and Band Ratio 35, to enhance urban areas and distinguish water from vegetation [57].

These indices have been selected based on previous studies that mapped mangroves [29,58] and the fact that mangrove ecosystems are a combination of coastal vegetation and tidal conditions; therefore, the vegetation and water indices are expected to help distinguish between mangroves and nonmangroves.

Table 2. Band combinations and formulas for spectral indices NDVI, NDMI, MNDWI, SR, Band Ratio 54, Band Ratio 35, GCVI for Optical and mRFDI Radar Images.

Index	Sensor	Bands	Formula
NDVI	Landsat-8	Band 4 and Band 5	$\frac{NIR-RED}{NIR+RED}$
	Sentinel-2	Band 4 and Band 8	
NDMI	Landsat-8	Band 7 and Band 3	$\frac{SWIR2-GREEN}{SWIR2+GREEN}$
	Sentinel-2	Band 12 and Band 3	
MNDWI	Landsat-8	Band 3 and Band 3	$\frac{SWIR1-GREEN}{SWIR1+GREEN}$
	Sentinel-2	Band 11 and Band 3	
SR	Landsat-8	Band 5 and Band 4	$\frac{NIR}{RED}$
	Sentinel-2	Band 8 and Band 4	
Band Ratio 54	Landsat-8	Band 6 and Band 5	$\frac{SWIR1}{NIR}$
	Sentinel-2	Band 11 and Band 8	
Band Ratio 35	Landsat-8	Band 4 and Band 6	$\frac{RED}{SWIR1}$
	Sentinel-2	Band 4 and Band 11	
GCVI	Landsat-8	Band 5 and Band 3	$\frac{NIR}{GREEN}-1$
	Sentinel-2	Band 8 and Band 3	
mRFDI	Sentinel-1	VV and VH	$\frac{\sigma^o_{VV}-\sigma^o_{VH}}{\sigma^o_{VV}+\sigma^o_{VH}}$

2.3. Existing Guyana Mangrove Extent Map Comparison

We compared our results with a 2018 mangrove extent map produced by Earth and Marine Environmental Consultants (EAME) acquired from a report by NAREI and CI. EAME used Sentinel-2 imagery to classify mangroves along Guyana's coastline. Their results were used to verify the spatial distribution of mangrove forests within our study area.

Machine Learning Algorithm

Random forest is an ML algorithm [59,60] that builds large numbers of decision trees when training the sample data and returns a particular class depending on the maximum number of occurrences. This reduces the chance of overtraining the sample data [59,61,62]. The standard principles for RF training are selecting a random sample with a replacement of the training dataset and fitting trees to the samples by creating a split in the sample data. This provides the algorithm with matching responses for the training data [62]. Predictions are then established after training the sample by designating a particular classification of the trees [62].

Mangrove experts at NAREI generated training polygons for both mangrove and nonmangrove land cover types by visually inspecting Landsat-8 and Sentinel-2 imagery in GEE. Nonmangrove samples comprised coastal seagrasses, agriculture, water, grassland, barren ground, and settlement. We then sampled the spectral and backscatter information from the Landsat-8 composited image only, Sentinel-2 composited image only, Sentinel-1 composited image only, and combined composited images (data fusion). These samples were used for training the algorithm and assessing model performance in GEE. The stratified random sampling method was used for this research to reduce sampling bias [62,63]. When choosing a reference sample location, we considered the sample sizes for the two land cover types, where the larger land cover type (nonmangrove) required more training samples than the smaller land cover type (mangrove). The collected reference samples were divided into 60% for training the ML model algorithm and 40% for validation for each image at 30 m resolution (Table 3).

To determine the optimal number of trees with which to parameterize the random forest classification model for each dataset, hyperparameter tuning was conducted. Hyperparameter tuning can result in statistically significant improvements in supervised classification model performance [64]. In this case, every RF classifier was trained with a range of 5 to 200 decision trees in increments of 5 to determine the number of trees that leads to the greatest accuracy for the least amount of computational power. The accuracy of model performance for each iteration was graphed (Figure 4). Notably, Sentinel-1 (C) had a lower overall accuracy than other model iterations, and an adjusted y-axis range was used accordingly.

Table 3. Pixel samples from each data source.

Sensor	Mangrove	Nonmangrove	Total
Landsat-8	546	5859	**6405**
Sentinel-2	556	6122	**6678**
Sentinel-1	557	6922	**7479**
Data Fusion	546	5648	**6194**

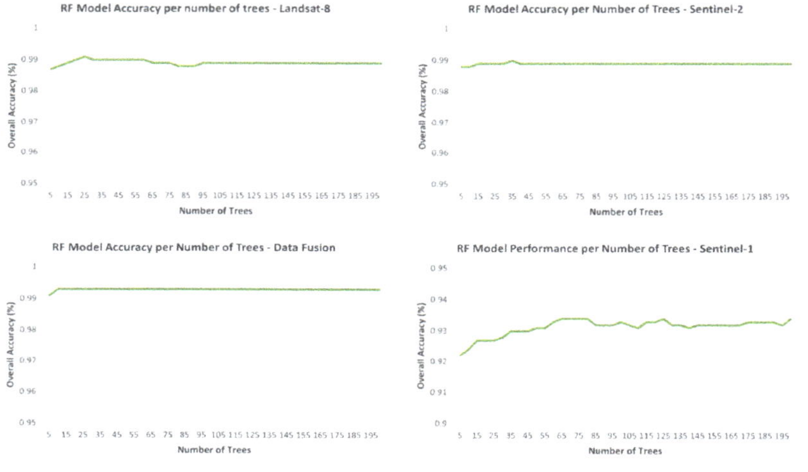

Figure 4. RF accuracy per number of decision trees for each dataset method.

For the validation, we used the F1-score (Equation (1)) as the metric for model performance. The F1-score is calculated using the precision and recall, also known as producer's and user's accuracy [65,66]. The precision is the ratio of correctly predicted positive observations to the total predicted positive observations. The recall represents the ratio of correctly predicted positive observations to all the observations in the class. The F1-score is the harmonic mean of precision and recall, considering both the false positives and false negatives.

$$F1 - score = \frac{2x(recall x precision)}{recall + precision} \quad (1)$$

2.4. Independent Accuracy Assessment and Unbiased Area Estimates

A systematic quantitative accuracy assessment is necessary to measure a map's accuracy and estimate each class's area based on reference sample data. We used methods described by Olofsson et al. (2014) [67] to extract the sample-based producer's accuracy, user's accuracy, overall accuracy (Equations (2)–(4)), and unbiased area estimates (±95% confidence interval).

$$\text{Producer's accuracy} = \frac{\Sigma \text{ correctly classified reference data for a class}}{\text{Total reference points for that class}} \quad (2)$$

$$\text{User's accuracy} = \frac{\Sigma \text{ correctly classified points for a class}}{\text{Total classified points for that class}} \quad (3)$$

$$\text{Overall accuracy} = \frac{\Sigma \text{ correctly classified points for all classes}}{\text{Total points}} \quad (4)$$

A total of 401 reference sample points within the study area were generated in GEE using a simple random sampling design [67]. The points were classified into mangrove and nonmangrove classes using the freely available Norway's International Climate and Forests Initiative (NICFI) Planet high-resolution (4.77 m spatial resolution) base maps. All monthly base maps for the year of classification (2020) were used in image interpretation. This process was completed using a visual interpretation sampling approach [68,69] in Collect Earth Online (CEO), a free and open-source interpretation system connected with high-resolution satellite image display functionality. CEO was created as a collaborative platform by the Food and Agriculture Organization (FAO) and SERVIR [70]. Image interpretation was conducted by remote sensing analysts at NAREI and the SERVIR Science Coordination Office, who have combined expertise in Guyana's mangroves and in interpreting remotely sensed imagery. Reference points in which the visual interpreter had low confidence were flagged and reviewed by multiple analysts for final classification.

3. Results

3.1. Mangrove Mapping Extent Results

The RF machine learning algorithm produced 2020 mangrove extent maps at 30 m for the study area using Landsat-8, Sentinel-2, Sentinel-1, and data fusion. Smaller mangrove stands within riverine environments and extensively fragmented patches can be detected and mapped in considerable detail using these high-resolution data and RF ML algorithm. Based on the unbiased area estimates, Landsat-8 only produced the largest mangrove extent, with a total of 3645.0 ± 836.7 hectares, accounting for 8.5% of the study area (Table 4). On the other hand, Sentinel-1 data produced the lowest mangrove extent with a total of 908.4 ± 579.7 hectares, accounting for 2.1% of the study area (Table 4).

Table 4. Comparison of total mangrove coverage (ha) and percentage coverage for optical and radar images at 30 m resolution, Landsat-8, Sentinel-2, Sentinel-1, and data fusion.

Sensor	Nonmangrove Coverage (ha)	Mangrove Coverage (ha)	95% Confidence Interval of Area (ha)	Mangrove Percentage Coverage
Landsat-8	39,360.8	3645.0	±836.7	8.5%
Sentinel-2	39,734.0	3271.8	±712.0	7.6%
Sentinel-1	42,097.3	908.4	±579.7	2.1%
Data fusion	39,817.0	3188.8	±707.0	7.4%

3.2. Mangrove Map Extent Comparison

The resulting mangrove extent map outputs were compared with the EAME 2018 mangrove baseline that was generated using only Sentinel-2 data. The visual comparisons demonstrated how closely our mangrove extent maps matched the 2018 baseline (Figure 5). The Landsat-8, Sentinel-2, and data fusion outputs were consistent with the 2018 baseline.

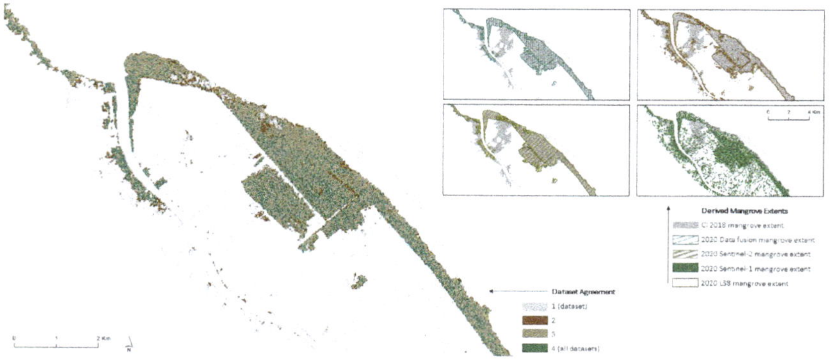

Figure 5. Mangrove extent map comparison for the study area based on Landsat-8, Sentinel-2, Sentinel-1 data, and data fusion. (Larger) Mangrove extent agreement between all datasets (Smaller); separated mangrove extents.

The EAME 2018 mangrove baseline displayed the spatial extent of mangroves at the estuary of the Pomeroon River, along both banks of the river, and along both sides of the Atlantic coast. The baseline also revealed that mangroves were expanding further inland of the Pomeroon estuary. While the distribution of mangroves between Landsat-8 and Sentinel-2 mangrove extents was similar to that of the 2018 baseline, Landsat-8 detected more mangroves in the Pomeroon riverine area and inland patches than Sentinel-2 (Figure 5). Sentinel-1's mangrove extent revealed sparse mangrove distribution both within mangrove areas highlighted by Landsat-8, Sentinel-2, data fusion, and baseline data, and within areas where mangroves were not highlighted by the other datasets and where mangroves may not be present with certainty (Figure 5). The mangrove extent produced using data fusion revealed a general distribution that was consistent with Landsat-8, Sentinel-2, and baseline data. Data fusion, Landsat-8, and Sentinel-2 extents intersected well, highlighting the areas where mangroves were likely to be found and coincided with the baseline data.

3.3. Accuracy

3.3.1. Model Performance

Using the 40% reference points dedicated to assessing the model performance (see Section 2.4), F1-scores were calculated based on confusion matrices of point counts for the four different products (Landsat-8 only, Sentinel-2 only, Sentinel-1 only, and data fusion). The results are summarized in Table 5.

Table 5. F1-scores describing ML algorithm performance (%) for classified mangrove extent using confusion matrix of point counts.

	F1-Score (%)	
	Nonmangrove	**Mangrove**
Landsat-8	99.5	93.5
Sentinel-2	99.5	94.2
Sentinel-1	96.0	28.6
Data Fusion	99.6	95.8

Model performance varied, with F1-scores ranging between 96.0% and 99.6% for the nonmangrove class and 28.6% and 95.8% for the mangrove class. The Sentinel-1 only product had the lowest performance for both mangrove and nonmangrove classes, while the optical and data fusion outputs performed better for the mangrove class. The data fusion approach showed the highest F1-score for both mangrove (95.8%) and nonmangrove

(99.6%), with the latter being a 0.1% difference compared with the Sentinel-2 only and Landsat-8 only outputs (99.5%). However, model performance results based exclusively on the reference data do not guarantee accurate maps. Additional results from an independent accuracy assessment are needed and summarized in the next section.

3.3.2. Independent Accuracy Assessment

The independent accuracy assessment we conducted compares the mangrove extent results with a random sample of points collected over monthly Planet base maps from 2020. Overall accuracy varied from 88% (Sentinel-1 only) to 95% (data fusion) (Tables 6–9). User's accuracy for the mangrove classification varied greatly (13% to 69%), as did producer's accuracy (50% to 87%). The poorest performance across these metrics was from the Sentinel-1 only classification. Performances across the Landsat-8 only, Sentinel-2 only, and data fusion approaches were more consistent, with the data fusion slightly outperforming the other model iterations in overall, user's, and producer's accuracy (Tables 6, 7 and 9). Tables 6–9 also report biased area totals (A_m) and class proportion (W_h).

Table 6. Landsat-8 only.

Pixel Count		Predicted		Total	A_m (ha)	W_h
		Nonmangrove	Mangrove			
Observed	Nonmangrove (NM)	344	9	353	39,030.9	91%
	Mangrove (M)	16	32	48	3974.8	9%
Total		360	41	401	43,005.8	100%
OA: 94%	M Users A: 67%	M Producers A: 78%	NM Users A: 97%	NM Producers A: 96%		

Table 7. Sentinel-2 only.

Pixel Count		Predicted		Total	A_m (ha)	W_h
		Nonmangrove	Mangrove			
Observed	Nonmangrove (NM)	348	5	353	39,051.4	91%
	Mangrove (M)	15	33	48	3954.4	9%
Total		363	38	401	43,005.8	100%
OA: 95%	M Users A: 69%	M Producers A: 87%	NM Users A: 99%	NM Producers A: 97%		

Table 8. Sentinel-1 only.

Pixel Count		Predicted		Total	A_m (ha)	W_h
		Nonmangrove	Mangrove			
Observed	Nonmangrove (NM)	347	6	353	41,362.7	96%
	Mangrove (M)	42	6	48	1643.1	4%
Total		389	12	401	43,005.8	100%
OA: 88%	M Users A: 13%	M Producers A: 50%	NM Users A: 98%	NM Producers A: 89%		

Table 9. Data fusion.

Pixel Count		Predicted		Total	A_m (ha)	W_h
		Nonmangrove	**Mangrove**			
Observed	Nonmangrove (NM)	348	5	353	39,174.6	91%
	Mangrove (M)	15	33	48	3831.1	9%
Total		363	38	401	43,005.8	100%
OA: 95%	M Users A: 69%	M Producers A: 87%	NM Users A: 99%	NM Producers A: 96%		

3.3.3. Field Verification Exercise

The ground truth fieldwork activity featured three targeted mangrove sites for cross validation, which NAREI conducted from 30 September to 1 October 2021. These sites were identified based on observation of visible changes between the 2018 and 2020 extent maps. In order to visualize the land cover and land use within the areas, the chosen target sites were studied using high-resolution still photography taken from a DJI Phantom 4 Advanced drone. The drone was first stationed in the Pomeroon River close to the estuary (target site 1), where agriculture was the predominant use of the surrounding land, along with dispersed settlements (Figure 6A). At the second target site of located La Union, mangroves were not visible in the photographed foreshore, but other tiny shrubs and seagrasses were observed in that location (Figure 6B). Target site 3 was located at Supenaam Stelling, where communities were observed more inland from the river bank to the north and south of the drone station (Figure 6C).

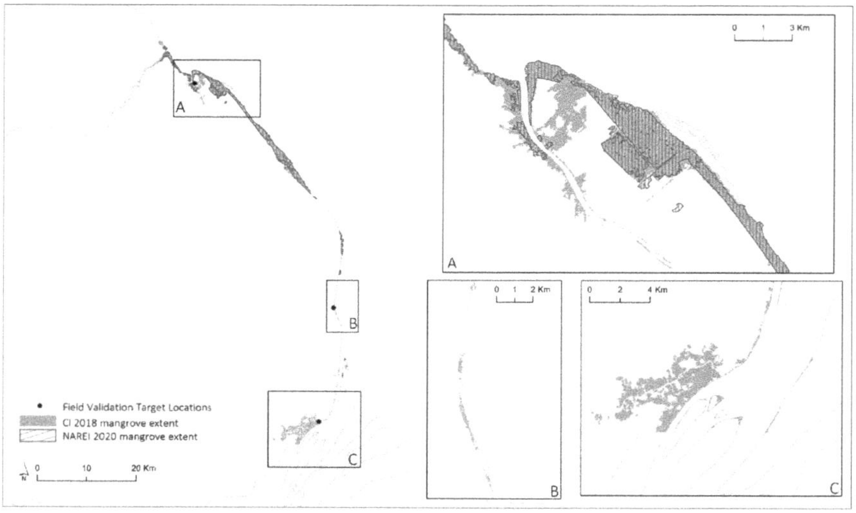

Figure 6. Target Sites for mangrove distribution validation and change detection. EAME 2018 mangrove baseline and 2020 data fusion and target sites identified for further field validation: (**A**) target site 1—Pomeroon Estuary, (**B**) target site 2—La Union Village, (**C**) target site 3—Supenaam Stelling.

Figure 6 illustrates 2018 and 2020 map extents, which highlighted notable changes within target areas. For target site 1, the 2020 mangrove extent showed significant mangrove losses as a direct result of agricultural expansion in the area compared with the 2018 mangrove extent. Figure 7A,B show a delineation of a small fringes of mangroves along the river banks and agricultural lands immediately adjacent to mangroves. Furthermore, evidence of mangrove displacement can be observed at the boundary between the mangrove fringe and agricultural lands Figure 7E. These changes were detected using our mangrove mapping method. At target site 2, the 2018 mangrove extent showed mangrove fringe located along the La Union Village foreshore; however, the 2020 mangrove extent and Figure 7C,D showed absence of mangrove. Nonetheless, instead of mangroves, seagrasses can be observed at the site. At target site 3, the 2018 mangrove extent showed mangrove inland; however, our method showed mangrove fringes on the river banks, with clustered settlements located in the proximity of mangroves (Figure 7E,F). Furthermore, the 2018 extent showed other vegetation species detected inland, while our model detected mangrove fringe, which was verified by the photographs.

Figure 7. (**A–F**) Land cover type within target areas depicted in Figure 6 captured using drone photography.

4. Discussion

4.1. Key Findings

Overall, the classification based on a combination of optical and radar data resulted in higher user's, producer's, and overall accuracies than models trained with a single dataset. Nevertheless, the performance of the model using only optical data was strong, likely due in part to the sufficient data coverage of Region 2 in 2020 from Landsat-8 and Sentinel-2. The Sentinel-1 product did not perform well alone, but results were enhanced when it was combined with the optical datasets.

When the map outputs from Sentinel-2, Landsat-8, and the data fusion in our study were compared with existing ones, results indicated that the classification of mangroves in the EAME 2018 output identified mangroves where they were not likely to be found. Ground truth activities corroborated this finding, since large swaths of mangroves were not observed along the banks of the Supenaam River; rather, very sparse fringes of mangroves were observed along the eastern bank of the Supenaam River and along the west bank of Essequibo. Nonetheless, the EAME 2018 extent managed to clearly detect patches of mangroves within the Pomeroon Riverine environment.

Moreover, a comparison between the 2018 and 2020 map products detected significant changes, particularly mangrove losses at the Pomeroon estuary and upriver (Figure 6). Results from ground truth activities showed significant mangrove forest loss due to the large expansion of coconut farms northeast of the estuary and the large clearing of mangroves for construction purposes (Figure 6).

We also compared the accuracies of our models with those of similar mangrove extent studies. We assessed GMW [21] in GEE and found that their overall, producer's, and user's accuracies were 91.8%, 70.8%, and 64.2%, respectively, for Region 2 using Advanced Land Observing Satellite (ALOS) Phased Array Type L-Band Synthetic Aperture Radar (PALSAR) L-Band and Landsat data, which are lower than the accuracies of our data fusion approach. Additionally, the GMW 2020 data misclassify a clearly deforested area of mangroves (see Figure 6A). We also assessed the European Space Agency's WorldCover Mangrove Layer [26] for 2020 in Region 2, and their accuracies were 89.0%, 79.2%, and 52.8%, respectively. This product is also produced with optical and SAR data, specifically from Sentinel-1 and 2. Our data fusion outperformed this product as well. Lastly, we assessed the mangrove extent in the MapBiomas Amazon 2020 product from Collection 5 [71] for Region 2 and found that the accuracies were 95.0%, 58.8%, and 71.4%, respectively. The overall accuracy is the same, and the producer's and user's accuracies are lower than our data fusion results. It should be noted that the MapBiomas product does not cover the entire Region 2 coastline that we assessed in our study.

4.2. Limitations and Future Work

While radar data have been an adequate data source for mangrove classification in other studies [29], our results demonstrate poor performance using only Sentinel-1 for classification relative to optical data. The poor performance may be due to the nature of C-band data, which may not be able to penetrate the mangrove forest canopy. A study by Simard et al. (2002) [72] successfully used L-band in the HH polarization to capitalize on the double-bounce backscatter mechanism and clearly discriminate flooded forests from other forest classes. In order to have a clearer double-bounce signal, L-band is required in order to penetrate vegetation and reach the ground [73]. If most of the C-band signal scattering is caused by canopy volume scattering rather than double-bounce scattering from interactions between tree trunks and the ground, the signal may not be as discernible [73]. In this case, the exclusive use of C-band data was likely a limiting factor in the effectiveness of SAR to map the mangrove extent [74]. Future work would involve employing free, open, and regularly available L-band data in this analysis, potentially from the new NISAR mission by NASA and the Indian Space Research Organisation set to launch in early 2024 [75].

There is potential for the future estimation of carbon stock in mangrove forests based on the maps produced by these methods. Coastal forests are one of the richest carbon sinks globally, sequestering more carbon than most inland rainforests. Mangroves are considered a blue carbon ecosystem due to their accumulation of rich soils and ability to store organic carbon both internally and externally [76]. Their complex root system traps sediment buildup, sequestering the carbon over a long period of time, while their dense, leafy canopy sequesters over a shorter period of time but throughout the living biomass of the forest [77].

Carbon reporting is a crucial component of the United Nations Framework Convention on Climate Change framework that guides country participants seeking to reduce deforestation and forest degradation emissions. These reports are frequently based on land cover datasets produced by the country or on global datasets produced with remotely sensed data. SERVIR's Science Coordination Office has begun implementing the SERVIR Carbon Pilot Project (S-CAP), which derives carbon emission estimates from a wide array of land cover extent datasets and compares them to better support countries in selecting the most efficient method for their applications. The S-CAP team plans to incorporate their methodologies into this mangrove monitoring service to better quantify the emission and sequestration of Guyana's mangroves. The S-CAP approach would employ the mangrove change extents generated through this study and apply various biomass datasets to estimate the emissions of cleared mangroves. The integration of these methods will assist NAREI in monitoring and understanding the climate impacts of these changes.

5. Conclusions

This study used Landsat-8, Sentinel-2, Sentinel-1, and a combination of those datasets to produce high-resolution mangrove forest data products for Pomeroon-Supenaam (Region 2) based on the RF ML algorithm in GEE. The resulting mangrove extent maps revealed that Sentinel-2, Landsat-8, and the data fusion product produced higher accuracies that were consistent with the general spatial distribution of mangrove forests in the EAME 2018 mangrove forest baseline, although smaller mangrove stands were additionally detected in the riverine environment. On the other hand, Sentinel-1 showed sparse mangrove forest distribution that was inconsistent with Landsat-8, Sentinel-2, the fusion dataset, and the EAME 2018 baseline.

Based on these results, a random forest model applied in a cloud computing platform like GEE is a suitable approach for mapping the mangrove extent along Guyana's coast. Time series analytics can require considerable computing power; therefore, a cloud-based approach is necessary, especially in regions with intermittent internet connection. Moreover, GEE provides quick access to free and openly available datasets across multiple sensors and data providers, facilitating these analyses. In terms of the computing power needed to run these models, we estimate the use of 48–53 EECU hours (Earth Engine Compute Unit) or USD 24–31 to replicate this same analysis to another year of study, proving that operationalizing this approach in a cloud-based environment is feasible and cost-effective.

This analysis was needed to identify the most appropriate technique to develop a mangrove monitoring service for Guyana that provides updated annual data on Guyana's mangrove extent, gains, losses, and the impact of NAREI's restoration efforts. It was also necessary for the data produced by the analysis to be freely accessible to a number of agencies, including the Sea and River Defence Division, Guyana Forestry Commission, Conservation International, World Wildlife Fund, and other stakeholders.

This paper provides the first comparison of mangrove extent mapping approaches in Guyana using remotely sensed data. We evaluate the performance of an ML algorithm to classify three different data combinations to determine the best approach for mapping the annual mangrove extent in a tropical, cloud-persistent, coastal ecosystem. Using only free and open data, this method empowers NAREI to produce mangrove extent datasets to their own specifications with the confidence that it will perform on par with other mangrove monitoring efforts and meet their mandate to monitor and protect this critical and vibrant ecosystem.

Author Contributions: Conceptualization, K.C.-B., K.E.H. and K.M.; data curation, K.C.-B., K.E.H., A.P.N. and V.M.-A.; formal analysis, K.C.-B. and K.E.H.; funding acquisition, B.Z.; investigation, K.C.-B. and K.E.H.; methodology, K.C.-B., K.E.H. and A.P.N.; resources, K.C.-B. and K.E.H.; software, K.C.-B., K.E.H. and A.P.N.; supervision, K.E.H., K.M. and B.Z.; validation, K.C.-B., K.E.H., A.P.N. and Z.N.; visualization, C.E.; writing—original draft, K.C.-B., K.E.H. and A.P.N.; writing—review and editing, A.P.N., V.M.-A., C.E. and H.P. All authors have read and agreed to the published version of the manuscript.

Funding: Funding for this work was provided through the cooperative agreement 80MSFC22N0004 between NASA and UAH. SERVIR is a joint NASA- and USAID-led program.

Data Availability Statement: The Landsat-8, Sentinel-2, and Sentinel-1 imageries used in this study are open and freely available in Google Earth Engine. The NICFI monthly base maps are available for noncommercial use through Planet and on Google Earth Engine. The training and testing datasets employed are available in the linked Google Earth Engine scripts.

Acknowledgments: Special thanks to Marc Simard, Abigail Barenblitt, Lola Fatoyinbo, and NAREI.

Conflicts of Interest: The authors declare no conflicts of interest.

Abbreviations

The following abbreviations are used in this manuscript:

ALOS	Advanced Land Observing Satellite
EECU	Earth Engine Compute Unit
FAO	Food and Agriculture Organization
GCVI	Green Chlorophyll Vegetation Index
GEE	Google Earth Engine
GMW	Global Mangrove Watch
GRD	Ground Range Detected
LULC	Land Use and Land Cover
ML	Machine Learning
MNDWI	Modified Normalized Difference Water Index
mRFDI	Modified Radar Forest Degradation Index
MSI	MultiSpectral Instrument
NAREI	National Agricultural and Research Extension Institute
NASA	National Aeronautics and Space Administration
NDMI	Normalized Difference Moisture Index
NDVI	Normalized Difference Vegetation Index
NICFI	Norway's International Climate and Forests Initiative
OLI	Operational Land Imager
PALSAR	Phased Array Type L-Band Synthetic Aperture Radar
RF	Random Forest
RTC	Radiometric Terrain Corrected
SAR	Synthetic Aperture Radar
S-CAP	SERVIR Carbon Pilot
SDG	Sustainable Development Goals
SEDAC	NASA Socioeconomic Data and Applications
SR	Simple Ratio
SRTM	Shuttle Radar Topography Mission
TM	Thematic Mapper

References

1. Tomlinson, P. *The Botany of Mangroves*; Cambridge University Press: London, UK, 1986.
2. Hutchings, P.; Saenger, P. *Ecology of Mangroves*; University of Queensland Press: St. Lucia, Australia, 1987.
3. Ricklefs, R.E.; Latham, R.E. Global patterns of diversity in mangrove floras. *Species Diversity in Ecological Communities: Historical and Geographical Perspectives*; University of Chicago Press: Chicago, IL, USA, 1993; pp. 215–229.
4. Eddy, S.; Milantara, N.; Sasmito, S.D.; Kajita, T.; Basyuni, M. Anthropogenic drivers of mangrove loss and associated carbon emissions in South Sumatra, Indonesia. *Forests* **2021**, *12*, 187. [CrossRef]
5. Kumari, P.; Singh, J.K.; Pathak, B. Potential contribution of multifunctional mangrove resources and its conservation. In *Biotechnological Utilization of Mangrove Resources*; Elsevier: Amsterdam, The Netherlands, 2020; pp. 1–26.
6. Tam, N.F.; Wong, Y.S. Mangrove soils as sinks for wastewater-borne pollutants. *Hydrobiologia* **1995**, *295*, 231–241. [CrossRef]
7. Alongi, D.M. Present state and future of the world's mangrove forests. *Environ. Conserv.* **2002**, *29*, 331–349. [CrossRef]
8. Walters, B.B.; Rönnbäck, P.; Kovacs, J.M.; Crona, B.; Hussain, S.A.; Badola, R.; Primavera, J.H.; Barbier, E.; Dahdouh-Guebas, F. Ethnobiology, socio-economics and management of mangrove forests: A review. *Aquat. Bot.* **2008**, *89*, 220–236. [CrossRef]
9. Giri, C.; Ochieng, E.; Tieszen, L.L.; Zhu, Z.; Singh, A.; Loveland, T.; Masek, J.; Duke, N. Status and distribution of manrgove forests of the world using earth oversvation satellite data. *Glob. Ecol. Biogeogr.* **2010**, *20*, 154–159. [CrossRef]
10. Sandilyan, S.; Kathiresan, K. Mangrove conservation: A global perspective. *Biodivers. Conserv.* **2012**, *21*, 3523–3542. [CrossRef]
11. Himes-Cornell, A.; Grose, S.O.; Pendleton, L. Mangrove ecosystem service values and methodological approaches to valuation: Where do we stand? *Front. Mar. Sci.* **2018**, *5*, 376. [CrossRef]
12. Spalding, M.; Parrett, C.L. Global patterns in mangrove recreation and tourism. *Mar. Policy* **2019**, *110*, 103540. [CrossRef]
13. Augustinus, P. The Changing Shoreline of SURINAM (South America). Ph.D. Thesis, University of Utrecht, Utrecht, The Netherlands, 1978.

14. Best, Ü.S.; van der Wegen, M.; Dijkstra, J.; Reyns, J.; van Prooijen, B.C.; Roelvink, D. Wave attenuation potential, sediment properties and mangrove growth dynamics data over Guyana's intertidal mudflats: Assessing the potential of mangrove restoration works. *Earth Syst. Sci. Data* **2022**, *14*, 2445–2462. [CrossRef]
15. Pastakia, C. A Preliminary Study of the Mangroves of Guyana. Final Report 1991. Available online: https://www.semanticscholar.org/paper/A-preliminary-study-of-the-mangroves-of-Guyana.-Pastakia/46dbc5766091240e0ea7a55f43736e2abac98b51 (accessed on 19 November 2023).
16. Carlson, R.R.; Evans, L.J.; Foo, S.A.; Grady, B.W.; Li, J.; Seeley, M.; Xu, Y.; Asner, G.P. Synergistic benefits of conserving land-sea ecosystems. *Glob. Ecol. Conserv.* **2021**, *28*, e01684. [CrossRef]
17. Van Doimen, T. Climate Change and Sea Level Rise in Guyana. 2013. Available online: https://www.ecojesuit.com/climate-change-and-sea-level-rise-in-guyana/ (accessed on 19 November 2023).
18. Lewis III, R.R. Ecological engineering for successful management and restoration of mangrove forests. *Ecol. Eng.* **2005**, *24*, 403–418. [CrossRef]
19. Maurya, K.; Mahajan, S.; Chaube, N. Remote sensing techniques: Mapping and monitoring of mangrove ecosystem—A review. *Complex Intell. Syst.* **2021**, *7*, 2797–2818. [CrossRef]
20. National Agricultural Research Extension Institute. Mangrove Restoration Project. Available online: https://narei.org.gy/mangrove/ (accessed on 20 September 2022).
21. Bunting, P.; Rosenqvist, A.; Lucas, R.M.; Rebelo, L.M.; Hilarides, L.; Thomas, N.; Hardy, A.; Itoh, T.; Shimada, M.; Finlayson, C.M. The Global Mangrove Watch—A New 2010 Global Baseline of Mangrove Extent. *Remote Sens.* **2018**, *10*, 1669. [CrossRef]
22. Spalding, M. *World Atlas of Mangroves*; Routledge: London, UK, 2010.
23. Hamilton, S.E.; Casey, D. Creation of a high spatio-temporal resolution global database of continuous mangrove forest cover for the 21st century (CGMFC-21). *Glob. Ecol. Biogeogr.* **2016**, *25*, 729–738. [CrossRef]
24. Thomas, N.; Lucas, R.; Bunting, P.; Hardy, A.; Rosenqvist, A.; Simard, M. Distribution and drivers of global mangrove forest change, 1996–2010. *PLoS ONE* **2017**, *12*, e0179302. [CrossRef] [PubMed]
25. Gumbricht, T.; RománCuesta. R.; Verchot, L.; Herold, M.; Wittmann, F ; Householder, E.; Herold, N.; Murdiyarso, D. Tropical and Subtropical Wetlands Distribution. 2017. Available online: https://data.cifor.org/dataset.xhtml?persistentId=doi:10.17528/CIFOR/DATA.00058 (accessed on 19 November 2023).
26. Zanaga, D.; Van De Kerchove, R.; Daems, D.; De Keersmaecker, W.; Brockmann, C.; Kirches, G.; Wevers, J.; Cartus, O.; Santoro, M.; Fritz, S.; et al. ESA WorldCover 10 m 2021 v200 2022. Available online: https://developers.google.com/earth-engine/datasets/catalog/ESA_WorldCover_v200 (accessed on 19 November 2023).
27. Giri, C.; Ochieng, E.; Tieszen, L.L.; Zhu, Z.; Singh, A.; Loveland, T.; Masek, J.; Duke, N. *Global Mangrove Forests Distribution, 2000*; NASA Socioeconomic Data and Applications Center (SEDAC): Palisades, NY, USA, 2005.
28. Goldberg, L.; Lagomasino, D ; Thomas, N.; Fatoyinbo, T. Global declines in human-driven mangrove loss. *Glob. Chang. Biol.* **2020**, *26*, 5844–5855. [CrossRef] [PubMed]
29. Nedd, G.A.; Oyedotun, T.D.T.; Simard, M. Evaluation of Spatio-Temporal Dynamics of Guyana's Mangroves Using SAR and GEE. *Earth Syst. Environ.* **2021**, *7*, 509–524. [CrossRef]
30. Ferriera, M.A.; Andrade, F.; Nogueira, R.M.; Paula, J. Use of satellite remote sensing for coastalc onservation in the Eastern African Coast: Advantages ad shortcomings. *Eur. J. Remote Sens.* **2012**, *45*, 293–304. [CrossRef]
31. Zhang, Y.; Woodcock, C.E.; Arévalo, P.; Olofsson, P.; Tang, X.; Stanimirova, R.; Bullock, E.; Tarrio, K.R.; Zhu, Z.; Friedl, M.A. A global analysis of the spatial and temporal variability of usable Landsat observations at the pixel scale. *Front. Remote Sens.* **2022**, *3*, 894618. [CrossRef]
32. Flores-Anderson, A.I.; Cardille, J.; Azad, K.; Cherrington, E.; Zhang, Y.; Wilson, S. Spatial and temporal availability of Cloud-free Optical Observations in the tropics to Monitor Deforestation. *Sci. Data* **2023**, *10*, 550. [CrossRef]
33. Flores-Anderson, A.I.; Herndon, K.; Thapa, R.; Cherrington, E.A.E. *The SAR Handbook: Comprehensive Methodologies for Forest Monitoring and Biomass Estimation*; U.S. National Aeronautics & Space Administration (NASA): Washington, DC, USA, 2019. [CrossRef]
34. Rocha de Souza Pereira, F.; Kampel, M.; Marilia, C.L. Mangrove vegetation structure in Southeast Brazil from phased array L-band synthetic aperture radar data. *J. Appl. Remote Sens.* **2016**, *10*, 036021. [CrossRef]
35. Zhu, W.; Chen, J.; Sun, Q.; Li, Z.; Tan, W.; Wei, Y. Reconstructing of High-Spatial-Resolution Three-Dimensional Electron Density by Ingesting SAR-Derived VTEC Into IRI Model. *IEEE Geosci. Remote. Sens. Lett.* **2022**, *19*, 1–5. [CrossRef]
36. Zhao, M.; Zhou, Y.; Li, X.; Cheng, W.; Zhou, C.; Ma, T.; Li, M.; Huang, K. Mapping urban dynamics (1992–2018) in Southeast Asia using consistent nighttime light data from DMSP and VIIRS. *Remote Sens. Environ.* **2020**, *248*, 111980. [CrossRef]
37. Tian, H.; Huang, N.; Niu, Z.; Qin, Y.; Pei, J.; Wang, J. Mapping winter crops in China with multi-source satellite imagery and phenology-based algorithm. *Remote Sens.* **2019**, *11*, 820. [CrossRef]
38. Tian, H.; Pei, J.; Huang, J.; Li, X.; Wang, J.; Zhou, B.; Qin, Y.; Wang, L. Garlic and winter wheat identification based on active and passive satellite imagery and the google earth engine in northern china. *Remote Sens.* **2020**, *12*, 3539. [CrossRef]
39. Yasir, M.; Liu, S.; Mingming, X.; Wan, J.; Pirasteh, S.; Dang, K.B. ShipGeoNet: SAR Image-Based Geometric Feature Extraction of Ships Using Convolutional Neural Networks. *IEEE Trans. Geosci. Remote. Sens.* **2024**, *62*, 5202613. [CrossRef]
40. Assembly, U.G. Transforming Our World: The 2030 Agenda for Sustainable Development; Technical Report, UN General Assembly; 2015. Available online: https://sdgs.un.org/2030agenda (accessed on 19 November 2023).

41. United Nations Children's Fund. Child-Friendly Regional Profile—Region Two (2) Pomeroon-Supenaam. 2017. Available online: https://www.unicef.org/lac/media/4581/file/PDF%20Portada%20Region%20two:%20Pomeroon-Supenaam.pdf (accessed on 19 November 2023).
42. Rama Rao, Y.; Alves, L.; Seulall, B.; Mitchell, Z.; Samaroo, K.; Cummings, G. Evaluation of the weather research and forecasting (WRF) model over Guyana. *Nat. Hazards* **2012**, *61*, 1243–1261. [CrossRef]
43. Gorelick, N.; Hancher, M.; Dixon, M.; Ilyushchenko, S.; Thau, D.; Moore, R. Google Earth Engine: Planetary-scale geospatial analysis for everyone. *Remote Sens. Environ.* **2017**, *202*, 18–27. [CrossRef]
44. Google. Sentinel-1 Algorithms | Google Earth Engine | Google for Developers, 2022. Available online: https://developers.google.com/earth-engine/guides/sentinel1 (accessed on 19 November 2023).
45. Mullissa, A.; Vollrath, A.; Odongo-Braun, C.; Slagter, B.; Balling, J.; Gou, Y.; Gorelick, N.; Reiche, J. Sentinel-1 SAR Backscatter Analysis Ready Data Preparation in Google Earth Engine. *Remote Sens.* **2021**, *13*, 1954. [CrossRef]
46. Lee, J.S.; Grunes, M.; de Grandi, G. Polarimetric SAR speckle filtering and its implication for classification. *IEEE Trans. Geosci. Remote. Sens.* **1999**, *37*, 2363–2373. [CrossRef]
47. Yommy, A.; Liu, R.; Wu, A. SAR Image Despeckling Using Refined Lee Filter. In Proceedings of the 2015 7th International Conference on Intelligent Human-Machine Systems and Cybernetics, Hangzhou, China, 26–27 August 2015.
48. Flores-Anderson, A.I.; Parache, H.B.; Martin-Arias, V.; Jiménez, S.A.; Herndon, K.; Mehlich, S.; Meyer, F.J.; Agarwal, S.; Ilyushchenko, S.; Agarwal, M.; et al. Evaluating SAR Radiometric Terrain Correction Products: Analysis-Ready Data for Users. *Remote Sens.* **2023**, *15*, 5110. [CrossRef]
49. Nicolau, A.P.; Flores-Anderson, A.; Griffin, R.; Herndon, K.; Meyer, F.J. Assessing SAR C-band data to effectively distinguish modified land uses in a heavily disturbed Amazon forest. *Int. J. Appl. Earth Obs. Geoinf.* **2021**, *94*, 102214. [CrossRef]
50. Hong, G.; Zhang, Y. A comparative study on radiometric normalization using high resolution satellite images. *Int. J. Remote Sens.* **2008**, *29*, 425–438. [CrossRef]
51. Rhew, I.C.; Vander Stoep, A.; Kearney, A.; Smith, N.L.; Dunbar, M.D. Validation of the normalized difference vegetation index as a measure of neighborhood greenness. *Ann. Epidemiol.* **2011**, *21*, 946–952. [CrossRef]
52. Taloor, A.K.; Manhas, D.S.; Kothyari, G.C. Retrieval of land surface temperature, normalized difference moisture index, normalized difference water index of the Ravi basin using Landsat data. *Appl. Comput. Geosci.* **2021**, *9*, 100051. [CrossRef]
53. Arreola-Esquivel, M.; Delgadillo-Herrera, M.; Toxqui-Quitl, C.; Padilla-Vivanco, A. Index-based methods for water body extraction in satellite data. In Proceedings of the Applications of Digital Image Processing XLII. SPIE, San Diego, CA, USA, 2–15 August 2019; Volume 11137, pp. 673–680.
54. Feng, S.; Zhao, J.; Liu, T.; Zhang, H.; Zhang, Z.; Guo, X. Crop type identification and mapping using machine learning algorithms and sentinel-2 time series data. *IEEE J. Sel. Top. Appl. Earth Obs. Remote Sens.* **2019**, *12*, 3295–3306. [CrossRef]
55. Brown, L.; Chen, J.M.; Leblanc, S.G.; Cihlar, J. A shortwave infrared modification to the simple ratio for LAI retrieval in boreal forests: An image and model analysis. *Remote Sens. Environ.* **2000**, *71*, 16–25. [CrossRef]
56. Ma, S.; Zhou, Y.; Gowda, P.H.; Dong, J.; Zhang, G.; Kakani, V.G.; Wagle, P.; Chen, L.; Flynn, K.C.; Jiang, W. Application of the water-related spectral reflectance indices: A review. *Ecol. Indic.* **2019**, *98*, 68–79. [CrossRef]
57. Yang, J.; Wang, Y.; Xiu, C.; Xiao, X.; Xia, J.; Jin, C. Optimizing local climate zones to mitigate urban heat island effect in human settlements. *J. Clean. Prod.* **2020**, *275*, 123767. [CrossRef]
58. Tran, T.V.; Reef, R.; Zhu, X. A Review of Spectral Indices for Mangrove Remote Sensing. *Remote Sens.* **2022**, *14*, 4868. [CrossRef]
59. Breiman, L. Random forests. *Mach. Learn.* **2001**, *45*, 5–32. [CrossRef]
60. Pal, M. Random forest classifier for remote sensing classification. *Int. J. Remote. Sens.* **2005**, *26*, 217–222. [CrossRef]
61. Shelestov, A.; Lavreniuk, M.; Kussul, N.; Novikov, A.; Skakun, S. Exploring Google Earth Engine platform for big data processing: Classification of multi-temporal satellite imagery for crop mapping. *Front. Earth Sci.* **2017**, *5*, 17. [CrossRef]
62. Liu, X.; Fatoyinbo, T.E.; Thomas, N.M.; Guan, W.W.; Zhan, Y.; Mondal, P.; Lagomasino, D.; Simard, M.; Trettin, C.C.; Deo, R.; et al. Large-scale high-resolution coastal mangrove forests mapping across West Africa with machine learning ensemble and satellite big data. *Front. Earth Sci.* **2021**, *8*, 560933. [CrossRef]
63. Taherdoost, H. Sampling methods in research methodology; how to choose a sampling technique for research. In *How to Choose a Sampling Technique for Research (10 April 2016)*; Elsevier: Amsterdam, The Netherlands, 2016. Available online: https://papers.ssrn.com/sol3/papers.cfm?abstract_id=3205035 (accessed on 19 November 2023).
64. Mantovani, R.; Horvath, T.; Cerri, R.; Vanschoren, J.; Carvalho, A.C.P.L.F.d. Hyper-Parameter Tuning of a Decision Tree Induction Algorithm. In Proceedings of the 2016 5th Brazilian Conference on Intelligent Systems (BRACIS), Recife, Brazil, 9–12 October 2016; pp. 37–42. [CrossRef]
65. Van Rijsbergen, C. Information retrieval: Theory and practice. In Proceedings of the Joint IBM/University of Newcastle upon Tyne Seminar on Data Base Systems, Callaghan, Australia, 4–7 September 1979; Volume 79.
66. Chicco, D.; Jurman, G. The advantages of the Matthews correlation coefficient (MCC) over F1 score and accuracy in binary classification evaluation. *BMC Genom.* **2020**, *21*, 6. [CrossRef]
67. Olofsson, P.; Foody, G.M.; Herold, M.; Stehman, S.V.; Woodcock, C.E.; Wulder, M.A. Good practices for estimating area and assessing accuracy of land change. *Remote Sens. Environ.* **2014**, *148*, 42–57. [CrossRef]
68. Lister, T.W.; Lister, A.J.; Alexander, E. Land use change monitoring in Maryland using a probabilistic sample and rapid photointerpretation. *Appl. Geogr.* **2014**, *51*, 1–7. [CrossRef]

69. Woodward, B.D.; Evangelista, P.H.; Young, N.E.; Vorster, A.G.; West, A.M.; Carroll, S.L.; Girma, R.K.; Hatcher, E.Z.; Anderson, R.; Vahsen, M.L.; et al. Co-Rip: A riparian vegetation and corridor extent dataset for Colorado river basin streams and rivers. *ISPRS Int. J. -Geo-Inf.* **2018**, *7*, 397. [CrossRef]
70. Saah, D.; Johnson, G.; Ashmall, B.; Tondapu, G.; Tenneson, K.; Patterson, M.; Poortinga, A.; Markert, K.; Quyen, N.H.; San Aung, K.; et al. Collect Earth: An online tool for systematic reference data collection in land cover and use applications. *Environ. Model. Softw.* **2019**, *118*, 166–171. [CrossRef]
71. Project, M.A. MapBiomas Amazonía Project—Collection [Version 5.0] of Annual Land Cover and Land Use Maps. 2024. Available online: https://amazonia.mapbiomas.org/en/ (accessed on 19 November 2023).
72. Simard, M.; Grandi, G.D.; Saatchi, S.; Mayaux, P. Mapping tropical coastal vegetation using JERS-1 and ERS-1 radar data with a decision tree classifier. *Int. J. Remote Sens.* **2002**, *23*, 1461–1474. [CrossRef]
73. Richards, J.A.; Sun, G.q.; Simonett, D.S. L-band radar backscatter modeling of forest stands. *IEEE Trans. Geosci. Remote Sens.* **1987**, pp. 487–498. [CrossRef]
74. Wang, Y.; Imhoff, L. Simulated and observed L-HH radar backscatter from tropical mangrove forests. *Int. J. Remote Sens.* **1993**, *14*, 2819–2828. [CrossRef]
75. Rajwi, T. NISAR mission on track for early 2024 launch says JPL official. *The Hindu*, 23 January 2024.
76. Chatting, M.; Al-Maslamani, I.; Walton, M.; Skov, M.W.; Kennedy, H.; Husrevoglu, Y.S.; Le Vay, L. Future mangrove carbon storage under climate change and deforestation. *Front. Mar. Sci.* **2022**, *9*, 781876. [CrossRef]
77. Mcleod, E.; Chmura, G.L.; Bouillon, S.; Salm, R.; Björk, M.; Duarte, C.M.; Lovelock, C.E.; Schlesinger, W.H.; Silliman, B.R. A blueprint for blue carbon: Toward an improved understanding of the role of vegetated coastal habitats in sequestering CO_2. *Front. Ecol. Environ.* **2011**, *9*, 552–560. [CrossRef] [PubMed]

remote sensing

MDPI

Article

Improving Land Use and Land Cover Information of Wunbaik Mangrove Area in Myanmar Using U-Net Model with Multisource Remote Sensing Datasets

Win Sithu Maung [1], Satoshi Tsuyuki [1,*] and Zhiling Guo [2,3]

1 Department of Global Agricultural Sciences, Graduate School of Agriculture and Life Sciences, The University of Tokyo, Tokyo 113-8654, Japan; winsithumaung@g.ecc.u-tokyo.ac.jp or sithumg1989@gmail.com
2 Center for Spatial Information Science, The University of Tokyo, Kashiwa 277-8568, Japan; guozhilingcc@csis.u-tokyo.ac.jp
3 Department of Building Environment and Energy Engineering, The Hong Kong Polytechnic University, Kowloon, Hong Kong, China
* Correspondence: tsuyuki@fr.a.u-tokyo.ac.jp

Abstract: Information regarding land use and land cover (LULC) is essential for regional land and forest management. The contribution of reliable LULC information remains a challenge depending on the use of remote sensing data and classification methods. This study conducted a multiclass LULC classification of an intricate mangrove ecosystem using the U-Net model with PlanetScope and Sentinel-2 imagery and compared it with an artificial neural network model. We mainly used the blue, green, red, and near-infrared bands, normalized difference vegetation index (NDVI), and normalized difference water index (NDWI) of each satellite image. The Digital Elevation Model (DEM) and Canopy Height Model (CHM) were also integrated to leverage the model performance in mixed ecosystems of mangrove and non-mangrove forest areas. Through a labeled image created from field ground truth points, the models were trained and evaluated using the metrics of overall accuracy, Intersection over Union, F1 score, precision, and recall of each class. The results demonstrated that the combination of PlanetScope bands, spectral indices, DEM, and CHM yielded superior performance for both the U-Net and ANN models, achieving a higher overall accuracy (94.05% and 92.82%), mean IoU (0.82 and 0.79), mean F1 scores (0.94 and 0.93), recall (0.94 and 0.93), and precision (0.94). In contrast, models utilizing the Sentinel-2 dataset showed lower overall accuracy (86.94% and 82.08%), mean IoU (0.71 and 0.63), mean F1 scores (0.87 and 0.81), recall (0.87 and 0.82), and precision (0.87 and 0.81). The best-classified image, which was produced by U-Net using the PlanetScope dataset, was exported to create an LULC map of the Wunbaik Mangrove Area in Myanmar.

Keywords: land use and land cover classification; U-Net; artificial neural network; PlanetScope; Sentinel-2; mangrove

1. Introduction

Mangrove ecosystems are among the most productive coastal habitats and are of great importance in contributing to the basic needs of local communities, such as wood, medicine, and food, and in protecting them from severe natural disasters [1,2]. Despite their significance, these ecosystems are experiencing a decline in status owing to the rapid increase in human population and developmental dynamics. It has been confronted with numerous anthropogenic disturbances, including the expansion of agriculture and aquaculture, urbanization, and the extensive exploitation of commercial fuelwood [3–5]. To compensate for mangrove deforestation, land use and land cover information for a region are essential components of an effective mangrove management regime. A reliable and up-to-date LULC map plays a crucial role in formulating management regulations, decision making, and restoration programs [6].

With the rapid development of Earth Observation, many researchers have been exploring remote sensing technology in environmental sciences. Remote sensing has become a powerful tool in mangrove conservation across various research areas such as mangrove extent mapping, species-level identification, LULC classification, and aboveground biomass estimation [7–9]. Acknowledging the contributions of space technology, numerous satellite missions, including Landsat and Sentinel, have provided invaluable insights into Earth observations, particularly for monitoring the condition of mangrove forests. Nguyen et al. employed Landsat and Sentinel imagery to map the extent of mangroves in Vietnam using a comparative analysis of unsupervised and supervised classification methods. Their results revealed that the highest overall accuracy (90.00%) was attained by employing Sentinel-2A images with a maximum-likelihood supervised classification method [10]. Manna and Raychaudhuri explored Sentinel-2 imagery to identify mangrove species and physiological conditions by integrating the Discriminant Normalized Vegetation index (DNVI) in the Sundarban mangroves [11]. The study underscored the effectiveness of Sentinel-2 coupled with its proposed spectral index, DNVI, achieving an impressive overall accuracy of 93.90% through a support vector machine-supervised classification approach. The research conducted by Zeng and Takeuchi demonstrated the performance of the integrated Advanced Land Observing Satellite (ALOS), Phased Array-type L-band Synthetic Aperture Radar 2 (PALSAR-2) and Landsat imagery to assess mangrove dynamics in China [12]. Some studies [13,14] have applied very high-resolution images such as World View-3, IKONOS and Unmanned Aerial Vehicle (UAV) images, and active light detection and ranging (LiDAR) data to improve the classification accuracy in mangrove species-level identification. However, the use of LiDAR data and UAV images for acquiring information about large-scale mangrove zones is relatively limited, and VHR images incur high financial costs. The resultant accuracy of mangrove classification varies based on the type of remote sensing data and the classification method employed, reflecting the benefits and drawbacks associated with different approaches [13,14].

Regarding classification methods, studies have explored LULC classification using both unsupervised and supervised approaches with different satellite images [15,16]. Currently, the application of machine and deep learning in LULC classification has proven to perform better than conventional approaches [17–21]. Unlike traditional machine learning classification methods, deep learning models exhibit the unique ability to automatically extract features and learn complex patterns from input features, thereby alleviating the necessity for extensive variable training. Numerous studies have demonstrated the superior performance of deep learning models over traditional ML methods in remote sensing applications [22–24]. Among state-of-the-art models, the U-Net segmentation model and artificial neural networks (ANNs) have consistently demonstrated superior performance in remote sensing image analysis [25–27]. However, limited studies have applied the U-Net and ANN models using different spatial resolution of satellite images for mangrove ecosystem [28–30]. Moreover, their research focused only on mangrove and non-mangrove classifications to highlight the distribution and dynamics of mangroves. Consequently, there is still a need to deploy U-Net and ANN models with various satellite images for LULC classification of an intricate mangrove ecosystem.

Despite the limitations of remote sensing technology, the mangrove community itself thrives in the intricate ecosystem of the intertidal zone of tropical and subtropical coastlines. Consequently, LULC classification in mangrove regions encounters unique challenges, such as spectral confusion between similar classes of mangrove status and other land use types. Therefore, the contribution of accurate information to LULC classification of mangrove ecosystems remains challenging. Furthermore, this study explored the integration of differences of Digital Elevation Models to leverage model performance in complex areas where mangrove and non-mangrove forests are mixed. The advent of Plant Scope imagery with 3 m resolution enables researchers to better understand and detail the acquisition of mangrove ecosystems for multiclass LULC classification. Through the Research and Education Program offered by PlanetScope, this study was granted access to high-resolution

PlanetScope imagery with 3 m resolution for LULC classification in the Wunbaik Mangrove region in Myanmar. It is worth examining the performance of DL models using different satellite images for the LULC classification of an intricate mangrove ecosystem.

The Wunbaik Mangrove Area (WMA) is one of the largest remnant mangrove biomes endowed with diverse mangrove species and provides invaluable ecosystem services to local communities in Myanmar. Owing to the rapid population explosion, low income, and scarcity of electricity, local people are highly dependent on mangrove forests for their fuelwood. Large areas of mangrove forests are cleared for main livelihoods, including shrimp ponds and paddy fields [31,32]. Being inaccessible and the least developed region in Myanmar, the Wunbaik Mangrove Forest is not well managed by officials, despite being an impressive natural mangrove area. Although mangrove restoration projects have been implemented by the Forest Department, few studies have been conducted to provide reliable information on the mangrove conservation of the WMA. A project report by the Food and Agricultural Organization in 2011 [31] has investigated the LULC classification of the Wunbaik Reserved Mangrove Forest (WRMF), a reserved forest. The doctoral research by Saw [33] investigated mangrove dynamics in the WRMF using Landsat images from 1990 to 2014. However, they only employed traditional classification approaches using medium-resolution satellite images. In 2020, Maung and Sasaki applied an ANN model with 10 m resolution Sentinel-2 imagery to detect the natural recovery of mangrove forests in the WMF, but their focus was also on mangrove and non-mangrove identification [34]. Therefore, there is a prominent research gap in updating the LULC information for such massive mangrove areas. In addition, it is challenging to determine how state-of-the-art deep learning models can accurately distinguish LULC classes in an intricate mangrove ecosystem.

Based on the detailed literature review above, this study aimed to (1) compare U-Net and ANN models using PlanetScope and Sentinel-2 images for LULC classification in a complex mangrove ecosystem and (2) contribute accurate and up-to-date LULC information to mangrove management regimes for Wunbaik Mangrove conservation. Furthermore, we explored the integration of topographic and canopy height information to improve the classification accuracy. By fulfilling these research objectives, the proposed method can improve the performance of multiclass LULC classification in an intricate mangrove environment.

2. Materials and Methods

2.1. Study Area

Wunbaik Mangrove Area (WMA) is one of large remnant mangrove assets covering a land area of about 55,000 ha, which is located between 19°05′00′′N–19°25′00′′N and 93°50′00′′E–94°05′00′′E in Rakhine State, Myanmar (Figure 1). Most mangrove forests in the study area have been reserved as forest areas under management by the Forest Department since 1931: Main Wunbaik Reserved Mangrove Forest (WRMF), Mingyaung Public Mangrove Forest, and Extended Mingyaung Mangrove Forest. According to a Food and Agricultural Organization (FAO) report, 34 mangrove species thrive in mangrove-reserved forests. The study area is endowed with diverse ecosystems of mangrove and non-mangrove forests, with different topographic features ranging from 0 to 95 m. The report of the Myanmar Information Management Unit (MIMU) (https://themimu.info/states_regions/rakhine, accessed on 12 June 2023) in 2019 mentions that temperature of the study area is between 27 °C and 33 °C and annual rainfall is 4860 mm. The main livelihoods of local people in the study area are rice production and shrimp pond aquaculture.

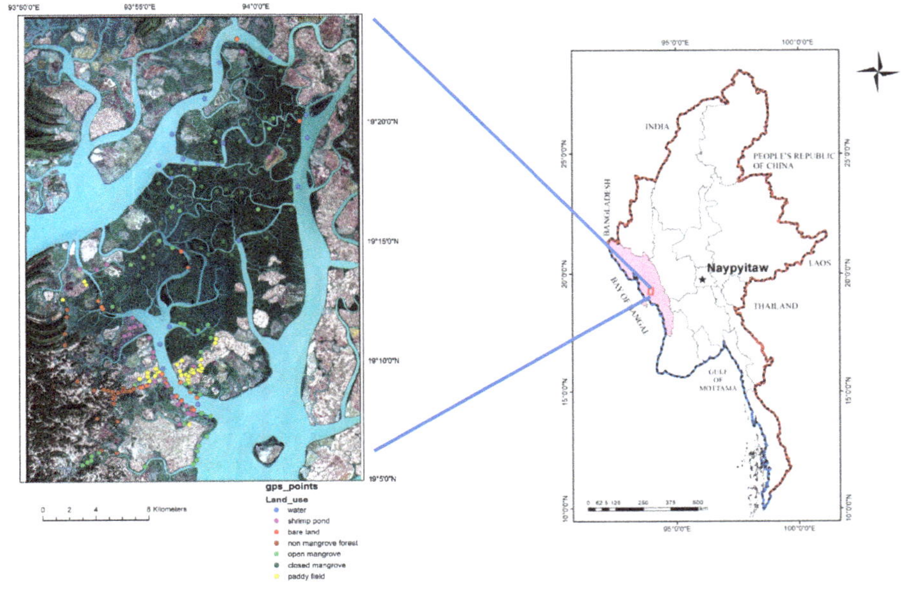

Figure 1. Location of field GPS points collected in Wunbaik Mangrove Area, Myanmar (PlanetScope image) (coastal border is in blue and terrestrial is in red).

2.2. Ground Truth Data Collection and Creating a Labeled Image

To create a robust deep learning model, training the model with accurately labeled data is necessary and challenging. Some studies have employed existing globally labeled datasets, but considering the unique characteristics of the mangrove ecosystem in our study region, we created a field-based fully labeled image for the classification. We conducted a field survey of ground-truth data collection for model training from 2 to 28 May 2023. Due to the diverse ecosystems and local livelihoods depending on the mangrove area, the study area contains various land use and land cover types, as shown in Figure 2. We used a handheld GARMIN GPS with 3 m accuracy, which is suitable for collecting land cover data. A total of 224 GPS points were recorded with latitude, longitude, and LULC information (Table 1) and then imported them into ArcGIS Pro 3.2.1 version for double checking using very high-resolution Google Earth and PlanetScope images.

To minimize the potential for misinterpretation, which is occasionally encountered in the conventional manual annotation of multiple LULC classes, we initially applied the ISODATA approach to the QGIS acquired on 12 February 2023, which was not used for model training [25,35]. Based on 224 Field GPS points and very high-resolution Google Earth and Planet Scope Dove images, the ISODATA output image was manually corrected using a raster editor tool in QGIS (QGIS Development Team (2009); Open-Source Geospatial Foundation Project, http://qgis.osgeo.org, accessed on 10 June 2023). We also referred to a classified image with 95% accuracy for mangrove and non-mangrove areas in 2020, which was provided by Maung and Sasaki [34]. Following careful visualization and validation by field experts, a reliable and accurately labeled image with seven LULC classes was obtained for training deep learning models.

Figure 2. Different LULC types of our classification in the study area. (**a**) Water body; (**b**) shrimp pond; (**c**) bare land; (**d**) non-mangrove forest; (**e**) open mangrove; (**f**) closed mangrove; (**g**) paddy field.

Table 1. Ground truth points collected in field survey.

Annotation Code	Description of LULC Classes	Number of GPS Points
0	Water: areas mainly covered by marine water.	20
1	Shrimp pond: man-made ponds used for shrimp farming, characterized by the presence of water and often geometric shapes.	31
2	Bare land: unvegetated areas such as tidal flat in mangrove areas or exposed soil in terrestrial areas.	22
3	Non-mangrove forest: forested areas consisting of tree cover that is not dominated by mangrove species.	25
4	Open mangrove: mangrove areas with a relatively sparse canopy, allowing more sunlight to reach the ground.	38
5	Closed mangrove: dense mangrove areas with a thick canopy, where mangrove trees form a dense cover.	58
6	Paddy field: agricultural fields used for cultivating rice, typically similar shapes to shrimp ponds.	30
	Total	224

2.3. *Earth Observation Data*

2.3.1. Satellite Band and Spectral Indices

This study mainly used two different satellite images of Planet Scope images with 3 m resolution and Sentinel-2 images with 10 m resolution. Planet launches an incredible support of education and research program, by which we could access very high-resolution PlanetScope images for our study area. Orthorectified and atmospherically corrected images were downloaded from the Planet portal. Sentinel-2 level 2A products were collected from the Copernicus Open Access Hub using the semi-automatic Classification Plugin (SCP) tool in the QGIS 3.8 version. Level 2A products have provided atmospherically corrected Surface Reflectance images since December 2018. Sentinel-2 provides 13 bands with resolutions of 10, 20, and 60 m. Our study specifically utilized RGB and NIR bands with a 10 m resolution. To ensure consistency in spatial resolution, these bands were resampled to match the 3 m resolution of the PlanetScope bands; more detailed information is reported in Table 2.

Table 2. Specification of satellite imageries used in this study.

Title 1	Acquisition Date	Cloud Coverage	Band Used	Spatial Resolution	Processing Level
PlanetScope	21 January 2023	0	RGB, NIR	3	-
Sentinel-2	14 February 2023	0	RGB, NIR	10	Level-2A

We also used two popular and effective spectral indices, NDVI and NDWI, derived from the preprocessed spectral bands of Planet Scope and Sentinel-2. NDVI was calculated from Red and Near-infrared (NIR) bands, while NDWI was created from Green and NIR bands, which are available in both satellite images, using Equations (1) and (2), respectively [36,37]. The NDWI with a visible green band and NIR can be used to distinguish mangroves from water bodies because of their sensitivity to vegetation [37].

$$\text{NDVI} = \frac{\text{NIR} - \text{RED}}{\text{NIR} + \text{RED}} \tag{1}$$

$$\text{NDWI} = \frac{\text{GREEN} - \text{NIR}}{\text{GREEN} + \text{NIR}} \tag{2}$$

2.3.2. Topographic and Canopy Height Information

Integration of topographic and canopy height information has been explored to show accuracy improvement in LULC classification [38–41]. This study exhibited Multi-Error-Removed-Improved-Terrain (MERIT) (https://hydro.iis.u-tokyo.ac.jp/~yamadai/MERIT_DEM/, accessed on 20 January 2020) and Global Canopy Height Model (GCHM) datasets (https://langnico.github.io/globalcanopyheight/, accessed on 17 July 2023) with the aim of leveraging classification accuracy in distinguishing mangrove and non-mangrove forest in a mixed ecosystem of the large study area. To the best of our knowledge, this study is the first attempt to use the GCHM as an input feature for the LULC classification of mangrove ecosystems. The mangrove forests in our study area thrive in specific topographic niches, which are flat and low-lying inland coastal regions. The MERIT DEM and GCHM can help highlight these topographic features and tree height information, making it easier to distinguish mangrove habitats from non-mangrove forests in mountainous regions. Yamazaki et al. improved the MERIT DEM by removing the vertical error from SRTM 3 arc (90 m resolution) imagery [42]. This information is accessible free of charge on the MERIT_DEM website. Lang et al. developed a GCHM by fusing the Global Ecosystem Dynamic Investigation (GEDI) space-borne LiDAR mission and Sentinel-2 imagery [43,44]. The GCHM dataset is available on the Google Earth Engine with a spatial resolution of

10 m. We then resampled MERIT and GCHM to a 3 m spatial resolution using the Nearest Neighbor approach to align them with other spectral information.

2.4. Deep Learning Models for LULC Classification

2.4.1. U-Net Model

The U-Net architecture proposed by Ronneberger et al. [45] is a deep learning model that has gained significant attention because of its exceptional performance in biomedical image segmentation tasks. With a unique U-shaped design, the U-Net architecture consists of an encoder and a decoder, facilitating the extraction of features and reconstruction of high-resolution segmentation maps using skip connections. Through hyperparameter tuning, we designed a U-Net architecture suitable for LULC classification (Figure 3). In the proposed U-Net design, the encoder processes the input features from 128 × 128 × 8 image patches. Using 3 × 3 2D convolutional layers with activation functions, hierarchical features were iteratively extracted, and the feature depth was progressively enhanced. To control the spatial dimensions and establish a contracting path that formed an efficient encoder for capturing contextual information, 2 × 2 max pooling layers were employed. In the bottleneck stage, the encoder converges to a central bottleneck using augmented filters, thereby capturing the most abstract features from the input data.

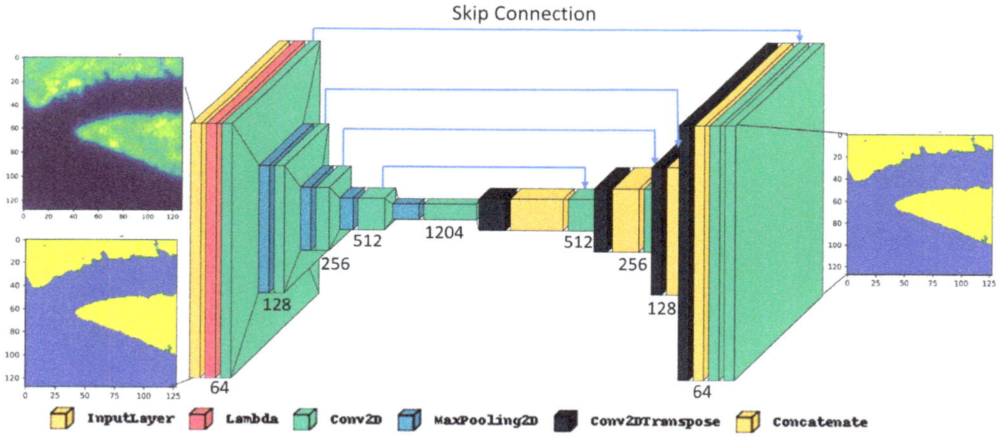

Figure 3. The proposed U-net architecture in this study.

The decoder, forming the expansive path, begins with 2 × 2 and 3 × 3 Conv2DTranspose upsampling layers to restore spatial resolution and reconstruct fine-grained details. Simultaneously, skip connections are established through concatenation, preserving low-level details and enabling seamless information transfer between paths. The conclusive layer generates the output, which is aligned with the input dimensions. The output layer mirrors the spatial dimensions of the input image, with each pixel predicting the LULC class for the corresponding region.

2.4.2. ANN Model

The ANN model was inspired by the design of a biological neural network of the human brain with a multilayer perceptron of input, hidden, and output layers [46]. In each layer, the nodes are interconnected to receive information, analyze it, and produce data. Referring to the basic ANN architecture, we developed the proposed deep neural network design with one input layer with eight variables, three hidden layers with 256,125, and 64 nodes, and one output layer with seven LULC classes, as shown in Figure 4.

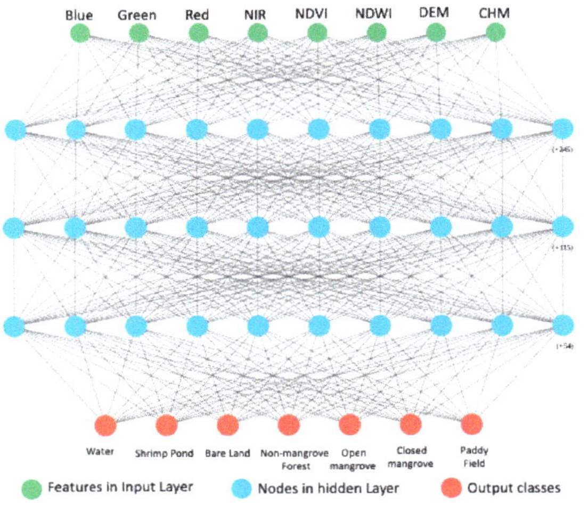

Figure 4. The proposed artificial neural network architecture in this study.

In both the U-net and ANN architectures, we employed an Exponential Linear Unit (ELU) in each layer as an activation function to analyze the features transferred from the input layer, as shown in Equation (3): ELU have shown faster and more precise learning in deep neural network computations than the commonly used Rectified Linear Unit (ReLU) [47]. Studies have already demonstrated the superior performance of ELU over ReLU through experimental analyses [48,49] and have effectively applied ELU in deep-learning segmentation tasks [50–52]. The softmax activation function was applied to the output layer to produce seven probabilistic LULC classes (Equation (4)).

$$f_{(y)} = \begin{cases} y & \text{if } y > 0 \\ \propto (\exp(y) - 1) & \text{if } y \leq 0 \end{cases} \tag{3}$$

$$f_{(y)} = \frac{\exp(y_i)}{\sum \exp(y_j)} \tag{4}$$

2.4.3. Assessment of Model Performance

Using a separate testing dataset, the performances of the models were evaluated using different metrics: overall accuracy (OA), IoU, F1 score, precision, and recall. First, we created a confusion matrix to provide a better understanding of the matching of the prediction results with the labeled data in each class. OA is one of the most widely used measures in LULC classification, showing the percentage of classification accuracy [53]. It was calculated based on the proportion of correctly classified pixels to the total number of pixels using Equation (5). A higher OA represents a better overall classification performance, but it sometimes leads to misleading results in an imbalanced dataset. Therefore, we employed other IoU metrics: F1 score, precision, and recall.

The IoU is used when the classes are imbalanced in a dataset. It measures the overlap between the predicted class and true class by calculating the ratio of intersection of the predicted and true classes to their union (Equation (6)) [54]. The F1 score is described as the harmonic mean of the precision and recall, and its value ranges from 0 to 1, indicating that higher F1 scores represent better classification results (Equation (7)). Precision is also expressed as the ratio of correctly classified positive classes to the total number of true and false positive classes, whereas recall focuses on the total number of true positive and false negative classes, as shown in Equations (8) and (9)) [55].

$$\text{OA} = \frac{\text{Number of correct prediction}}{\text{Total number of prediction}} \tag{5}$$

$$\text{IoU} = \frac{\text{True postive}}{\text{True positive} + \text{False positive} + \text{False negative}} \tag{6}$$

$$\text{F1} = 2 \times \frac{\text{Precision} \times \text{Recall}}{\text{Precision} + \text{Recall}} \tag{7}$$

$$\text{Precision} = \frac{\text{True positive}}{\text{True positive} + \text{False positive}} \tag{8}$$

$$\text{Recall} = \frac{\text{True positive}}{\text{True positive} + \text{False negative}} \tag{9}$$

2.5. LULC Classification Workflow

The main workflow of this study is illustrated in the flowchart shown in Figure 5. First, we acquired the necessary remote sensing data, as mentioned in Section 2.3. During the pre-processing stage, the images were mosaiced, and the study area was extracted from each image. All data were resampled to a 3 m resolution based on PlantScope imagery using the Nearest Neighborhood resampling approach. We then created two main datasets, PlanetScope and Sentinel-2, each integrated with their spectral indices, DEM, and GCHM data. The following processes were performed separately for the Planet and Sentinel-2 datasets.

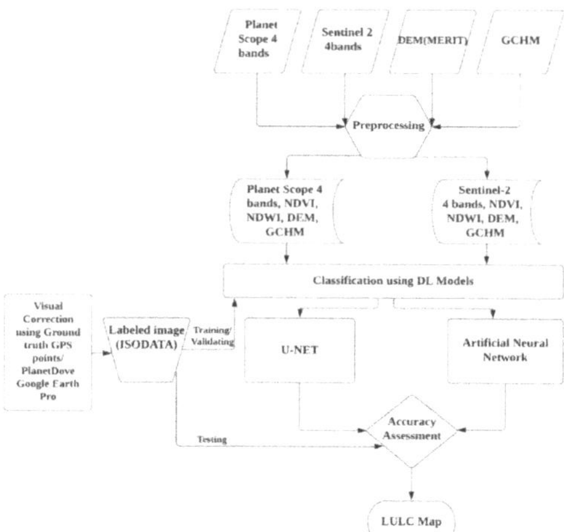

Figure 5. Workflow of LULC classification in this study.

All remote sensing images were sliced into 128 × 128 patches. A total of 3700 images with 128 × 128 patches were created and aligned with the same number of labeled images encoding seven classes of LULC information. Sliced remote sensing images with 128 × 128 patches were concatenated into a dataset of 128 × 128 × 8 patches. We then split the datasets into a 75% training dataset with 2389 images and 15% for validation with 500 images. The remaining 10% of the 388 images were reserved as the testing dataset to evaluate the performance of the models. Our proposed U-Net and ANN models are then trained in 200 epochs and accessed using TensorFlow version 2.1.5 in python 3.10 on the Wisteria/BDEC-01 supercomputer system of the University of Tokyo.

3. Results

In this study, LULC classification was explored using the U-Net and ANN models with two different datasets: Planet Scope and Sentinel-2. Each dataset was also combined with DEM and GCHM data to leverage the classification performance. The models were trained using a training dataset and evaluated using a testing dataset to compare their performances for complex LULC classes in the Wunbaik Mangrove Area.

3.1. Model Performance with PlanetScope Imagery

Using the RGB and NIR bands of Planet Scope, NDVI, NDWI, DEM, and GCHM, both U-Net and ANN models demonstrated promising results in classifying various LULC categories. The U-Net model exhibits an overall accuracy (OA) of 94.88% for the training dataset and 94.05% for the testing dataset. The U-Net model also produced an Intersection over Union (IoU) of 0.82, and average values of precision, recall, and F1 score of 0.94. The ANN model, on the other hand, achieved an OA of 93.61% with the training dataset and 92.82% when the test dataset was predicted, with an IoU of 0.79, an F1 score of 0.93, a precision of 0.94 and a recall of 0.93.

We also investigated all matrices in each class to analyze the model performance in each LULC category in detail. The results showed that both the U-Net and ANN models classified water, open mangrove, and closed mangrove areas with the highest scores in all metrics. The U-Net consistently outperformed the ANN in most LULC classes. However, notable strengths of the ANN model were observed in some classes, which are in bold in Table 3. Both models identified bare land with the lowest scores in most matrices among other classes. In the context of non-mangrove classes, which represent the smallest portion within the dataset, U-Net demonstrated a relatively strong performance in distinguishing them from mangroves, whereas ANN exhibited notably lower scores for this specific task, except for the precision score. In our study area, the paddy fields had a shape similar to shrimp ponds and similar spectral information to that of bare land during the post-harvest season. U-Net, however, could extract paddy fields from confusing classes with slightly higher IoU and F1 scores compared to the ANN's performance. The images classified by each model were randomly extracted, and their differences are shown with the labeled images in Figure 6.

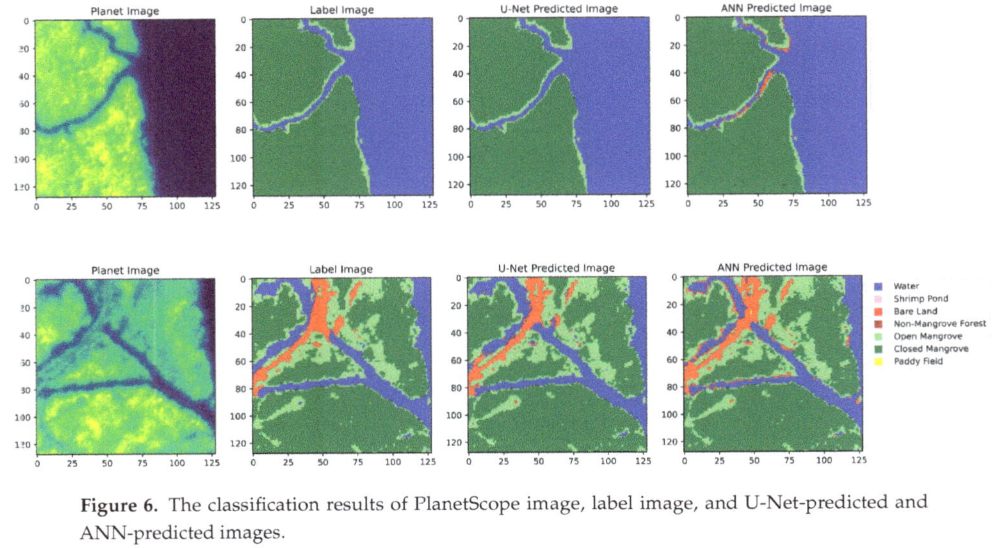

Figure 6. The classification results of PlanetScope image, label image, and U-Net-predicted and ANN-predicted images.

Table 3. Classification results of U-Net and ANN using Planet Scope dataset.

LULC Class	IoU		F1 Score		Precision		Recall	
	U-Net	**ANN**	**U-Net**	**ANN**	**U-Net**	**ANN**	**U-Net**	**ANN**
Water	0.97	0.96	0.97	0.97	0.97	**0.98**	0.97	0.96
Shrimp pond	0.82	0.72	0.84	0.79	0.85	**0.88**	0.82	0.72
Bare land	0.76	**0.84**	0.73	0.70	0.70	0.60	0.76	**0.84**
Non-mangrove forest	0.92	0.83	0.91	0.85	0.89	0.87	0.92	0.83
Open mangrove	0.91	**0.95**	0.93	**0.94**	0.95	0.94	0.91	**0.94**
Closed mangrove	0.99	0.99	0.99	0.99	0.99	0.98	0.99	0.98
Paddy field	0.92	0.89	0.92	0.90	0.92	0.92	0.92	0.92
Average Score	0.82	0.79	0.94	0.93	0.94	0.94	0.94	0.93
Overall accuracy on testing dataset (%)							94.05	92.82

Note: Notable strengths of the ANN model were observed in some classes, which are in bold.

3.2. Model Performance with Sentinel-2 Imagery

We also employed U-Net and ANN on the Sentinel-2 dataset, including the same spectral bands, indices, DEM, and GCHM. The U-Net model exhibited a higher OA of 98.15% in the training phase; however, the accuracy decreased up to 86.94% for the testing datasets. However, the ANN model showed steady performance, resulting in 82.03% and 82.08% performance on both the training and testing datasets, respectively. A mean IoU of 0.71 and an average value of 0.87 for the F1 score, precision, and recall were obtained with the U-Net model. In sharp contrast, the ANN model classified LULC classes with lower scores: mean IoU, 0.63; average F1 score, 0.81; precision, 0.81; and recall, 0.82. With a detailed analysis of each class, the U-Net model outperformed the ANN model, with higher scores, except for IoU, recall of shrimp ponds, and recall of closed mangroves highlighted in bold in Table 4. The results indicated that when utilizing the Sentinel-2 dataset, the models demonstrated lower overall performance scores than when using the PlanetScope dataset. However, they exhibited slightly higher precision and recall scores for shrimp pond classifications. Both models produced unsatisfactory results for the bare land and open mangrove classes, whereas a higher performance was achieved when applying the PlanetScope dataset. The models exhibited noticeable and steady scores in identifying paddy fields using both datasets. The differences in classification are shown with the labeled images in Figure 7.

Table 4. Classification results of U-Net and ANN using Sentinel-2 dataset.

LULC Class	IoU		F1 Score		Precision		Recall	
	U-Net	**ANN**	**U-Net**	**ANN**	**U-Net**	**ANN**	**U-Net**	**ANN**
Water	0.94	0.89	0.94	0.91	0.94	0.94	0.94	0.89
Shrimp pond	0.83	**0.91**	0.82	0.77	0.81	0.67	0.83	**0.91**
Bare land	0.53	0.22	0.54	0.34	0.55	0.55	0.53	0.22
Non-mangrove forest	0.89	0.85	0.91	0.87	0.93	0.89	0.91	0.85
Open mangrove	0.60	0.43	0.64	0.49	0.68	0.57	0.64	0.43
Closed mangrove	0.96	0.94	0.93	0.89	0.91	0.84	0.93	**0.94**
Paddy field	0.92	0.91	0.92	0.91	0.92	0.91	0.92	0.91
Average Score	0.71	0.63	0.87	0.81	0.87	0.81	0.87	0.82
Overall accuracy on testing dataset (%)							86.94	82.08

Note: With a detailed analysis of each class, the U-Net model outperformed the ANN model, with higher scores, except for IoU, recall of shrimp ponds, and recall of closed mangroves highlighted in bold.

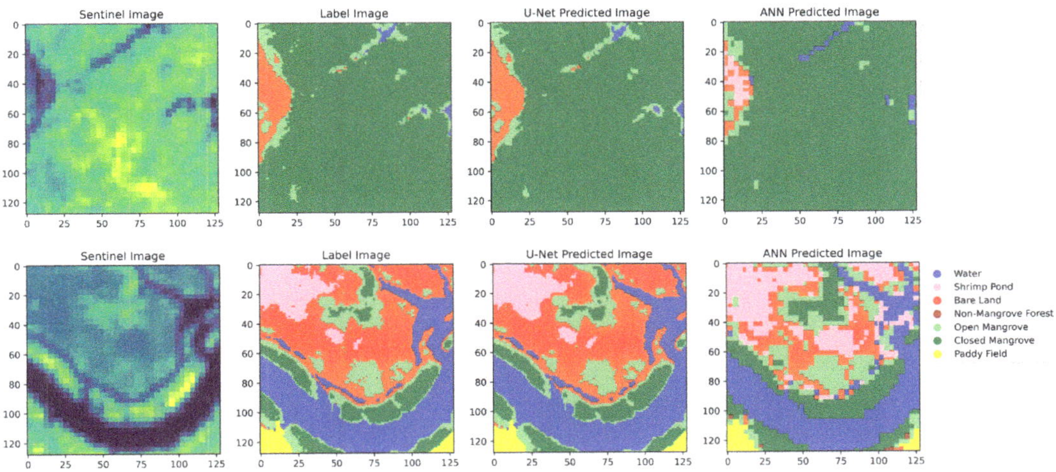

Figure 7. The classification results of Sentinel-2 image, label image, and U-Net-predicted and ANN-predicted images.

3.3. LULC Classification Map for the Whole Study Area

After evaluating the models with different RS datasets, they were applied to produce an LULC map of the Wunbaik Mangrove Area. The LULC maps of the entire study area were visualized using ArcGIS Pro by comparing true color images and labeled ground truth images (Figure 8). The results showed that the LULC map produced by U-Net with Planet Scope had a strong correspondence with the labeled ground truth images in all classes. However, despite the promising scores in the training and testing phases, the ANN model misclassified bare land in most paddy field areas.

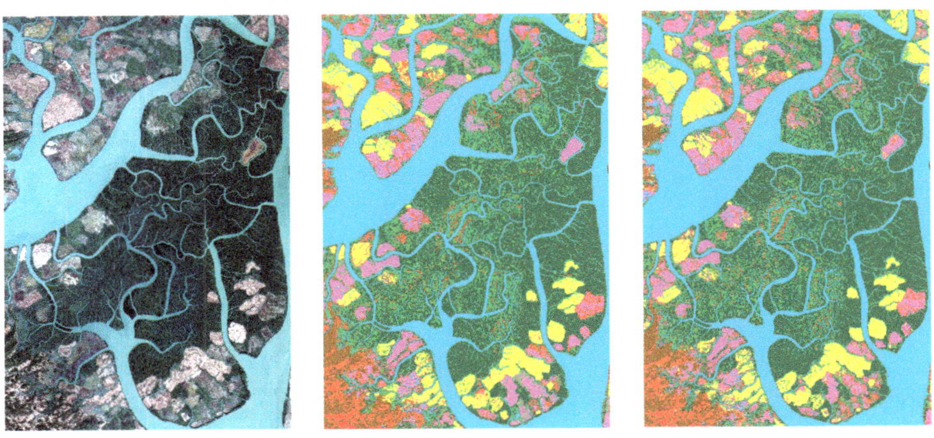

(**a**) (**b**) (**c**)

Figure 8. *Cont.*

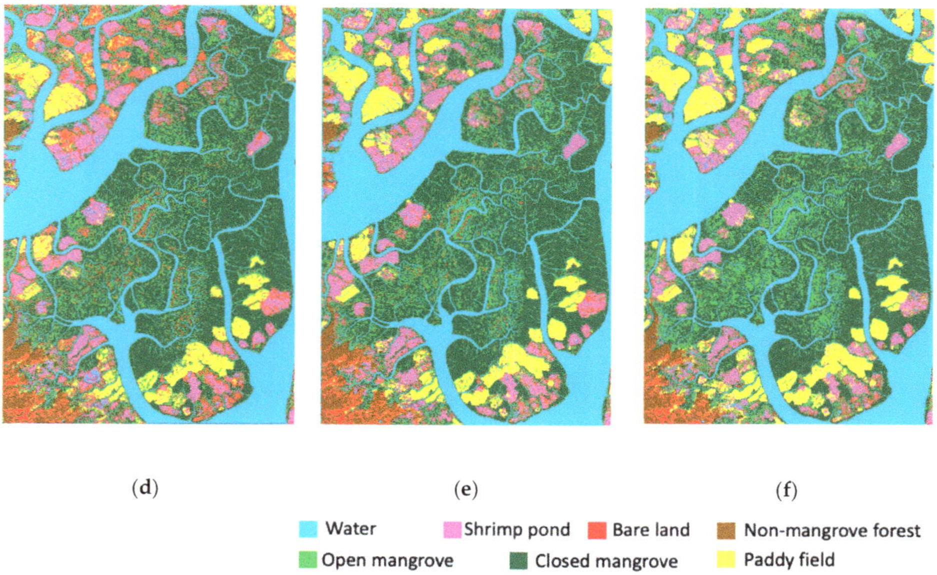

Figure 8. (**a**) True color image of PlanetScope, (**b**) labeled image, LULC classification maps classified by (**c**) U-Net model with PlanetScope, (**d**) ANN model with PlanetScope, (**e**) U-Net model with Sentinel-2, and (**f**) ANN model with Sentinel-2 Dataset.

We also exported and visually analyzed the LULC classification maps from the predictions of the U-Net and ANN models using the Sentinel-2 dataset. Using Sentinel-2 data, the U-Net model produced a classification map that was as accurate as the U-Net model using Planet Scope. However, there was some noise and misclassified pixels when identifying bare land in non-mangrove areas. In addition, both the U-Net and ANN models misclassified bare land areas and open mangroves with unsatisfactory results. Similarly, using the PlanetScope data, the ANN model could not distinguish between bare land and paddy fields.

3.4. Estimating LULC Area for Wunbaik Mangrove Forest Managment

Because the U-Net model with Planet Scope produced the highest scores in all metrics and predicted a promising classification map, we calculated LULC areas using the U-Net prediction model. According to our prediction result, closed and open mangrove area cover 187.5 km^2 and 57.00 km^2 while water body cover 149.96 km^2 in our study area. As our study area is a complex ecosystem, we also identified 15.15 km^2 of non-mangrove forests. Our model detected 44.89 km^2 of shrimp ponds and 45.36 km^2 of paddy fields, which are the main livelihoods of nearby local communities (Table 5).

With the aim of contributing up-to-date LULC information to sustainable mangrove forest management in the Wunbaik area, we extracted used (WRMF) data from the whole classification map (Figure 9). According to our classification results for 2023, the WRM comprises 167.38 km^2 of closed mangrove, 35.52 km^2 of open mangroves, and 72.43 km^2 of water bodies. Although there was a small portion of non-mangrove forests on small mounds inside the WRM, we mainly focused on mangrove forest management. Despite the large portion of remnant mangrove, due to human disturbances and their effects, we also detected 20.45 km^2 of shrimp pond, 17.77 km^2 of paddy fields, and 9.41 km^2 of bare land.

Table 5. LULC information of the whole study area and Wunbaik Reserved Mangrove Forest.

LULC Class	Whole Study Area		Wunbaik Reserved Mangrove Forest	
	Area (km^2)	Portion (%)	Area (km^2)	Portion (%)
Water	149.96	28.14	72.94	22.51
Shrimp pond	44.89	8.42	17.99	5.55
Bare land	32.98	6.19	11.84	3.65
Non-mangrove forest	15.15	2.84	0.29	0.09
Open mangrove	57.00	10.70	35.52	10.96
Closed mangrove	187.50	35.19	167.38	51.65
Paddy field	45.36	8.51	18.12	5.59

Figure 9. (**a**) LULC map and (**b**) LULC information of Wunbaik Reserved Mangrove Forest.

4. Discussion

For the effective management of mangrove areas, comprehensive LULC information is pivotal for systematic planning and successful restoration programs. This study conducted an LULC classification of an intricate ecosystem by comparing two deep learning models, U-Net and ANN, with confusing classes such as closed and open mangroves, non-mangrove forests, shrimp ponds, paddy fields, bare land, and water bodies.

4.1. Complexity of Land Use Compared to Land Cover

Some studies have shown promising results in binary classification in which mangrove are distinguished from non-mangrove classes such as water, human settlement, and agricultural field [28,29]. However, this study identified multiple classes of intricate mangrove regions that contributed detailed LULC information for the management of mangroves. In the LULC classification of mangrove regions, it is straightforward to classify a binary class, such as mangrove and non-mangrove, or simple land cover classes, such as mangrove, water, and bare land. However, the complexity of different land uses and cover types makes the classification process challenging when land use is considered. As a complicated land use system in our study area, the integration of shrimp ponds and paddy fields requires more attention when selecting the classification methods and RS datasets. For instance, shrimp ponds are a type of land use but it possesses water and bare land as land cover. Shrimp pond areas are sometimes inundated during high tide and have similar spectral information as water, while being confused with bare land during low tide. Similarly, the land use and paddy fields after harvesting confused the models with bare land. Zabaar et al. employed CNN models and compared them with different machine learning algorithms using Pléiades VHSR and Sentinel-2A for multiclass LULC mapping of the Ain Témouchent coastal area situated in western Algeria. Their results showed that the CNN model outperformed them with a higher OA of 93.5% for Pléiades data and 83.4% for Sentinel-2A data [56]. Alteraz et al. also demonstrated the superior performance of U-Net with Sentinel 1 and 2 fusion in multiclass LULC classification compared with other machine learning models [57]. Therefore, the LULC classification of multiple classes with similar land covers could lead to misclassification when using traditional approaches. However, the U-Net model, using PlanetScope with topographic and canopy height information, produced better performance for the intricate coastal land use system in our study.

Due to an imbalanced class distribution in this research, the water and closed mangrove classes represented over half of the dataset, while other classes represented less than 10 percent of the study area (Figure 10). Therefore, the U-Net and ANN models produced noticeably higher scores than the other classes for both datasets. Considering the non-mangrove forest class with the smallest portion of the dataset, the proposed U-Net model outperformed both PlanetScope and Sentinel-2 datasets. In the case of bare land, which was the second smallest portion, the ANN model overpredicted the PlanetScope dataset, whereas the model could not be identified in most areas, particularly inside the mangrove forest with the Sentinel-2 dataset.

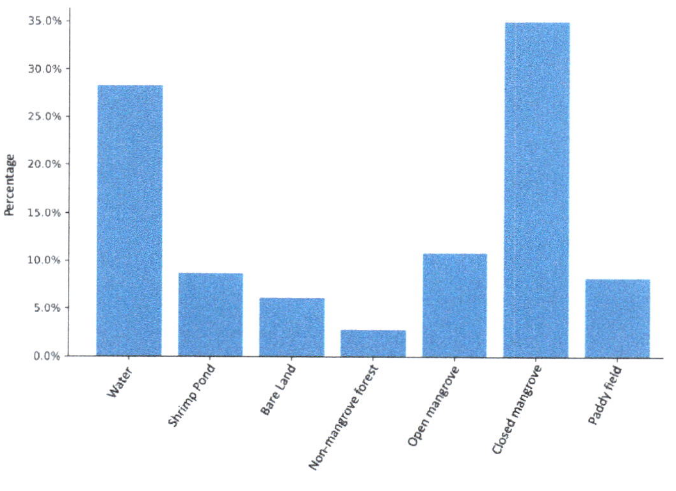

Figure 10. Area of LULC classes in our dataset of the study area.

4.2. Integration of DEM and CHM

Paying attention to non-mangrove forested areas have similar spectral information to mangroves despite ecological differences. Scholars have also developed mangrove-specific vegetation indices, such as the Combined Mangrove Recognition Index (CMRI) [58] the Enhanced Mangrove Vegetation Index (EMVI) [59], to distinguish mangroves from other forest types. Paying careful attention to the ecological conditions of the study area, we applied DEM and CHM instead of using additional indices. According to our fieldwork, non-mangrove forests grow in terrestrial regions with higher elevations than mangrove regions. Canopy height information derived from Unmanned Aerial Vehicle (UAV) images, light detection, and airborne radar (Lidar) data has been applied as a complementary feature in mangrove classification of specific and small areas. The integration of DEM improved the distinction between these two forest types, whereas CHM provided more information for the case of forested and non-forested classes. Maung and Sasaki [34] introduced a Canopy Height Model derived from SRTM and MERIT DEMs to identify mangrove distribution in the Wunbaik mangrove region. Their experimental research highlighted that using differences of SRTM and MERIT DEMs improved the accuracy compared to CMRI in mangrove regions. However, they also considered the binary classification of mangroves and non-mangroves using only an ANN model and Sentinel-2 imagery for intricate coastal ecosystems. Considering seven LULC classes in the same study area, we explored the U-Net segmentation model and high-resolution PlanetScope imagery and compared them with the ANN and Sentinel-2 image used in the binary classification of Maung and Sasaki's previous study. Our results show that U-Net was impressively classified using the PlanetScope dataset, whereas the ANN model produced some misclassifications between paddy fields and bare land, as shown in Figure 8.

4.3. Model Performance and Labeled Data Requirement

The deep learning models proposed in our study were trained using the training and validation datasets separately with the same hyperparameters and input shapes of 128 × 128 × 8. The models were monitored during the learning phase using a validation dataset. As shown in Figure 11, the U-Net learned well with PlanetScope dataset, which means no significance difference between the training and validating loss. When the U-Net model was introduced to the Sentinel-2 dataset, there was overfitting with different trends of training and validation losses after 25 epochs of learning phases. However, the ANN model significantly reduced the loss during early epochs in both datasets and, also, a remarkable loss reduction occurred when using PlanetScope as compared to the Sentinel-2 dataset. The overfitting problem of the U-Net model with Sentinel-2 dataset was considered a reason for the resampling approach of Sentinel-2 from 10 m to 3 m. Being a segmentation model, U-Net is sensitive for spatial information of a dataset. In response to these overfitting challenges, the fusion of PlanetScope and Sentinel-2 data was not considered in our analysis, as the U-Net model with the Planet dataset had already yielded satisfactory results in improving LULC information. In our future studies, we are aiming to explore various state of the art deep learning approaches to compare with our proposed method.

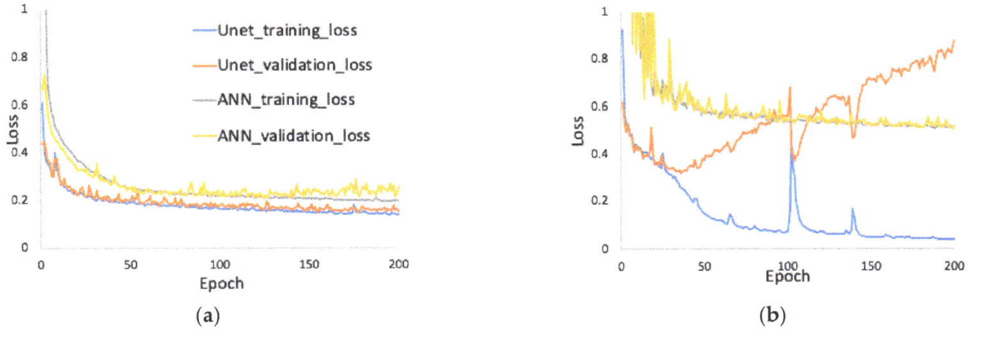

Figure 11. Learning process of the U-Net and ANN models using (**a**) PlanetScope and (**b**) Sentinel-2.

Aside from the overfitting, a notable discrepancy in the spatial resolution between the U-Net and ANN-predicted images occurred when utilizing the Sentinel dataset, as shown in Figure 7. This discrepancy can be attributed to the resampling process applied to the Sentinel images, which affected all corresponding predictions. This assumption finds support in the ANN-predicted image of the PlanetScope dataset, illustrated in Figure 6, which maintains its original 3 m resolution. In contrast, the U-Net model demonstrated a notable ability to preserve fine spatial details in both datasets, which were closely aligned with the resolution of the labeled images. Leveraging its semantic segmentation approach and skip connections, U-Net showcased superior performance in terms of both classification accuracy and spatial resolution of the predicted images. However, it is worth trying original spatial resolution or different resampling approaches for the Sentinel-2 imagery in future research. Through a detailed quantitative and qualitative assessment of the model learning and prediction, we assumed that the U-Net model with PlanetScope bands, NDVI, NDWI, MERIT DEM, and GCHM, is the best candidates for the LULC classification of the Wunbaik Mangrove Area.

Semantic segmentation models such as U-Net are pivotal for accurately categorizing objects in images, particularly in remote sensing, where access to fully labeled images is common. With full labels, the model can capture intricate patterns, leading to a more precise and nuanced segmentation. By contrast, partial labels may hinder the model's ability to distinguish between similar classes or subtle distinctions. However, it is important to acknowledge the challenge of acquiring substantial amounts of labeled data, which is resource intensive and time consuming. In some cases, partially labeled datasets may be more practical, and exploring techniques such as semi-supervised learning or active learning can optimize the labeling resources. Marsocci and Scardapane investigated continual self-supervised learning for remote sensing semantic segmentation by employing heterogeneous remote sensing sensors, resolutions, acquisitions, and scenes with varying levels of labeled data (10%, 50%, and 100%) [60]. Their findings revealed that the proposed model, Continual Barlow Twins (CBTs), achieved satisfactory IoU scores with 10% and 50% partially labeled images from the US3D dataset derived from the WorldView 3 satellite, whereas using a 100% labeled dataset produced the highest IoU score. To address the challenge of labeled data dependence in remote sensing image semantic segmentation, Li et al. introduced a semi-supervised segmentation network called SegMind, tested with partially labeled datasets (5%, 10%, and 20%) [61]. Their experimental results demonstrated competitive IoU scores compared with other state-of-the-art deep learning models. Saha et al. conducted unsupervised single-scene segmentation with different input datasets and no labeled data [62]. Their proposed method outperformed other models, despite low metric scores and limitations in segmenting small classes. Given the rapid developments in computer science and remote sensing applications, it is essential to consider both model performance and resource effectiveness in future studies.

4.4. Challenges and Opportunities of Sustainable Wunbaik Mangrove Mangement

To better manage mangrove forests, other land use information from a reliable LULC map needs to be considered. Some researchers have explored mangrove mapping on a global and large scale to better understand mangrove distribution and dynamic changes as the foundation of international collaboration [63]. Detailed, accurate, and up-to-date regional information is of utmost importance when considering practical strategies for implementing regional mangrove restoration projects. Despite a massive remnant mangrove track, the Wunbaik Mangrove Area is also facing human disturbances such as the extension of shrimp ponds and paddy fields, and illegal firewood extraction due to growth in population with low income and lack of job opportunities. Over the years, conversion into shrimp ponds and paddy fields has led to mangrove deforestation, whereas overexploitation of fuelwood has downgraded the mangrove status. Other studies have also pointed out mangrove loss in Wunbaik due to human disturbances, such as shrimp ponds and agricultural expansion, except for mangrove plantation, but there are no particular programs in which local communities can participate in mangrove restoration projects [34]. According to our LULC map, shrimp ponds and paddy fields still exist, even in the Wunbaik Reserved Mangrove Forest. Being a vulnerable mangrove area due to many interactions with local livelihoods, it would be impossible to implement mangrove restoration programs without the active participation of local people. Therefore, based on our findings of LULC information, we would recommend mangrove projects, such as community forestry, mangrove-friendly aquaculture systems other than plantations, and the constitution of reserved forests. In pursuit of a long-term mangrove management strategy for the Wunbaik Reserved Forest, policymakers, conservationists, and local communities can leverage the LULC information to formulate a collective land use policy across all government sectors. The insights derived from our LULC information lay the groundwork for a comprehensive forest management plan, in which the creation of buffer zones between existing mangrove areas and anthropogenically disturbed zones is recommended to curtail the expansion of shrimp ponds and paddy fields. With the goal of mitigating the impact of climate change, our accurate and up-to-date LULC data serve as a benchmark for estimating mangrove carbon sequestration, thereby contributing to improved accuracy of global carbon accounting efforts.

5. Conclusions

This study provides reliable and up-to-date LULC information for an intricate mangrove ecosystem using the U-Net model in conjunction with multisource remote sensing datasets of PlanetScope, Sentinel-2, MERIT DEM, and Global Canopy Height Model. Through different scenarios of the U-Net and ANN models using different dataset combinations, the U-Net model using PlanetScope, topography, and canopy height information achieved the highest OA (94.05%), IoU (0.82), and a mean F1 score, recall, and precision of 0.94. Our proposed model could predict LULC classes in the entire study area and produce an accurate LULC map for the mangrove management regime. According to the LULC results, Wunbaik Reserved Mangrove Forest is endowed with varying states of closed and open mangroves. However, disturbances due to human activities in shrimp ponds and paddy fields are still occurring in reserved forests, thereby altering for the need of conservation. To safeguard this remnant mangrove area, immediate attention should be paid to restoration approaches in which local communities can participate under expert guidance. In summary, our study not only proposes a promising classification method but also an urgent call for conservation measures for this unique mangrove ecosystem.

Author Contributions: Conceptualization, methodology, W.S.M., S.T. and Z.G.; data curation, formal analysis, and writing—original draft preparation, W.S.M.; project administration, and funding acquisition, S.T.; writing—review and editing, supervision, Conceptualization, S.T. and Z.G. All authors have read and agreed to the published version of the manuscript.

Funding: This research received no external funding.

Data Availability Statement: The remote sensing data used in this study are available on their websites as mentioned in the related section. Other data presented in the study are available on request from the first author.

Acknowledgments: We express our sincere thanks to Planet (https://www.planet.com/, accessed on 20 April 2022) for granting us access to the high-resolution PlanetScope imagery through the Education and Research Program. Deep learning analysis was performed using the supercomputer resources of International Technological Center (ITC) supercomputer, the University of Tokyo. The authors would like to express their gratitude to the anonymous reviewers for their insightful comments and suggestions to improve the quality of this article.

Conflicts of Interest: The authors declare no conflicts of interest.

References

1. Brander, L.M.; Wagtendonk, A.J.; Hussain, S.S.; McVittie, A.; Verburg, P.H.; de Groot, R.S.; van der Ploeg, S. Ecosystem service values for mangroves in Southeast Asia: A meta-analysis and value transfer application. *Ecosyst. Serv.* **2012**, *1*, 62–69. [CrossRef]
2. Hawkins, S.J.; Evans, A.J.; Dale, A.C.; Firth, L.B.; Hughes, D.J.; Smith, I.P. Ecosystem Services and Disservices of Mangrove Forests and Salt Marshes. In *Oceanography and Marine Biology*; Taylor & Francis: Hoboken, NJ, USA, 2017; Volume 55, ISBN 9781351987592.
3. Richards, D.R.; Friess, D.A. Rates and drivers of mangrove deforestation in Southeast Asia, 2000–2012. *Proc. Natl. Acad. Sci. USA* **2016**, *113*, 344–349. [CrossRef] [PubMed]
4. Friess, D.A.; Rogers, K.; Lovelock, C.E.; Krauss, K.W.; Hamilton, S.E.; Lee, S.Y.; Lucas, R.; Primavera, J.; Rajkaran, A.; Shi, S. The State of the World's Mangrove Forests: Past, Present, and Future. *Annu. Rev. Environ. Resour.* **2019**, *44*, 89–115. [CrossRef]
5. Awuku-Sowah, E.M.; Graham, N.A.J.; Watson, N.M. Investigating mangrove-human health relationships: A review of recently reported physiological benefits. *Dialogues Health* **2022**, *1*, 100059. [CrossRef]
6. Chen, R.; Yang, H.; Yang, G.; Liu, Y.; Zhang, C.; Long, H.; Xu, H.; Meng, Y.; Feng, H. Land-Use Mapping with Multi-Temporal Sentinel Images Based on Google Earth Engine in Southern Xinjiang Uygur Autonomous Region, China. *Remote Sens.* **2023**, *15*, 3958. [CrossRef]

7. Wang, L.; Jia, M.; Yin, D.; Tian, J. A review of remote sensing for mangrove forests: 1956–2018. *Remote Sens. Environ.* **2019**, *231*, 111223. [CrossRef]
8. Pham, T.; Yokoya, N.; Yoshino, K.; Bui, D.; Friess, D. Remote Sensing Approaches for Monitoring Mangrove Species, Structure, and Biomass: Opportunities and Challenges. *Remote Sens.* **2019**, *11*, 230. [CrossRef]
9. Giri, C. Recent advancement in mangrove forests mapping and monitoring of the world using earth observation satellite data. *Remote Sens.* **2021**, *13*, 563. [CrossRef]
10. Nguyen, H.H.; Nghia, N.H.; Nguyen, H.T.T.; Le, A.T.; Tran, L.T.N.; Duong, L.V.K.; Bohm, S.; Furniss, M.J. Classification methods for mapping mangrove extents and drivers of change in Thanh Hoa province, Vietnam during 2005–2018. *For. Soc.* **2020**, *4*, 225–242. [CrossRef]
11. Manna, S.; Raychaudhuri, B. Mapping distribution of Sundarban mangroves using Sentinel-2 data and new spectral metric for detecting their health condition. *Geocarto Int.* **2020**, *35*, 434–452. [CrossRef]
12. Zheng, Y.; Takeuchi, W. Quantitative assessment and driving force analysis of mangrove forest changes in china from 1985 to 2018 by integrating optical and radar imagery. *ISPRS Int. J. Geo-Inf.* **2020**, *9*, 513. [CrossRef]
13. Lu, D.; Weng, Q. A survey of image classification methods and techniques for improving classification performance. *Int. J. Remote Sens.* **2007**, *28*, 823–870. [CrossRef]
14. Thakur, S.; Mondal, I.; Ghosh, P.B.; Das, P.; De, T.K. A review of the application of multispectral remote sensing in the study of mangrove ecosystems with special emphasis on image processing techniques. *Spat. Inf. Res.* **2020**, *28*, 39–51. [CrossRef]
15. Green, E.P.; Clark, C.D.; Mumby, P.J.; Edwards, A.J.; Ellis, A.C. Remote sensing techniques for mangrove mapping. *Int. J. Remote Sens.* **1998**, *19*, 935–956. [CrossRef]
16. Ahmad, A.; Quegan, S. Comparative analysis of supervised and unsupervised classification on multispectral data. *Appl. Math. Sci.* **2013**, *7*, 3681–3694. [CrossRef]
17. Toosi, N.B.; Soffianian, A.R.; Fakheran, S.; Pourmanafi, S.; Ginzler, C.; Waser, L.T. Comparing different classification algorithms for monitoring mangrove cover changes in southern Iran. *Glob. Ecol. Conserv.* **2019**, *19*, e00662. [CrossRef]
18. Ge, G.; Shi, Z.; Zhu, Y.; Yang, X.; Hao, Y. Land use/cover classification in an arid desert-oasis mosaic landscape of China using remote sensed imagery: Performance assessment of four machine learning algorithms. *Glob. Ecol. Conserv.* **2020**, *22*, e00971. [CrossRef]
19. Campomanes, F.; Pada, A.V.; Silapan, J. Mangrove classification using support vector machines and random forest algorithm: A comparative study. In Proceedings of the GEOBIA 2016: Solutions and Synergies, Enschede, The Netherlands, 14–16 September 2016; pp. 2–6.
20. Castelo-Cabay, M.; Piedra-Fernandez, J.A.; Ayala, R. Deep learning for land use and land cover classification from the Ecuadorian Paramo. *Int. J. Digit. Earth* **2022**, *15*, 1001–1017. [CrossRef]
21. Han, W.; Zhang, X.; Wang, Y.; Wang, L.; Huang, X.; Li, J.; Wang, S.; Chen, W.; Li, X.; Feng, R.; et al. A survey of machine learning and deep learning in remote sensing of geological environment: Challenges, advances, and opportunities. *ISPRS J. Photogramm. Remote Sens.* **2023**, *202*, 87–113. [CrossRef]
22. Zhang, C.; Sargent, I.; Pan, X.; Li, H.; Gardiner, A.; Hare, J.; Atkinson, P.M. Joint Deep Learning for land cover and land use classification. *Remote Sens. Environ.* **2019**, *221*, 173–187. [CrossRef]
23. Yuan, Q.; Shen, H.; Li, T.; Li, Z.; Li, S.; Jiang, Y.; Xu, H.; Tan, W.; Yang, Q.; Wang, J.; et al. Deep learning in environmental remote sensing: Achievements and challenges. *Remote Sens. Environ.* **2020**, *241*, 111716. [CrossRef]
24. Fan, Z.; Zhan, T.; Gao, Z.; Li, R.; Liu, Y.; Zhang, L.; Jin, Z.; Xu, S. Land Cover Classification of Resources Survey Remote Sensing Images Based on Segmentation Model. *IEEE Access* **2022**, *10*, 56267–56281. [CrossRef]
25. Chen, J.; Sasaki, J. Mapping of subtidal and intertidal seagrass meadows via application of the feature pyramid network to unmanned aerial vehicle orthophotos. *Remote Sens.* **2021**, *13*, 4880. [CrossRef]
26. Mas, J.F.; Flores, J.J. The application of artificial neural networks to the analysis of remotely sensed data. *Int. J. Remote Sens.* **2008**, *29*, 617–663. [CrossRef]
27. Dai, X.L.; Khorram, S. Remotely sensed change detection based on artificial neural networks. *Photogramm. Eng. Remote Sens.* **1999**, *65*, 1187–1194.
28. Wei, Y.; Cheng, Y.; Yin, X.; Xu, Q.; Ke, J.; Li, X. Deep Learning-Based Classification of High-Resolution Satellite Images for Mangrove Mapping. *Appl. Sci.* **2023**, *13*, 8526. [CrossRef]
29. Islam, M.D.; Di, L.; Mia, M.R.; Sithi, M.S. Deforestation Mapping of Sundarbans Using Multi-Temporal Sentinel-2 Data and Transfer Learning. In Proceedings of the 2022 10th International Conference on Agro-geoinformatics (Agro-Geoinformatics), Quebec City, QC, Canada, 11–14 July 2022; pp. 1–4.
30. Chun, B.B.; Mat Jafri, M.Z.; San, L.H. Comparison of remote sensing approach for mangrove mapping over Penang Island. In Proceedings of the 2012 International Conference on Computer and Communication Engineering, Kuala Lumpur, Malaysia, 3–5 July 2012; pp. 258–262.
31. Stanley, D.O.; Broadhead, J.; Aung Aung, M. *The Atlas and Guidelines for Mangrove Management in Wunbaik Reserved Forest*; Forest Department; Food and Agriculture Organization of the United Nations FAO-UN: Yangon, Myanmar, 2011; 132p.
32. Saw, A.A.; Kanzaki, M. Local Livelihoods and Encroachment into a Mangrove Forest Reserve: A Case Study of the Wunbaik Reserved Mangrove Forest, Myanmar. *Procedia Environ. Sci.* **2015**, *28*, 483–492. [CrossRef]

33. Saw, A.A. Deforestation and Local Livelihood Strategy: A Case of Encroachment into the Wunbaik Reserved Mangrove Forest, Myanmar. Ph.D. Thesis, Kyoto University, Kyoto, Japan, 2017.
34. Maung, W.S.; Sasaki, J. Assessing the natural recovery of mangroves after human disturbance using neural network classification and sentinel-2 imagery in wunbaik mangrove forest, Myanmar. *Remote Sens.* **2021**, *13*, 52. [CrossRef]
35. Xu, C.; Wang, J.; Sang, Y.; Li, K.; Liu, J.; Yang, G. An Effective Deep Learning Model for Monitoring Mangroves: A Case Study of the Indus Delta. *Remote Sens.* **2023**, *15*, 2220. [CrossRef]
36. Teillet, P.M.; Staenz, K.; Williams, D.J. Effects of spectral, spatial, and radiometric characteristics on remote sensing vegetation indices of forested regions. *Remote Sens. Environ.* **1997**, *61*, 139–149. [CrossRef]
37. McFeeters, S.K. The use of the Normalized Difference Water Index (NDWI) in the delineation of open water features. *Int. J. Remote Sens.* **1996**, *17*, 1425–1432. [CrossRef]
38. Alsaaideh, B.; Al-Hanbali, A.; Tateishi, R.; Kobayashi, T.; Hoan, N.T. Mangrove Forests Mapping in the Southern Part of Japan Using Landsat ETM+ with DEM. *J. Geogr. Inf. Syst.* **2013**, *5*, 369–377. [CrossRef]
39. Eiumnoh, A.; Shrestha, R.P. Application of DEM data to Landsat image classification: Evaluation in a tropical wet-dry landscape of Thailand. *Photogramm. Eng. Remote Sens.* **2000**, *66*, 297–304.
40. Alsaaideh, B.; Al-Hanbali, A.; Tateishi, R.; Nguyen Thanh, H. The integration of spectral analyses of Landsat ETM+ with the DEM data for mapping mangrove forests. In Proceedings of the 2011 IEEE International Geoscience and Remote Sensing Symposium, Vancouver, BC, Canada, 24–29 July 2011; pp. 1914–1917.
41. Ou, J.; Tian, Y.; Zhang, Q.; Xie, X.; Zhang, Y.; Tao, J.; Lin, J. Coupling UAV Hyperspectral and LiDAR Data for Mangrove Classification Using XGBoost in China's Pinglu Canal Estuary. *Forests* **2023**, *14*, 1838. [CrossRef]
42. Yamazaki, D.; Ikeshima, D.; Neal, J.C.; O'Loughlin, F.; Sampson, C.C.; Kanae, S.; Bates, P.D. MERIT DEM: A new high-accuracy global digital elevation model and its merit to global hydrodynamic modeling. *AGUFM* **2017**, *2017*, H12C-04.
43. Lang, N.; Jetz, W.; Schindler, K.; Wegner, J.D. A high-resolution canopy height model of the Earth. *arXiv* **2022**, arXiv:2204.08322. [CrossRef] [PubMed]
44. Lang, N.; Schindler, K.; Wegner, J.D. Country-wide high-resolution vegetation height mapping with Sentinel-2. *Remote Sens. Environ.* **2019**, *233*, 111347. [CrossRef]
45. Ronneberger, O.; Fischer, P.; Brox, T. *U-Net: Convolutional Networks for Biomedical Image Segmentation*; Lecture Notes in Computer Science; Springer: Cham, Switzerland, 2015; Volume 9351, pp. 234–241.
46. Vanneschi, L.; Silva, S. Artificial Neural Networks. Natural Computing Series. 2023. Available online: https://link.springer.com/chapter/10.1007/978-3-031-17922-8_7 (accessed on 13 November 2023).
47. Clevert, D.A.; Unterthiner, T.; Hochreiter, S. Fast and accurate deep network learning by exponential linear units (ELUs). In Proceedings of the 4th International Conference on Learning Representations, ICLR 2016, San Juan, Puerto Rico, 2–4 May 2016; pp. 1–14.
48. Cococcioni, M.; Rossi, F.; Ruffaldi, E.; Saponara, S. A Novel Posit-based Fast Approximation of ELU Activation Function for Deep Neural Networks. In Proceedings of the 2020 IEEE International Conference on Smart Computing (SMARTCOMP), Bologna, Italy, 14–17 September 2020; pp. 244–246.
49. Devi, T.; Deepa, N. A novel intervention method for aspect-based emotion Using Exponential Linear Unit (ELU) activation function in a Deep Neural Network. In Proceedings of the 2021 5th International Conference on Intelligent Computing and Control Systems (ICICCS), Madurai, India, 6–8 May 2021; pp. 1671–1675.
50. Zhang, Y.; Wu, J.; Chen, W.; Chen, Y.; Tang, X. Prostate segmentation using z-net. In Proceedings of the 2019 IEEE 16th International Symposium on Biomedical Imaging, Venice, Italy, 8–11 April 2019; pp. 11–14.
51. Pan, X.; Liu, Z.; He, C.; Huang, Q. Modeling urban expansion by integrating a convolutional neural network and a recurrent neural network. *Int. J. Appl. Earth Obs. Geoinf.* **2022**, *112*, 102977. [CrossRef]
52. Wang, X.; Jing, S.; Dai, H.; Shi, A. High-resolution remote sensing images semantic segmentation using improved UNet and SegNet. *Comput. Electr. Eng.* **2023**, *108*, 108734. [CrossRef]
53. Rwanga, S.S.; Ndambuki, J.M. Accuracy Assessment of Land Use/Land Cover Classification Using Remote Sensing and GIS. *Int. J. Geosci.* **2017**, *8*, 611–622. [CrossRef]
54. Van Beers, F.; Lindström, A.; Okafor, E.; Wiering, M.A. Deep neural networks with intersection over union loss for binary image segmentation. In Proceedings of the ICPRAM 2019—8th International Conference on Pattern Recognition Applications and Methods, Prague, Czech Republic, 19–21 February 2019; pp. 438–445.
55. Yacouby, R.; Axman, D. Probabilistic Extension of Precision, Recall, and F1 Score for More Thorough Evaluation of Classification Models. In Proceedings of the First Workshop on Evaluation and Comparison ofNLP Systems, Eval4NLP 2020, Online, 20 November 2020; pp. 79–91. Available online: https://aclanthology.org/2020.eval4nlp-1.9/ (accessed on 13 November 2023).
56. Zaabar, N.; Niculescu, S.; Kamel, M.M. Application of Convolutional Neural Networks with Object-Based Image Analysis for Land Cover and Land Use Mapping in Coastal Areas: A Case Study in Ain Témouchent, Algeria. *IEEE J. Sel. Top. Appl. Earth Obs. Remote Sens.* **2022**, *15*, 5177–5189. [CrossRef]
57. Altarez, R.D.D.; Apan, A.; Maraseni, T. Deep learning U-Net classification of Sentinel-1 and 2 fusions effectively demarcates tropical montane forest's deforestation. *Remote Sens. Appl. Soc. Environ.* **2023**, *29*, 100887. [CrossRef]

58. Gupta, K.; Mukhopadhyay, A.; Giri, S.; Chanda, A.; Datta Majumdar, S.; Samanta, S.; Mitra, D.; Samal, R.N.; Pattnaik, A.K.; Hazra, S. An index for discrimination of mangroves from non-mangroves using LANDSAT 8 OLI imagery. *MethodsX* **2018**, *5*, 1129–1139. [CrossRef] [PubMed]
59. Yang, G.; Huang, K.; Sun, W.; Meng, X.; Mao, D.; Ge, Y. Enhanced mangrove vegetation index based on hyperspectral images for mapping mangrove. *ISPRS J. Photogramm. Remote Sens.* **2022**, *189*, 236–254. [CrossRef]
60. Marsocci, V.; Scardapane, S. Continual Barlow Twins: Continual Self-Supervised Learning for Remote Sensing Semantic Segmentation. *IEEE J. Sel. Top. Appl. Earth Obs. Remote Sens.* **2023**, *16*, 5049–5060. [CrossRef]
61. Li, Z.; Chen, H.; Wu, J.; Li, J.; Jing, N. SegMind: Semisupervised Remote Sensing Image Semantic Segmentation with Masked Image Modeling and Contrastive Learning Method. *IEEE Trans. Geosci. Remote Sens.* **2023**, *61*, 1–17. [CrossRef]
62. Saha, S.; Shahzad, M.; Mou, L.; Song, Q.; Zhu, X.X. Unsupervised Single-Scene Semantic Segmentation for Earth Observation. *IEEE Trans. Geosci. Remote Sens.* **2022**, *60*, 1–11. [CrossRef]
63. Giri, C.; Ochieng, E.; Tieszen, L.L.; Zhu, Z.; Singh, A.; Loveland, T.; Masek, J.; Duke, N. Status and distribution of mangrove forests of the world using earth observation satellite data. *Glob. Ecol. Biogeogr.* **2011**, *20*, 154–159. [CrossRef]

remote sensing MDPI

Article

Using Multisource High-Resolution Remote Sensing Data (2 m) with a Habitat–Tide–Semantic Segmentation Approach for Mangrove Mapping

Ziyu Sun [1,2], Weiguo Jiang [1,2,*], Ziyan Ling [1,2,3], Shiquan Zhong [3], Ze Zhang [1,2], Jie Song [1,2] and Zhijie Xiao [1,2]

1 State Key Laboratory of Remote Sensing Science, Faculty of Geographical Sciences, Beijing Normal University, Beijing 100875, China; sunzy@mail.bnu.edu.cn (Z.S.); lingziyan@nnnu.edu.cn (Z.L.); 202231051043@mail.bnu.edu.cn (Z.Z.); songjie@mail.bnu.edu.cn (J.S.); xiaozhijie@mail.bnu.edu.cn (Z.X.)
2 Beijing Key Laboratory for Remote Sensing of Environment and Digital Cities, Faculty of Geographical Science, Beijing Normal University, Beijing 100875, China
3 Key Laboratory of Environmental Change and Resource Use in Beibu Gulf (Ministry of Education), School of Geography and Planning, Nanning Normal University, Nanning 530001, China; zhongshq@nnnu.edu.cn
* Correspondence: jiangweiguo@bnu.edu.cn; Tel.: +86-010-5880-4142

Abstract: Mangrove wetlands are hotspots of global biodiversity and blue carbon reserves in coastal wetlands, with unique ecological functions and significant socioeconomic value. Annual fine-scale monitoring of mangroves is crucial for evaluating national conservation programs and implementing sustainable mangrove management strategies. However, annual fine-scale mapping of mangroves over large areas using remote sensing remains a challenge due to spectral similarities with coastal vegetation, tidal periodic fluctuations, and the need for consistent and dependable samples across different years. In previous research, there has been a lack of strategies that simultaneously consider spatial, temporal, and methodological aspects of mangrove extraction. Therefore, based on an approach that considers mangrove habitat, tides, and a semantic segmentation approach, we propose a method for fine-scale mangrove mapping suitable for long time-series data. This is an optimized hybrid model that integrates spatial, temporal, and methodological considerations. The model uses five sensors (GF-1, GF-2, GF-6, ZY-301, ZY-302) to combine deep learning U-Net models with mangrove habitat information and algorithms during low-tide periods. This method produces a mangrove map with a spatial resolution of 2 m. We applied this algorithm to three typical mangrove regions in the Beibu Gulf of Guangxi Province. The results showed the following: (1) The model scored above 0.9 in terms of its F1-score in all three study areas at the time of training, with an average accuracy of 92.54% for mangrove extraction. (2) The average overall accuracy (OA) for the extraction of mangrove distribution in three typical areas in the Beibu Gulf was 93.29%. When comparing the validation of different regions and years, the overall OA accuracy exceeded 89.84% and the Kappa coefficient exceeded 0.74. (3) The model results are reliable for extracting sparse and slow-growing young mangroves and narrow mangrove belts along roadsides. In some areas where tidal flooding occurs, the existing dataset underestimates mangrove extraction to a certain extent. The fine-scale mangrove extraction method provides a foundation for the implementation of fine-scale management of mangrove ecosystems, support for species diversity conservation, blue carbon recovery, and sustainable development goals related to coastal development.

Keywords: mangrove forest; U-Net; multisource high resolution; remote sensing; mangrove habitat

1. Introduction

Mangroves are mainly distributed in tropical and subtropical regions between 32°N and 38°S around the world, and they play important roles in coastal environmental protection [1,2], water purification [3,4], carbon sequestration [5–7], biodiversity conservation [8,9], and other ecological functions [4,10,11]. Due to human activities and climate change, the mangrove ecosystem is one of the most fragile ecosystems on Earth and has unique habitat conditions. Over the past 50 years, global mangrove coverage has decreased by approximately 20–35% [12]. In China, mangroves have experienced a gradual increase recently after an initial decrease, but the current stock is still only approximately 40% of that in 2000 [13]. The severe degradation of mangroves has attracted the attention of the Chinese government. Through the efforts of various sectors in China, mangroves have slowly begun to recover since 2015 [14]. Reliable, high-precision, and regular monitoring of mangroves is crucial for formulating and implementing sustainable mangrove management strategies.

Remote sensing plays a unique role in monitoring inaccessible coastal ecosystems [11]. With the development of satellite remote sensing and intelligent classification algorithms, significant progress has been made in the extraction and monitoring of mangrove wetland areas [15–18]. In recent years, high-resolution GF and ZY series satellites have been widely used for mangrove monitoring tasks [14,19]. With characteristics such as high image resolution and short revisit intervals, these satellites have great potential for detailed and effective mangrove monitoring.

To date, based on existing remote sensing datasets, researchers have conducted a large amount of mangrove monitoring work on local to global scales. Using Landsat datasets with a spatial resolution of 30 m, researchers used machine learning methods such as random forests to create a global mangrove map with a resolution of 30 m from 2000 to 2016 to capture human and natural impacts on mangroves [12]. They have also generated long-term (1973–2015) dynamics of Chinese mangroves using image thresholding methods and analyzed the effectiveness of Chinese mangrove protection measures [13]. This research strategy takes into account temporal optimization by identifying the low-tide periods for mangroves using corresponding ecological indices. This method is more suitable for global-scale mapping of mangroves at a 30 m resolution. However, such methods have a higher demand for samples and still pose challenges for fine-scale annual monitoring of long-term mangrove changes. Using Sentinel and Landsat datasets with spatial resolutions ranging from 10 m to 30 m, researchers evaluated the applicability of different indicators and selected the interannual changes of EVI and NDVI indices in mangroves time-series analysis [20]. And Theil–Sen median trend analysis and the Mann–Kendall test were used to explore the spatiotemporal variation trends in mangroves in the Qinglan Harbor Nature Reserve in Hainan, China from 1987 to 2020 [20]. Researchers generated mangrove maps of China using the Otsu method for six time periods from 1973 to 2015, and analyzed national- and local-scale changes in mangroves over a 42-year period [13]. Studies based on UAV (unmanned aerial vehicle) data, hyperspectral data, and multispectral data have also used methods such as XGBoost (extreme gradient boosting) for mangrove extraction at a small spatial scale [21,22], SVMs (support vector machines) for mangrove species classification [23–25], CNN (convolutional neural networks) to overcome manual leaf sample identification during image recognition process for mangrove extraction and species classification [26]. This research strategy considers spatial optimization by combining deep learning with mangrove habitat information to identify potential growth areas of mangroves and then extract mangroves based on this information. This approach helps avoid confusion between mangroves and other terrestrial vegetation due to spectral similarity. However, this research strategy often overlooks the uncertainty of tides, leading to underestimation of extracted mangroves compared with the actual extent of the mangroves. Additionally, these methods require high-quality samples and are still not suitable for long-term mangrove monitoring. These studies indicate that Landsat 30 m spatial resolution and Sentinel 10 m spatial resolution images are mainly used for large-scale mangrove monitoring, but there is a gap in large-scale fine spatial resolution monitoring (2 m) for long time series. Since the policy for protecting mangroves was established, the fluctuation in mangrove areas has increased. In this situation, high spatial resolution and dense time-series mangrove monitoring pose even more rigorous challenges in evaluating national protection plans.

Machine learning algorithms are the most widely used methods for extracting mangroves. For example, machine learning algorithms rely on rich and unbiased training samples from which they can learn to obtain accurate mappings [27]. In addition, machine learning algorithms require setting the weights of features to balance prior probabilities and fitting quality [28–30]. CNNs achieve high scores in automatic classification tasks, but the results of traditional CNNs for mangrove extraction rely on stable sample quality [31]. Researchers have utilized attention mechanisms and CNNs to extract mangroves in the Dongzhai Port National Nature Reserve in Hainan, China, further improving mangrove extraction accuracy [32]. However, obtaining stable, long-term mangrove samples remains a challenge. Semantic segmentation models like U-Net provide an opportunity for full-sample mapping, requiring only one year of stable samples for training; the model results can be applied to other years [33]. The U-Net deep learning semantic segmentation model network can reduce the dependence on sample points and achieve good classification results (OA = 81%) based on Landsat images [34]. In addition, some improved U-Net models have demonstrated excellent performance in mangrove extraction [35,36]. Semantic segmentation models have shown excellent performance in binary extraction tasks and have been repeatedly proven to be more suitable for fine-scale, long-term mangrove extraction. However, mangrove extraction tasks still present challenges, such as errors caused by spectral similarity confusion leading to an overestimation of mangrove area, and errors resulting from tidal uncertainty leading to an underestimation of mangrove area. Similarly, an independent study in China showed that farmlands, inland shrublands, bushes, and aquatic plants have very similar spectral characteristics to mangroves, and thus misclassifications are common [37]. This indicates that obtaining stable samples for mangrove extraction tasks is highly challenging.

Fine-scale annual monitoring of mangroves still presents three major challenges: (1) Spectral similarity: Mangroves share a high spectral similarity with coastal vegetation. Relying solely on spectral data often results in the misclassification of a significant amount of other vegetation as mangroves. (2) Tidal influence: The appearance of mangroves in imagery is greatly influenced by tidal conditions. Existing extraction strategies struggle to obtain accurate mangrove boundaries due to these tidal fluctuations. (3) Sample availability: Current mangrove extraction strategies require a substantial number of reliable samples. Obtaining reliable samples from noncurrent years is challenging, making it difficult to monitor long-term changes in mangroves.

The aim is to solve these challenges. The objective of this research is to develop and evaluate hybrid U-Net deep learning models for mapping mangroves from multisource high-resolution data. This research includes the following aspects: (1) developing a hybrid U-Net model based on a habitat–tide-semantic segmentation approach; (2) extracting the maximum extent of annual mangrove data using multisource high-resolution imagery; (3) qualitatively and quantitatively evaluating the model results through visual interpretation and validation by multiple experts; and (4) comparing the model results with existing mangrove datasets.

The rest of this paper is organized as follows: Section 2 provides detailed information about the study area, data sources, and auxiliary data. Section 3 describes the hybrid U-Net deep learning model for mapping mangroves from multisource high-resolution data. Section 4 provides an evaluation of the performance of the hybrid U-Net model. Section 5 provides a discussion of the advantages and limitations of the hybrid U-Net model and emphasizes the advantages of multisource high score data in mangrove extraction. Finally, the major findings of this research are presented in Section 6.

2. Study Areas and Data Sources

2.1. Study Areas

The study areas are located in three typical regions of Guangxi Province in southern China (Figure 1): (1) Zhenzhu Har bor–Fangcheng Harbor in Guangxi (ZZW); (2) Dafeng River in Guangxi (DFJ); and (3) Tieshan Harbor in Guangxi (TSG). The study areas have two national nature reserves (Guangxi Beilun Estuary National Nature Reserve and Guangxi Shankou Mangrove Ecological and National Nature Reserve) and one autonomous region nature reserve (Guangxi Maowei Sea-Dafeng River Mangrove Forest Autonomous Region Nature Reserve). ZZW belongs to the Guangxi Beilun Estuary National Nature Reserve, which is the largest and most typical bay-type mangrove in China. The mangroves of DFJ have the typical characteristics of estuary-type mangroves. The major species of mangroves include *Aegiceras corniculatum*, *Kandelia obovata*, *Avicennia Marina*, *Acanthus ilicifolius* L., and *Sonneratia apetala* [38]. The predominant species in the mangrove communities of TSG, which span the low, mid-, and high intertidal zones, include *Avicennia Marina*, *Aegiceras corniculatum*, *Kandelia obovata*, *Rhizophora stylosa Griff*, and *Bruguiera gymnorrhiza (Linn.) Savigny* [39]. The mangrove data presented in Figure 1 are openly available at [Science Data Bank (ScienceDB)] at [doi:10.11922/sciencedb.00449], reference number [40].

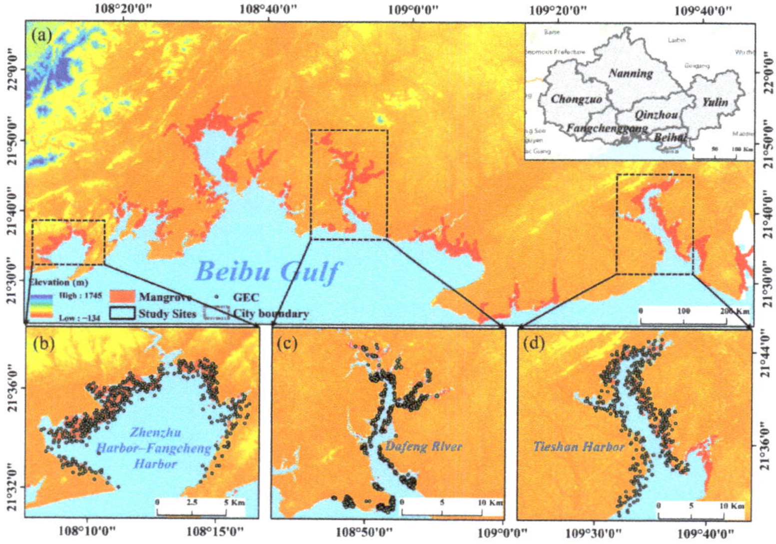

Figure 1. Locations of study areas: (**a**) the Guangxi Beibu Gulf, (**b**) Zhenzhu Har-bor–Fangcheng Harbor in Guangxi (ZZW), (**c**) Dafeng River in Guangxi (DFJ), (**d**) Tieshan Harbor in Guangxi (TSG).

2.2. Data Preprocessing

This study used five types of multispectral satellites, namely GF-1, GF-2, GF-6, ZY3-01, and ZY3-02. The main parameters are shown in Table 1.

Table 1. Parameters of multispectral data used in this study (m for meters).

Satellite/Sensor	GF-1	GF-2	GF-6	ZY3-01	ZY3-02
Study area	ZZW, DFJ, TSG	ZZW, DFJ, TSG	ZZW, DFJ, TSG	TSG	ZZW, DFJ, TSG
Data collection time	15 January 2015 2 December 2021	29 February 2016 15 August 2020	23 November 2019 4 December 2021	24 August 2015 4 December 2015	18 December 2018 9 November 2021
Number of images	101	79	9	4	33
Number of bands	5	5	5	8	8
Multispectral band resolution/m	8	3.2	8	5.8	5.8
Pansharpened band resolution/m	2	0.8	2	2.1	2.1
After image fusion resolution /m	2	0.8	2	2.1	2.1

For the preprocessing of the images, the preprocessing workflows for GF-1, GF-2, GF-6, ZY3-01, and ZY3-02 were generally the same. The processes included (1) orthorectification of the panchromatic and multispectral images, (2) radiometric calibration, (3) atmospheric correction, (4) image fusion and pansharpening, and (5) data cropping and subset extraction. First, all of the images were orthorectified to the World Geodetic System 1984 (WGS 84) reference system, with a root mean square geometric error of less than one pixel. The changes in solar angle and Earth–sun distance was normalized by converting the digital values to top-of-atmosphere reflectance for each image. Then, the radiometric calibration tool of ENVI5.3 and the FLAASH atmospheric correction tool were used to process the multispectral images. The calibration coefficients and atmospheric correction parameters were obtained from the ENVI5.3 lookup table (GMTED2010). Finally, the pansharpening algorithm [41] was used to fuse the orthorectified panchromatic and multispectral images. Pansharpening is a radiation transformation that combines high-resolution panchromatic images (or raster bands) with lower-resolution multiband raster datasets. Combining the high spatial resolution information of panchromatic imagery with the low spatial resolution information of multispectral imagery allows us to obtain fused images with both high spatial and spectral resolution. The data fusion resolution was the same as the panchromatic band spatial resolution. The fused images were resampled to a spatial resolution of 2 m (m for meters). The processed images were then divided into separate image sets by year for subsequent mangrove extraction. In this study, the five types of high-resolution satellite images with a cloud cover of less than 15% in the Beibu Gulf from 2015 to 2021 were selected [42]. A total of 226 images were selected for the extraction of the mangrove forests in the Beibu Gulf.

2.3. *Auxiliary Data*

2.3.1. Potential Growth Areas of Mangroves

Using information about the mangrove habitat, mangroves are classified into a specific ecological niche, which eliminates most areas where mangroves are not found [23]. In this study, potential mangrove habitats were determined based on common mangrove habitat characteristics. First, a 20 km buffer zone was set along the Beibu Gulf coastline. Second, Guangxi has 12 species of true mangroves, with the tallest being smooth cordgrass (*Sonneratia apetala*), which can grow up to 20 m high [43]. The areas where the DSM (Digital Surface Model) value was less than 20 m were extracted from the buffer zone to obtain the mangrove habitat. The DSM data were downloaded from http://www.eorc.jaxa.jp/ALOS/en/aw3d30/data/index.html, accessed on 21 March 2023.

2.3.2. Reference Data Collection

The sample dataset of this study was mainly used to evaluate the effectiveness of the proposed method for mangrove extraction. The sample discrimination was based on visual interpretation of Google Earth high-resolution imagery and field investigations. The sample dataset consisted of three parts: (1) For the first part, ArcGIS 10.8 was used to generate random points in the mangrove and its adjacent areas. These points were determined to be mangroves by multiple geographers and mangrove experts. Each sampling point's category attribute was manually determined based on high-resolution imagery from Google Earth, and uncertain sampling points were removed. (2) The second part came from the area of change in the mangrove patches, primarily focusing on areas where differences existed between existing datasets for the same year. These areas were initially assessed by an expert system and ultimately visually interpreted using high-resolution imagery from Google Earth. (3) The third part of the sample points comes from the 2015 field investigation results. The number and sources of the sampling points in each region are shown in Table 2.

Table 2. Number of sample points in each area and their sources. (Note: GER stands for Google Earth high-resolution image interpretation of random points; GEC stands for Google Earth high-resolution image interpretation of area change over two consecutive years; and FSs stands for field surveys).

Study Areas	Years	Number of Sample Points (Statistics/Validation)							Source
		Mangrove			Non-Mangrove			Total	
		GER	GEC	FSs	GER	GEC	FSs		
ZZW	2015	412	50	/	106	72	/	640	GER + GEC
	2018	434	14	/	83	100	/	631	GER + GEC
	2019	355	37	/	67	41	/	500	GER + GEC
	2020	372	59	/	87	83	/	601	GER + GEC
DFJ	2015	398	1885	18	99	3	/	706	GER + GEC + FSs
	2018	332	186	/	98	4	/	620	GER + GEC
	2019	361	188	/	89	32	/	670	GER + GEC
	2020	424	156	/	101	15	/	696	GER + GEC
TSG	2015	468	11	34	98	80	/	691	GER + GEC + FSs
	2018	342	6	/	97	54	/	499	GER + GEC
	2019	341	14	/	74	136	/	565	GER + GEC
	2020	375	15	/	101	71	/	562	GER + GEC

2.3.3. Existing Data Productions

In this study, we compared our research model's results with five publicly available mangrove datasets. To ensure the reliability of data comparison, we only selected the baseline years of these datasets for comparison. These five datasets consist of one set at 2 m resolution (XCina, 2018) [23], two sets at 10 m resolution (TW, LREIS) [44,45] (2019, 2020), and two sets at 30 m resolution (GMW, CAS) (2015, 2018, 2019, 2020). The 10 m resolution products were mainly extracted based on Sentinel images, and then underwent field investigations and postprocessing, with an accuracy of over 90%. The 30 m resolution products were mainly extracted based on Landsat images, and after postprocessing, the results were published. The most representative products are the GMW data, which have formed annual time-series data since 2015 [15].

3. Hybrid U-Net for Extracting Mangrove Forests and Evaluation Criteria

3.1. Overview of the Mangrove Extraction Process

The overview of the workflow for mangrove extraction and evaluation is outlined in Figure 2 and mainly consists of four parts: (1) image preprocessing, (2) mangrove maximum growth area assessment, (3) U-Net mangrove extraction, and (4) accuracy assessment. Image preprocessing: the satellite images were preprocessed to obtain a high-resolution multisource image set with a spatial resolution of 2 m (m for meters) for 2015–2021. Mangrove maximum growth area assessment: Based on the habitat characteristics and NDVI, we acquired cloud-free images during the low-tide period for mangroves, representing the maximum growth area of mangroves. Extracting the maximum mangrove growth area, which is used for U-Net mangrove extraction, is crucial for the "full mapping" of the semantic segmentation model. U-Net mangrove extraction: Accurate determination of the maximum mangrove growth area contributes to achieving a "full mapping" relationship between stable samples (2021) and the spatial scale of remote sensing images. On this basis, image detection and segmentation were performed on the annual cloud-free mangrove images. Multiple-source image segmentation results were extracted using a blended U-Net model, which was used to derive mangrove maps for multiple years (2015–2020). Accuracy assessment: the accuracy of the mangrove maps was verified based on validation sample sites and field survey results, and quantitative and qualitative analyses were performed via comparing multi-person visual interpretations with existing data.

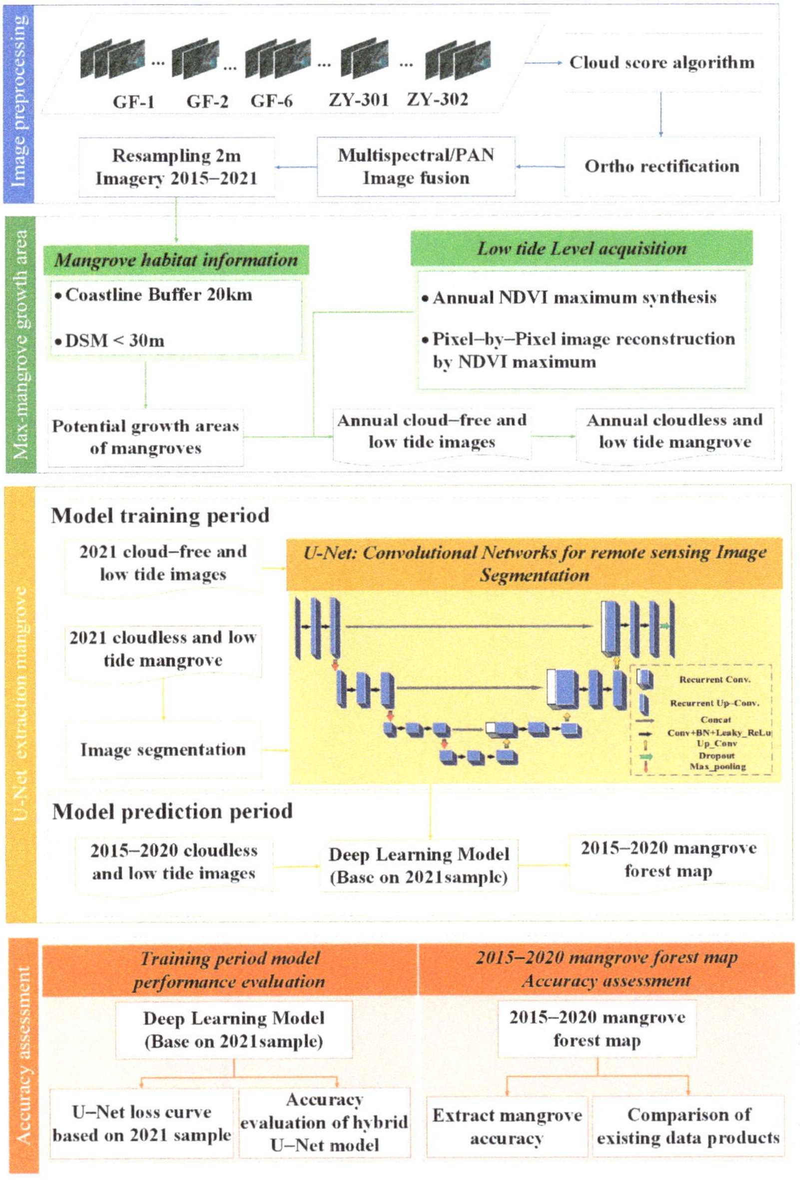

Figure 2. Overview of the workflow for the production of 2 m spatial resolution mangrove datasets.

3.2. Principle of the U-Net Model

The U-Net model is the most widely used model in semantic segmentation projects and has also been extensively applied to remote sensing image classification [46]. The input image first passes through the encoder to generate downsampling feature maps.

Then, the decoder part implements upsampling of the feature maps to make them the same size as the original image. U-Net is composed of stackable highway units, allowing deeper networks to be trained using a limited number of samples [34]. To address the problem of gradient degradation caused by excessively deep neural networks in deep learning tasks, this study used ResNet-34 as the kernel function to train the deep learning model to enhance the model's feature representation capability [47,48]. To optimize the update of network parameters, the Adam optimizer [49] was used. The specific equation, Equation (1), is as follows:

$$\theta_t = \theta_{t-1} - \alpha * \hat{m} / \left(\sqrt{\hat{v}_t + \varepsilon}\right) \tag{1}$$

In Equation (1), t represents the number of training iterations, α is the learning rate, m is the exponential moving average of the gradient, and v is the exponential moving average of the squared gradient. ε is usually a constant value of 1×10^{-8}.

The model loss was calculated during training using the average cross-entropy loss function, as shown in Equation (2):

$$loss = -\frac{1}{n}\sum\nolimits_{i=1}^{n} y_i loga_i + (1 - y_i)\log(1 - a_i) \tag{2}$$

where n represents the batch size and y_i and a_i are the predicted value and true value of the i-th sample in each batch, respectively. The U-Net network model has better spatial accuracy due to adding skip connections between the encoder and decoder layers.

3.3. Hybrid U-Net Based on the Habitat–Tide Semantic Segmentation Approach

Based on mangrove habitat characteristics and the collected low-tide data, a hybrid strategy combining the U-Net model was proposed for mangrove extraction based on the habitat data method. The hybrid U-Net model consists of three steps: (1) mangrove habitat information, (2) low tide level acquisition, and (3) U-Net model mangrove extraction (Figure 3).

The habitat extraction module identifies the potential growth areas of mangroves. The low-tide data collection module extracts the maximum mangrove extent based on the current data. The U-Net module for mangrove extraction constructs the relationship between the maximum mangrove extent and multisource high-resolution remote sensing imagery to build a mangrove extraction model. This model can be used for mangrove extraction in other years within the same area. In the U-Net module for mangrove extraction, the encoder and decoder parts are the same as those of the standard U-Net model. However, the classifier part is enhanced via integrating the habitat information module and the low-tide data collection module, improving the accuracy of the extraction process.

Step 1: Mangrove habitat information

According to the content in Section 2.3.1, mangroves grow on intertidal flats and the buffer zone is designed as 20 km from the coastline. Areas with a digital surface model (DSM) value less than 20 m are considered potential growth areas to reduce interference from other coastal vegetation.

This step involves using the unique habitat of mangroves to identify potential mangrove habitat areas and avoid interference from shoreline vegetation in the inversion results. This is an optimization in terms of the model's spatial aspects. The inputs for this step include GF series and ZY series remote sensing images, DSM (digital surface model) data, and coastline data. The output is the extent of the mangrove habitat.

Step 2: Low tide level acquisition

The minimum water level in the study area was based on the lowest tidal period of the year [50]. For the normalized difference vegetation index (NDVI), the NDVI values of non-water pixels (vegetation and tidal flats) are higher than those of water pixels, as vegetation absorbs more strongly in the green band. Maximum value compositing of the NDVI for the preprocessed 2 m resolution images from 2021 can be used to extract the minimum water surface [51]. This composite image not only enhances the difference between vegetation and water but also removes the periodic flooding effect of seawater on the mangrove forest, effectively separating the mangrove forest from the water. In this study, 51 images were ultimately used to track the low-tide period (GF-1: 13; GF-2: 17; GF-6: 4; ZY3-01: 5; ZY3-02: 12). We list the relevant information for the remote sensing data in the Supplementary Materials.

This step involves using NDVIs and multisource high-resolution data to determine the low-tide moments for mangroves, obtaining the maximum mangrove area. Based on this, multilevel interpretation is carried out, and the interpretation results serve as inputs to the U-Net model for training, achieving a "full mapping" with remote sensing imagery. The training results can then be applied to mangrove extraction for other years. This is an optimization in terms of the model's temporal aspects. The inputs for the second part include NDVIs constructed from GF series and ZY series remote sensing images. The output is the maximum extent of the mangrove areas.

Step 3: U-Net model mangrove extraction

First, multilevel interactive verification of the maximum range of mangrove areas (2021).Then, each detected mangrove patch was adjusted to obtain the best mangrove patch samples. Finally, the mangrove patch samples were trained to obtain a mangrove extraction model. The cloud-free low-tide images from other years (2015–2020) were used as inputs to the mangrove extraction model to obtain the mangrove map for the corresponding year.

This study was compiled using the Python language based on the TensorFlow deep learning framework. The hardware configuration of the operating platform includes a Lenovo Ren-7000 i7-12700 (F)/NVIDIA Quadro RTX3070Ti 16G GPU. The slice size was set to 256 × 256 pixels, the batch size was set to 32, the kernel function was ResNet-34, and the other parameters were set to the default values of ResNet-34.

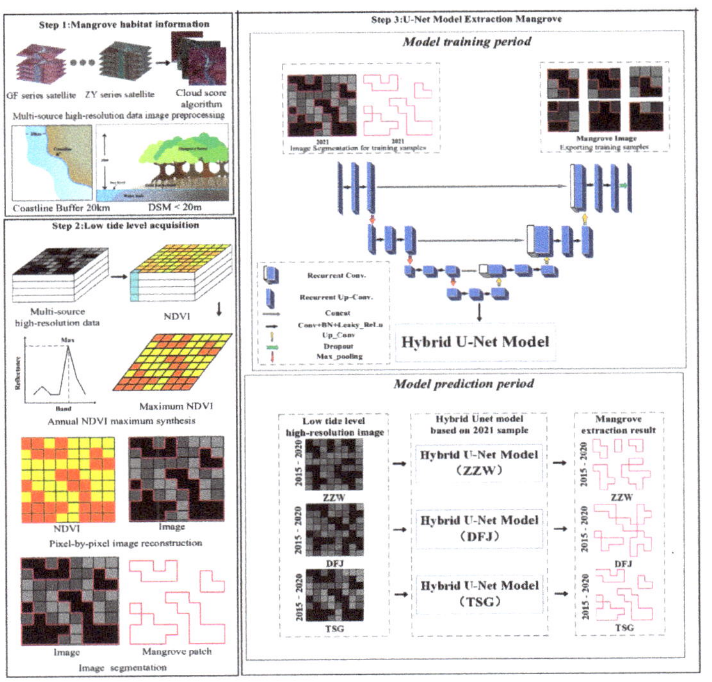

Figure 3. Hybrid U-Net model mangrove extraction process.

3.4. Accuracy Assessment

We evaluated the classification accuracy of the proposed hybrid U-Net model using three metrics: Kappa coefficient, overall accuracy (OA), and F1-scores [52]. The kappa coefficient, OA, and F1-scores are all based on confusion matrix calculations, which provide the number of correctly classified and misclassified samples [53]. Specifically, the Kappa coefficient is often used to measure the consistency of test results and the effectiveness of classification, while OA is the ratio of correctly classified categories to the total number of categories [34]. The F1-score reflects the overall comprehensive index, independent of the number of categories; it is a means of coordinating precision and recall. Precision is defined as the ratio of true positive (TP) predictions to sample predictions of the same type of land cover. It refers to the proportion of mangrove and non-mangrove types in the model results that is the same as in the validation point results. Recall is defined as the ratio of target type samples to total type samples. It refers to the proportion of model results defined as mangroves and the validation points (wild survey results and visual interpretation results) that are classified as mangroves [50].

$$OA = \frac{\sum_{i=1}^{2} a_{ii}}{N} \tag{3}$$

$$Kappa = \frac{OA - \sum_{i=1}^{2} a_i + {}^*a{+}_i / N^2}{1 - OA}, a_{i+} = \sum_i a_{ij}, a_{+i} = \sum_j a_{ij} \tag{4}$$

$$precision = \frac{TP}{TP + FP'}, \ recall = \frac{TP}{TP + FN'} \tag{5}$$

$$F1 - scores = \frac{2 \times (precision \times recall)}{precision + recall} \tag{6}$$

In Equations (3) and (4), a_{ii} represents the number of samples correctly predicted as type i, a_{ij} represents the number of samples of type i predicted as type j, and N is the total number of samples. In Equation (5), FN represents the number of samples for which the classification model predicts mangrove forests but the ground truth shows non-mangrove forests.

3.5. Multilevel Interactive Verification

In this study, the multilevel interactive verification was divided into four parts: preliminary interpretation, comparison of interpretation results, issue resolution and correction, and final review and confirmation. Preliminary interpretation: We engaged multiple experts in geographic information or experts in mangrove studies. Based on their respective professional knowledge and experience, they independently interpreted and identified mangroves in the same remote sensing image, generating interpretation logs. These interpretation logs include the basis for decisions, data sources, and key findings. Comparison of interpretation results: Interpretation results from multiple individuals were compared to identify stable mangrove areas and areas with discrepancies. Issue resolution and correction: Referring to the existing dataset, identify areas with differences in mangrove interpretation to determine whether they are mangroves. For areas that still could not be definitively determined, in-depth analysis of the imagery, additional data searches, or reference to ground validation data were conducted, and the final determination was made by local mangrove experts. Final review and confirmation: The stable mangrove areas obtained from the preliminary interpretation were overlaid with areas where there were discrepancies as determined by experts. The final visual interpretation results were ultimately confirmed.

4. Results

4.1. Performance Evaluation of the Hybrid U-Net Model during the Training Period

The training results (2021 samples) based on the hybrid U-Net model in three typical mangrove areas in the Beibu Gulf are shown in Figure 4. During the training period, the output of the hybrid U-Net's divided into two categories: mangrove and non-mangrove. The application effect of the U-Net deep learning model was evaluated using classic evaluation metrics such as accuracy, recall, and F1-score in the three study areas. As seen in Table 3, the F1-scores of the three areas were all above 0.9, indicating that the hybrid U-Net model theoretically performed well in precision and recall. Precision is an indicator from the perspective of model prediction, and there was an average mangrove prediction precision of 92.54% in the three areas. Recall is an indicator from the perspective of real results, and there was an average mangrove prediction recall of 91.51% in the three areas. Overall, the model performs similarly in the three areas, with slightly better results in TSG than in ZZW and DFJ. The lowest F1-score for non-mangrove areas during training in the three regions was 0.97. This demonstrates that very few mangrove patches were misinterpreted as non-mangrove patches. This further indicates that the boundaries of mangrove patches used for model training are quite accurate.

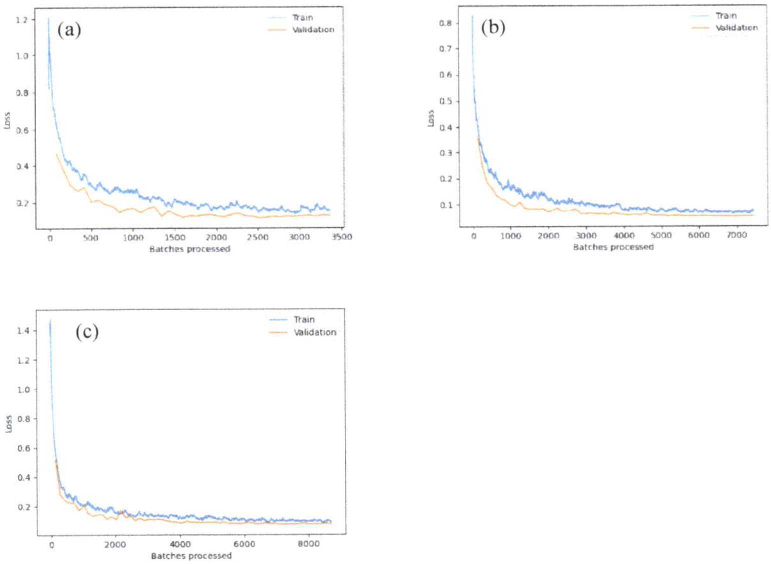

Figure 4. U-Net loss curves based on 2021 samples: (**a**) loss curve at ZZW, (**b**) loss curve at DFJ, (**c**) loss curve at TSG.

Table 3. Accuracy evaluation of hybrid U-Net model for various types (mangrove, non-mangrove).

Study Area	Ground Truth	Precision	Recall	F1-Score
ZZW	Mangrove	89.51%	90.04%	0.90
	Non-Mangrove	97.12%	96.95%	0.97
DFJ	Mangrove	93.78%	89.99%	0.92
	Non-Mangrove	98.63%	99.18%	0.99
TSG	Mangrove	94.34%	94.49%	0.94
	Non-Mangrove	98.62%	98.58%	0.99

The hybrid U-Net model effectively inverted the mangroves in the three areas. The training accuracy–verification curve of the hybrid U-Net is shown in Figure 4. The training accuracy curve was used to measure the performance of the model, and the loss curve was used to further optimize the model. After approximately 7000 iterations in the three typical areas, the model reached a convergence state, at which point the model weights were determined. After model convergence, the loss of three regions is around 0.05.

To avoid overfitting, the early stopping strategy was employed during the model training process [54]. This strategy involves monitoring the training progress and automatically stopping the training when the model has reached a point of convergence and the training loss on the training dataset no longer decreases significantly. This allows the model to be saved with the optimal parameter settings without continuing training until after the specified number of epochs.

4.2. Evaluation of the Hybrid U-Net Model's Mangrove Extraction Accuracy

The extraction results of the hybrid U-Net model can be seen in Figure 5 and Table 4 (the extraction results for DFJ and TSG are shown in the Supplementary Materials (Figures S1 and S2)). The PA, UA, OA, and Kappa coefficients were selected as indicators to validate the accuracy of the hybrid U-Net model in extracting the mangroves. The average OA value of the mangrove distribution extraction in the typical area of the Beibu Gulf was 93.29%. When comparing the validation results of the different regions and different years (Table 4), the overall OA accuracy exceeded 89.84%, and the Kappa coefficient exceeded 0.74. The lowest OA was 93.31% (kappa = 0.8) in the DFJ area, and the lowest OA was 91.68% (kappa = 0.82) in the TSG area. The best extraction accuracy of the mangroves in the DFJ area was in 2015 (OA = 96.58%, Kappa = 0.87), and the lowest accuracy was in the ZZW area in 2020 (OA = 89.84%, Kappa = 0.77). Compared with the results for DFJ, the extraction accuracy for ZZW and TSG in the four years averaged 94.74%. With the hybrid U-Net model, there were very few errors in mangrove extraction (Figure 5), resulting in high PA values for all the mangrove categories (DFJ reached a maximum value of 96.64% in 2015). The PA and UA values of the mangroves in the three typical areas were both greater than 90.11%.

Table 4. The Hybrid U-Net results and the producer's accuracy (PA), user's accuracy (UA), overall accuracy (OA), and Kappa coefficients (Kappa) of three typical mangrove forest areas in the Beibu Gulf. (1) ZZW; (2) DFJ; (3) TSG.

Study Areas	Years	Ground Truth	PA	UA	OA	Kappa
ZZW	2015	Mangrove	93.90%	98.93%	94.81%	0.88
		Non-Mangrove	97.27%	85.58%		
	2018	Mangrove	94.12%	95.52%	92.79%	0.83
		Non-Mangrove	89.71%	86.73%		
	2019	Mangrove	90.11%	96.79%	89.93%	0.74
		Non-Mangrove	89.27%	71.52%		
	2020	Mangrove	91.31%	94.10%	89.84%	0.77
		Non-Mangrove	86.29%	80.57%		
DFJ	2015	Mangrove	96.64%	99.34%	96.58%	0.87
		Non-Mangrove	96.23%	82.93%		
	2018	Mangrove	94.18%	99.23%	94.51%	0.82
		Non-Mangrove	96.23%	76.12%		
	2019	Mangrove	93.58%	97.86%	93.31%	0.80
		Non-Mangrove	90.98%	77.07%		
	2020	Mangrove	95.24%	98.14%	94.57%	0.82
		Non-Mangrove	91.34%	80%		

Table 4. *Cont.*

Study Areas	Years	Ground Truth	PA	UA	OA	Kappa
TSG	2015	Mangrove	91.44%	99.03%	92.88%	0.83
		Non-Mangrove	97.27%	78.76%		
	2018	Mangrove	92.80%	99.71%	94.69%	0.88
		Non-Mangrove	99.34%	84.83%		
	2019	Mangrove	93.67%	96.47%	93.85%	0.87
		Non-Mangrove	94.17%	89.74%		
	2020	Mangrove	90.49%	97.50%	91.68%	0.82
		Non-Mangrove	94.51%	80.75%		

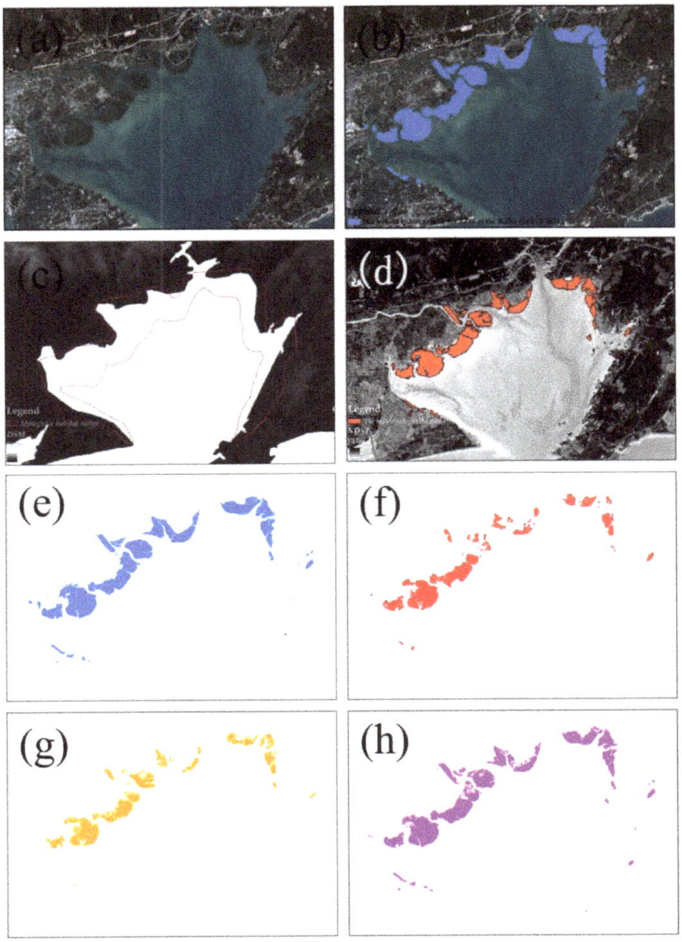

Figure 5. The hybrid U-Net results for ZZW at the Beibu Gulf. (**a**,**b**) The satellite image of mangrove area in 2021, (**c**) mangrove habitat information results, (**d**) low tide level acquisition results, (**e**) the hybrid U-Net results for 2020, (**f**) the hybrid U-Net results for 2019, (**g**) the hybrid U-Net results for 2018, (**h**) the hybrid U-Net results for 2015.

4.3. Multilevel Interactive Verification of Mangrove Extraction

To further validate and confirm the reliability of the mangrove extraction results, this study employed a multilevel interactive interpretation approach on remote sensing imagery to extract mangrove information. The extracted results were then compared with the findings of this study. In the context of multilevel interactive verification of mangrove extraction, multiple experts collaborated to review and validate the extracted mangrove maps (Figure 6). (A comparison of the results of hybrid U-Net extraction of mangroves with multi-person interactive validation for DFJ and TSG is shown in the Supplementary Materials (Figures S3 and S4)). The key steps involved in this interactive verification process included data sharing, a collaboration platform, comparative analysis, discussion and consensus, expert guidance, and iterative refinement. The verification process was carried out independently to ensure unbiased results.

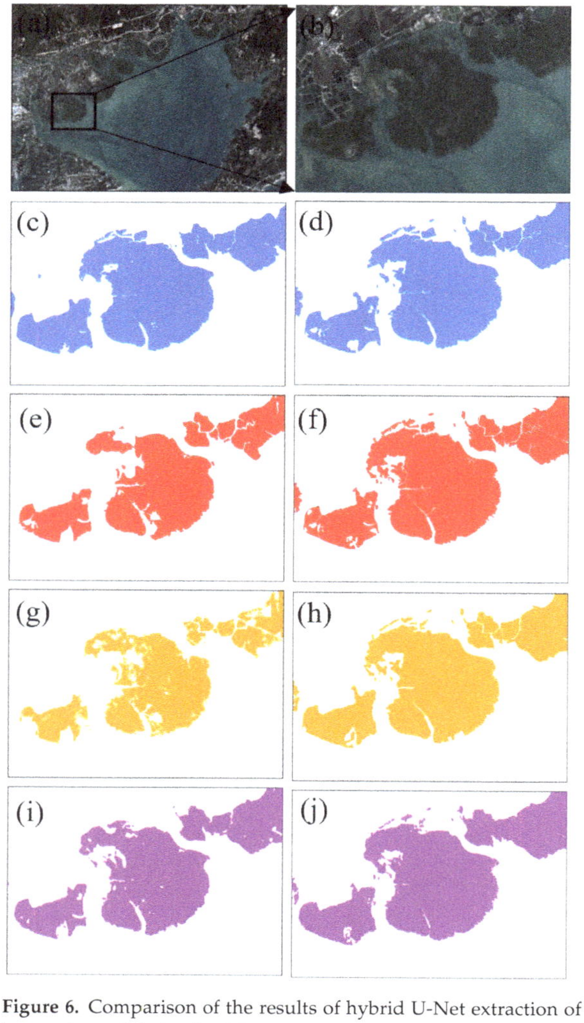

Figure 6. Comparison of the results of hybrid U-Net extraction of mangroves with multi-person interactive validation for ZZW: (**a**,**b**) The satellite image of mangrove area at ZZW in 2021 October. (**c**,**e**,**g**,**i**) The hybrid U-Net results for 2020, 2019, 2018, 2015. (**d**,**f**,**h**,**j**) multilevel interactive validation for 2020, 2019, 2018, 2015.

4.4. Evaluation of the Hybrid U-Net Model with Existing Data Products

We compared the model results with the baseline years of the existing dataset separately (Figures 7 and 8). Overall, the results of this study were more consistent with the 2/10 m resolution dataset, and in some areas, the results of this study were more reliable compared with the existing dataset, especially in the extraction of sparse and low-growth young forests and narrow mangrove belts along embankments (Figures 7 and 8). (The hybrid U-Net mangrove extraction results compared with existing data products for DFJ and TSG are shown in the Supplementary Materials (Figures S5 and S6)). In our Supplementary Materials, more details on the comparison are provided (Figure S7).

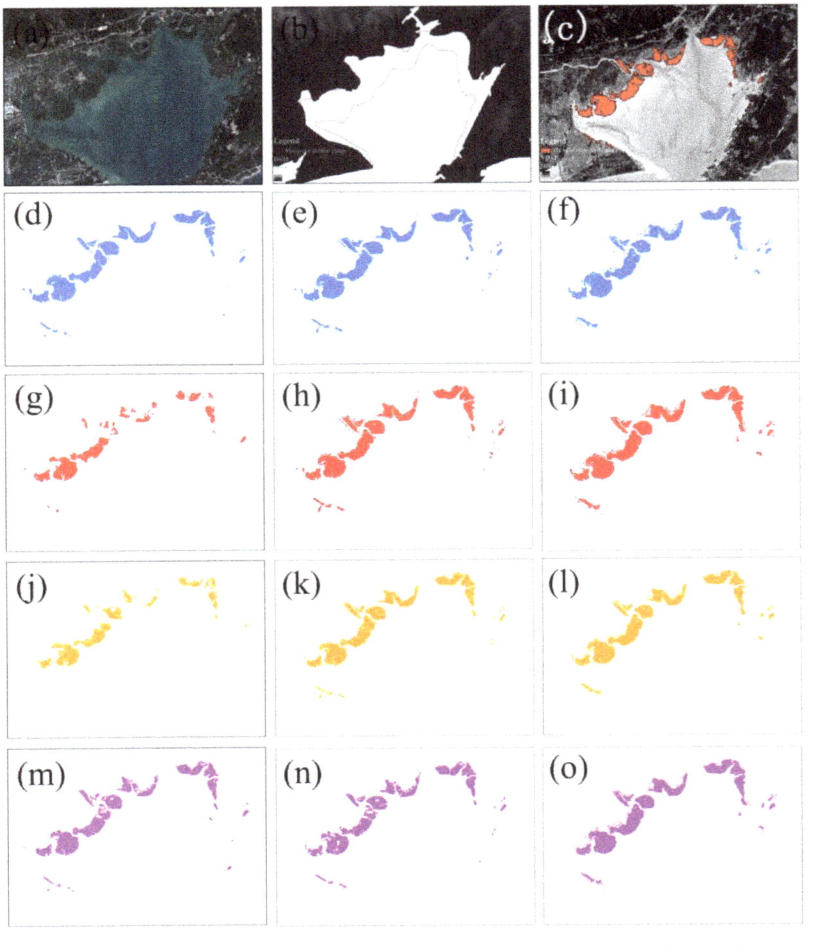

Figure 7. Hybrid U-Net mangrove extraction results compared with existing data products. (**a**) The satellite image of mangrove area at ZZW, (**b**) mangrove habitat information results, (**c**) low tide level acquisition results, (**d**,**g**,**j**,**m**) the hybrid U-Net results in 2020, 2019, 2018, 2015, (**f**,**i**,**l**,**o**) the Global Mangrove Watch (GMW) in 2020, 2019, 2018, 2015, (**e**) LREIS in 2020, (**h**) TW in 2019, (**k**) MCina in 2018, (**n**) CAS in 2015.

The 10/30 m data product did not provide sufficient detailed spatial information. Factors such as the spatial resolution of remote sensing images, tidal inundation, and information extraction strategies can affect the accuracy of mangrove extraction. Previous studies have shown that the area error measured with remote sensing images increases gradually with improvements in spatial resolution [55]. Therefore, for narrow and fragmented mangrove areas, the extraction effect of medium-to-low-resolution remote sensing images is poor.

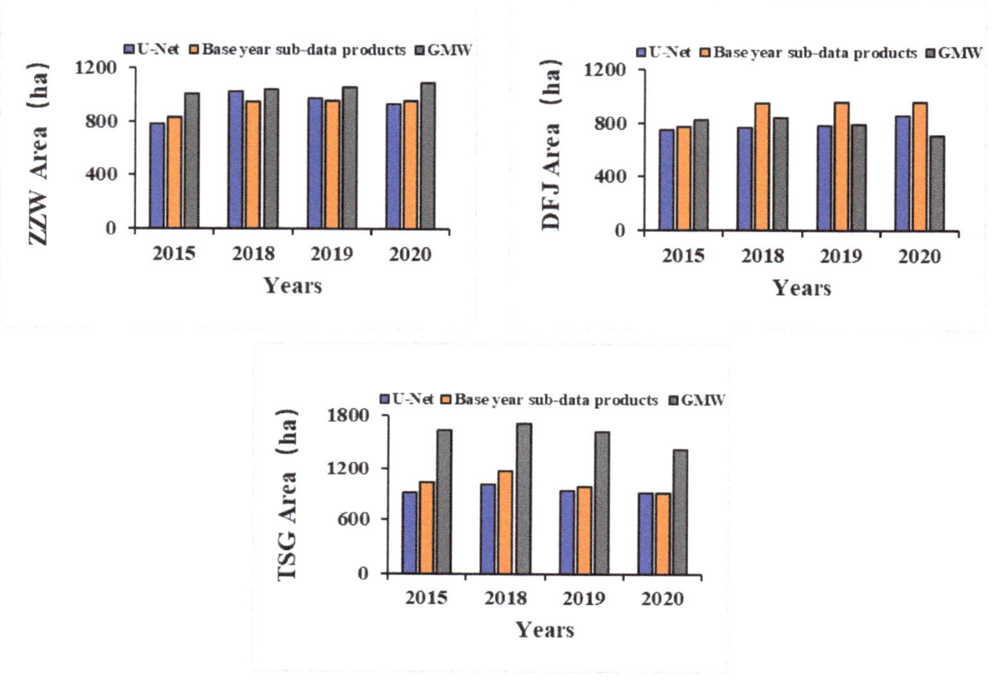

Figure 8. The hybrid U-Net results for mangrove area changes in the ZZW, DFJ, and TSG areas for 2015, 2018, 2019, 2020 with the base year sub-data products.

According to the base year sub-data products, the average area values of mangroves in ZZW, DFJ, and TSG were 927.36 ha, 787.55 ha, and 941.58 ha, respectively (Figures 7 and 8 and Table S1). Among them, ZZW and TSG showed a trend of first increasing and then decreasing in mangrove area at the four time points, while the mangrove area in DFJ showed a continuous increasing trend. From 2015 to 2020, the overall mangrove area in ZZW increased by 147.79 ha (18.9%), and that in DFJ increased by 104.66 ha (13.94%), while the mangrove area in TSG remained almost unchanged.

A comparison with 10/30 m data products in this study revealed significant differences in the changes in mangrove area in TSG and ZZW for 30 m data products. The large discrepancy of 693.96 ha in TSG from 2018 (30 m reference) may be because China has many small mangrove patches, as well as narrow mangroves along levees, which were not captured by the 30 m resolution product due to the lack of fine-scale information and different pixel sizes. In contrast, this study's results exhibit better performance in terms of extracting more detailed mangrove patches. In addition, differences in some areas may be due to different methods used for image acquisition and mangrove mapping in different years. From the time series of different mangrove datasets in Table S1, it can be seen that China's mangrove area has remained stable since 2015, with fluctuations in some regions, which is consistent with the results of [13]. The results of this study were generated using 2 m high-resolution images, with smooth edges of patches and complete structures of small patches. In contrast, the mangrove patches in the other five datasets showed serious saw-toothed edge effects and insufficient accuracy for small patches.

The results of this study are closer to the 10 m product in terms of mangrove area. Considering the pixel scale effect, the 2 m resolution result should be smaller than the 10 m resolution, which is consistent with the results of this study. For the Pearl River Estuary, where there are mainly large independent mangroves, the differences between the three products were not significant. For TSG and DFJ, there are narrow mangrove belts along the embankments and large independent mangroves. From a spatial perspective, the 10/30 m data have significant omissions for such mangroves. Considering the influence of the scale effect, although the overall changes in the mangrove area for DFJ among the three datasets are not significant, the 2/10 m results are much better than the 30 m result in terms of spatial information. For DFJ, the inversion of the western mangroves using 10/30 m data is disastrous, while the 2 m data can clearly show the spatial distribution of the western mangroves in DFJ.

5. Discussion

5.1. Stability of the Hybrid U-Net Network Performance

This article proposes an integrated spatial, temporal, and methodological (hybrid U-Net model) optimization for mangrove extraction. In this study, the spatial resolution of the mangroves extracted is 2 m, allowing for the capture of finer mangrove patches, especially small and narrow ones. Mangroves are primarily found near estuaries, and small and narrow mangrove patches are quite common. This study enhances the quality of training samples based on mangrove habitat and tidal information, utilizing a semantic segmentation model like U-Net that requires only one year of samples to map mangroves for multiple years. Compared with other machine learning methods that require repeated acquisition of stable samples, this approach enables rapid detection of mangrove changes.

The output of the hybrid U-Net network model has only two categories, mangrove and non-mangrove, making it a binary classification model. Compared with other models, this approach takes into account the potential growth areas of mangroves and the impact of tides on mangrove extraction, and combines the advantages of semantic segmentation models (reducing dependence on samples). This method is more suitable for large-scale mangrove extraction tasks [56]. Additionally, the combination of GF and ZY series satellite sensors compensates for the shortcomings of individual satellite image information, ensuring high-resolution mapping at a 2 m spatial resolution.

The Hybrid U-Net network model has three unique advantages in fine-scale mangrove mapping: (1) It excludes non-vegetation cover before mangrove extraction, avoiding the impact of non-vegetation on sparse mangroves and saving post-processing work; (2) It utilizes vegetation indices to determine the low tide moments, ensuring the accuracy of extracting the maximum extent of mangroves; (3) The Hybrid U-Net model requires only one year of samples for training and the training model can be applied to infer mangroves for other years within the same area. This reduces the traditional deep learning models' reliance on samples.

5.2. Advantages of High-Resolution Data Extraction for Mangrove Mapping

Currently, most of the published mangrove distribution datasets use medium-resolution satellite data with a spatial resolution of 30 m or synthetic aperture radar data at the L-band with a resolution of 25 m (Figure 9). In these medium-resolution images, the vegetation texture and spatial details are missing, leading to limitations in accurately identifying mangrove forests. In addition, small and irregularly shaped mangrove areas are difficult to accurately identify and extract.

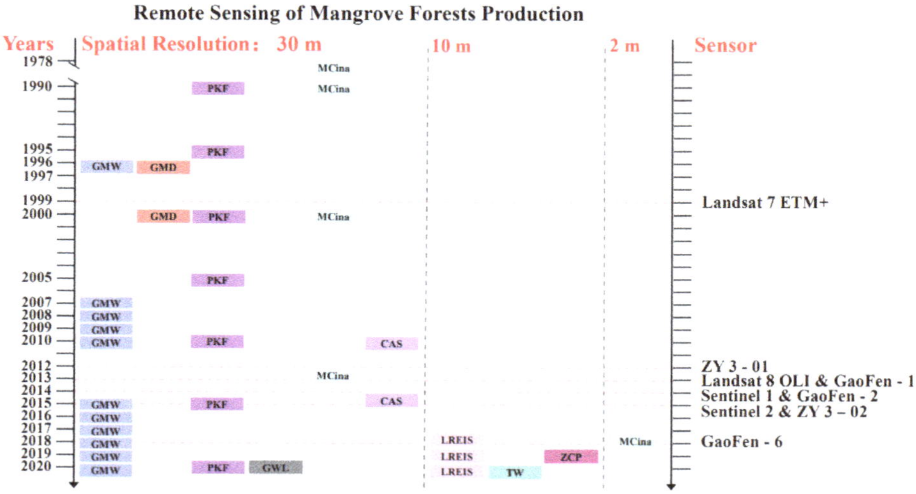

Figure 9. Time-series distribution of existing mangrove data products (30 m spatial resolution mangrove dataset: the Global Mangrove Watch (GMW [15]), Global Mangrove Distribution (GMD [57]), PKF [58], GWL_FCS30 (GWL [59]), (MCina [40] (1978, 1990, 2000, 2013)), (CAS [13]); 10 m spatial resolution mangrove dataset: Mangrove China_LASAC (LREIS [23]), Tidal Wetlands in East Asia (TW [44]), ZCP [18,40]; 2 m spatial resolution mangrove dataset: (XCina [40] (2018))).

This study uses satellite remote sensing images with a resolution of 2 m from GF and ZY series sensors for mangrove mapping. The finer spatial resolution allows for the delineation of small and scattered mangrove patches and narrow strips along the coastline. This is particularly important for mangrove mapping in China. There are numerous small and narrow patches of mangroves in China. Young plants that were newly planted but not forested were not included in existing dataset [23]. The use of high-spatial-resolution remote sensing images enables the extraction of more detailed mangrove distributions at a finer spatial scale and allows for rapid detection of subtle changes in the information. A detailed understanding of the distribution of and changes in these mangroves is of great significance for allowing local management authorities to carry out mangrove ecological protection and restoration and to coordinate coastal economic development with ecological protection.

5.3. The Limitations of the Hybrid U-Net Network Model

In general, remote sensing-based surveys of mangrove forests only include areas with a tree crown density of 20% or more [23]. Seedlings and mangrove forests with an area of less than 4 m^2 are currently unable to be extracted due to the limitations in sensor accuracy, which may result in an underestimation of model results compared with actual conditions.

There are other types of vegetation growing on the tidal flats where mangrove forests grow along the coast of the Beibu Gulf in Guangxi, China. These vegetation types are easily confused with mangrove forests in image interpretation. The most common types of vegetation are *Spartina alterniflora* and reeds (Figure 10) [51]. Many studies have provided evidence that *Spartina alterniflora* can entirely supplant mangrove forests within a short span of 1 to 5 years [60]. The latest research indicates that the invasion of *Spartina alterniflora* threatens the survival and reproduction of mangroves, as *Spartina alterniflora* inhibits the growth and expansion of mangroves [20,61]. There is spectral confusion between *Spartina alterniflora* and mangroves, which may lead to an overestimation of mangrove areas by the hybrid U-Net [62]. In future research, it will be possible to improve the accuracy of mangrove mapping further by overlaying thematic maps of misclassified species.

Figure 10. Field investigation: mangrove (**a**,**d**), *Spartina alterniflora* (**b**), reed (**c**).

6. Conclusions

This paper presents a spatial–temporal method optimization approach (hybrid U-Net) based on mangrove habitats, low tide, and semantic segmentation models to address challenges in satellite remote sensing of mangrove wetlands, including spectral similarity confusion, tidal uncertainty, and difficulties in obtaining stable samples. The paper demonstrates the advantages of combining tidal water levels and mangrove habitat information in mangrove extraction and mapping.

The hybrid U-Net model overcomes the dependency of traditional machine learning algorithms on large sample datasets. The use of mangrove habitats and low-tide periods helps reduce the interference of surrounding vegetation in the extraction of the mangroves. This method was applied to extract mangroves in three typical areas of the Beibu Gulf in 2015, 2018, 2019, and 2020. This method is less sample-dependent and more accurate and is more suitable for the rapid detection of subtle changes in large-scale mangroves. The results of the hybrid U-Net model were reliable for extracting sparse, slow-growing young mangroves and narrow mangrove strips along roadsides.

Future research will focus on using one or a set of indicators or rules to approximate classification results, attempting to reduce the impact of spectral uncertainty on model accuracy under different climatic conditions. This high-precision mangrove extraction method will provide an important basis for refined calculations and assessments of mangrove ecosystem changes, mangrove species classification, marine ecosystem management, blue carbon restoration, and biodiversity protection in the future.

Supplementary Materials: The following supporting information can be downloaded at https://www.mdpi.com/article/10.3390/rs15225271/s1. (Figure S1. The Hybrid U-Net result of DFJ at the Beibu Gulf. (a, b) The satellite image of mangrove area in 2021, (c) Mangrove habitat information, (d) Low tide level acquisition, (e) The Hybrid U-Net result in 2020, (f) The Hybrid U-Net result in 2019, (g) The Hybrid U-Net result in 2018, (h) The Hybrid U-Net result in 2015, Figure S2. The Hybrid U-Net result of TSG at the Beibu Gulf. (a, b) The satellite image of mangrove area in 2021, (c) Mangrove habitat information, (d) Low tide level acquisition, (e) The Hybrid U-Net result in 2020, (f) The Hybrid U-Net result in 2019, (g) The Hybrid U-Net result in 2018, (h) The Hybrid U-Net result in 2015. Figure S3: Comparison of the results of hybrid U-Net extraction of mangroves with multi-person interactive validation at DFJ: (a, b) The satellite image of mangrove area at DFJ in 2021 October, (c, e, g, I) The Hybrid U-Net result in 2020, 2019, 2018, 2015. (d, f, h, j) Multiplayer interactive validation in 2020, 2019, 2018, 2015. Figure S4. Comparison of the results of hybrid U-Net extraction of mangroves with multi-person interactive validation at TSG: (a, b) The satellite image of mangrove area at TSG in 2021 February, (c, e, g, I) The Hybrid U-Net result in 2020, 2019, 2018, 2015. (d, f, h, j) Multiplayer interactive validation in 2020, 2019, 2018, 2015. Figure S5. Hybrid U-Net extract mangrove results compare with existing data products. (a) The satellite image of mangrove area at DFJ, (b) Mangrove habitat information, (c) Low tide level acquisition, (d, g, j, m) The Hybrid U-Net result in 2020, 2019, 2018, 2015, (f, I, l, o) The Global Mangrove Watch (GMW) in 2020, 2019, 2018, 2015, (e) LREIS in 2020, (h) TW in 2019, (k) MCina in 2018, (n) CAS in 2015. Figure S6. Hybrid U-Net extract mangrove results compare with existing data products. (a) The satellite image of mangrove area at TSG, (b) Mangrove habitat information, (c) Low tide level acquisition, (d, g, j, m) The Hybrid U-Net result in 2020, 2019, 2018, 2015, (f, I, l, o) The Global Mangrove Watch (GMW) in 2020, 2019, 2018, 2015, (e) LREIS in 2020, (h) TW in 2019, (k) MCina in 2018, (n) CAS in 2015. Figure S7. Hybrid U-Net extract mangrove results compare with existing data products. (TSG details). (a) The GF-1 satellite image(2m) of mangrove area at TSG, (b) The GF-1 satellite image(2m) of mangrove area at TSG (details), (c) The Sentinal-2 satellite image(10m) of mangrove area at TSG (details), (d) The Landsat 8 OLI satellite image(30m) of mangrove area at TSG (details), (e, h, k, n) The Hybrid U-Net result in 2020, 2019, 2018, 2015, (g, j, m, p), The Global Mangrove Watch (GMW) in 2020, 2019, 2018, 2015, (f) LREIS in 2020, (i) TW in 2019, (l) MCina in 2018, (o) CAS in 2015. Table S1. Areal changes (ha) of mangrove forest with hybrid U-Net in the ZZW, DFJ, and TSG area for the 2015, 2018, 2019, 2020 with the base year sub-data products.)

Author Contributions: Z.S.: conceptualization, methodology, software, investigation, formal analysis, writing—original draft. W.J.: Conceptualization, funding acquisition, resources, data curation, investigation, writing—review and editing, supervision. Z.L.: funding acquisition, supervision. S.Z.: resources, investigation. Z.Z.: visualization, supervision. J.S.: validation, supervision. Z.X.: writing—review and editing. All authors have read and agreed to the published version of the manuscript.

Funding: This work was supported by the National Natural Science Foundation of China under grant no. U21A2022, grant no. 42101369, and grant no. 2023GXNSFBA026278. We also appreciate the anonymous reviewers for their comments and suggestions, which helped to improve the quality of this manuscript.

Data Availability Statement: The data presented in this study are available on request from the corresponding author. The mangrove datasets can be downloaded at https://zenodo.org/record/8041161, accessed on 26 October 2022.

Conflicts of Interest: The authors declare no conflict of interest.

References

1. Morris, R.L.; Fest, B.; Stokes, D.; Jenkins, C.; Swearer, S.E. The coastal protection and blue carbon benefits of hybrid mangrove living shorelines. *J. Environ. Manag.* **2023**, *331*, 117310. [CrossRef] [PubMed]
2. Shahbudin, S.; Zuhairi, A.; Kamaruzzaman, B.Y. Impact of coastal development on mangrove cover in Kilim river, Langkawi Island, Malaysia. *J. For. Res.* **2011**, *23*, 185–190. [CrossRef]
3. Zhou, H.Z.; Wang, J.F.; Jiang, H.M.; Cai, Z.X.; Tang, G.H.; Ding, S.; Liu, S.T.; Xu, Z.M. Distribution fractions and potential ecological risk assessment of heavy metals in mangrove sediments of the Greater Bay Area. *Environ. Sci. Pollut. Res.* **2023**, *30*, 45859–45871. [CrossRef] [PubMed]
4. Giri, C.; Long, J.; Abbas, S.; Murali, R.M.; Qamer, F.M.; Pengra, B.; Thau, D. Distribution and dynamics of mangrove forests of South Asia. *J. Environ. Manag.* **2015**, *148*, 101–111. [CrossRef]
5. Uddin, M.M.; Aziz, A.A.; Lovelock, C.E. Importance of mangrove plantations for climate change mitigation in Bangladesh. *Glob. Chang. Biol.* **2023**, *29*, 3331–3346. [CrossRef]
6. Pham, T.D.; Xia, J.; Ha, N.T.; Bui, D.T.; Le, N.N.; Tekeuchi, W. A Review of Remote Sensing Approaches for Monitoring Blue Carbon Ecosystems: Mangroves, Seagrasses and Salt Marshes during 2010–2018. *Sensors* **2019**, *19*, 1933. [CrossRef]
7. Tang, W.; Zheng, M.; Zhao, X.; Shi, J.; Yang, J.; Trettin, C. Big Geospatial Data Analytics for Global Mangrove Biomass and Carbon Estimation. *Sustainability* **2018**, *10*, 472. [CrossRef]
8. Come, J.; Peer, N.; Nhamussua, J.L.; Miranda, N.A.; Macamo, C.C.; Cabral, A.S.; Madivadua, H.; Zacarias, D.; Narciso, J.; Snow, B. A socio-ecological survey in Inhambane Bay mangrove ecosystems: Biodiversity, livelihoods, and conservation. *Ocean Coast. Manag.* **2023**, *244*, 106813. [CrossRef]

9. Rull, V. Rise and fall of Caribbean mangroves. *Sci. Total Environ.* **2023**, *885*, 163851. [CrossRef]
10. Malik, A.; Fensholt, R.; Mertz, O. Mangrove exploitation effects on biodiversity and ecosystem services. *Biodivers. Conserv.* **2015**, *24*, 3543–3557. [CrossRef]
11. Wang, L.; Jia, M.; Yin, D.; Tian, J. A review of remote sensing for mangrove forests: 1956–2018. *Remote Sens. Environ.* **2019**, *231*, 111223. [CrossRef]
12. Goldberg, L.; Lagomasino, D.; Thomas, N.; Fatoyinbo, T. Global declines in human-driven mangrove loss. *Glob. Chang. Biol.* **2020**, *26*, 5844–5855. [CrossRef] [PubMed]
13. Jia, M.; Wang, Z.; Zhang, Y.; Mao, D.; Wang, C. Monitoring loss and recovery of mangrove forests during 42 years: The achievements of mangrove conservation in China. *Int. J. Appl. Earth Obs. Geoinf.* **2018**, *73*, 535–545. [CrossRef]
14. Su, J.; Friess, D.A.; Gasparatos, A. A meta-analysis of the ecological and economic outcomes of mangrove restoration. *Nat. Commun.* **2021**, *12*, 5050. [CrossRef] [PubMed]
15. Bunting, P.; Rosenqvist, A.; Hilarides, L.; Lucas, R.M.; Thomas, N. Global Mangrove Watch: Updated 2010 Mangrove Forest Extent (v2.5). *Remote Sens.* **2022**, *14*, 1034. [CrossRef]
16. Lymburner, L.; Bunting, P.; Lucas, R.; Scarth, P.; Alam, I.; Phillips, C.; Ticehurst, C.; Held, A. Mapping the multi-decadal mangrove dynamics of the Australian coastline. *Remote Sens. Environ.* **2020**, *238*, 111185. [CrossRef]
17. Fu, B.L.; Liang, Y.Y.; Lao, Z.A.; Sun, X.D.; Li, S.Z.; He, H.C.; Sun, W.W.; Fan, D.L. Quantifying scattering characteristics of mangrove species from Optuna-based optimal machine learning classification using multi-scale feature selection and SAR image time series. *Int. J. Appl. Earth Obs. Geoinf.* **2023**, *122*, 103446. [CrossRef]
18. Zhao, C.; Qin, C.-Z. 10-m-resolution mangrove maps of China derived from multi-source and multi-temporal satellite observations. *ISPRS J. Photogramm. Remote Sens.* **2020**, *169*, 389–405. [CrossRef]
19. Xia, Q.; Qin, C.Z.; Li, H.; Huang, C.; Su, F.Z.; Jia, M.M. Evaluation of submerged mangrove recognition index using multi-tidal remote sensing data. *Ecol. Indic.* **2020**, *113*, 106196. [CrossRef]
20. Zhu, B.; Liao, J.; Shen, G. Combining time series and land cover data for analyzing spatio-temporal changes in mangrove forests: A case study of Qinglangang Nature Reserve, Hainan, China. *Ecol. Indic.* **2021**, *131*, 108135. [CrossRef]
21. Cao, J.; Liu, K.; Zhuo, L.; Liu, L.; Zhu, Y.; Peng, L. Combining UAV-based hyperspectral and LiDAR data for mangrove species classification using the rotation forest algorithm. *Int. J. Appl. Earth Obs. Geoinf.* **2021**, *102*, 102414. [CrossRef]
22. Zhao, D.M.; Zhen, J.N.; Zhang, Y.H.; Miao, J.; Shen, Z.; Jiang, X.P.; Wang, J.J.; Jiang, J.C.; Tang, Y.Z.; Wu, G.F. Mapping mangrove leaf area index (LAI) by combining remote sensing images with PROSAIL-D and XGBoost methods. *Remote Sens. Ecol. Conserv.* **2023**, *9*, 370–389. [CrossRef]
23. Zhang, T.; Hu, S.; He, Y.; You, S.; Yang, X.; Gan, Y.; Liu, A. A Fine-Scale Mangrove Map of China Derived from 2-Meter Resolution Satellite Observations and Field Data. *ISPRS Int. J. Geo-Inf.* **2021**, *10*, 92. [CrossRef]
24. Su, X.; Geng, J.; Ma, X.; Wang, H.; Wang, X. Mangrove species classification based on multiple vegetation index extraction and joint sparse representation. *Mar. Environ. Sci.* **2017**, *36*, 114–120.
25. Zhen, J.; Liao, J.; Shen, G. Remote Sensing Monitoring and Analysis on the Dynamics of Mangrove Forests in Qinglan Habor of Hainan Province since 1987. *Wetl. Sci.* **2019**, *17*, 44–51.
26. Mondal, P.; Liu, X.; Fatoyinbo, T.E.; Lagomasino, D. Evaluating Combinations of Sentinel-2 Data and Machine-Learning Algorithms for Mangrove Mapping in West Africa. *Remote Sens.* **2019**, *11*, 2928. [CrossRef]
27. Sheykhmousa, M.; Mahdianpari, M.; Ghanbari, H.; Mohammadimanesh, F.; Ghamisi, P.; Homayouni, S. Support Vector Machine Versus Random Forest for Remote Sensing Image Classification: A Meta-Analysis and Systematic Review. *IEEE J. Sel. Top. Appl. Earth Obs. Remote Sens.* **2020**, *13*, 6308–6325. [CrossRef]
28. Belgiu, M.; Dragut, L. Random forest in remote sensing: A review of applications and future directions. *ISPRS J. Photogramm. Remote Sens.* **2016**, *114*, 24–31. [CrossRef]
29. Tsai, Y.L.S. Monitoring 23-year of shoreline changes of the Zengwun Estuary in Southern Taiwan using time-series Landsat data and edge detection techniques. *Sci. Total Environ.* **2022**, *839*, 156310. [CrossRef]
30. Iovan, C.; Kulbicki, M.; Mermet, E. Deep convolutional neural network for mangrove mapping. In Proceedings of the IEEE International Geoscience and Remote Sensing Symposium (IGARSS), Waikoloa, HI, USA, 26 September–2 October 2020; pp. 1969–1972.
31. Guo, M.Q.; Yu, Z.Y.; Xu, Y.Y.; Huang, Y.; Li, C.F. ME-Net: A Deep Convolutional Neural Network for Extracting Mangrove Using Sentinel-2A Data. *Remote Sens.* **2021**, *13*, 1292. [CrossRef]
32. Kalinaki, K.; Malik, O.A.; Lai, D.T.C. FCD-AttResU-Net: An improved forest change detection in Sentinel-2 satellite images using attention residual U-Net. *Int. J. Appl. Earth Obs. Geoinf.* **2023**, *122*, 103453. [CrossRef]
33. Guo, Y.; Liao, J.; Shen, G. Mapping Large-Scale Mangroves along the Maritime Silk Road from 1990 to 2015 Using a Novel Deep Learning Model and Landsat Data. *Remote Sens.* **2021**, *13*, 245. [CrossRef]
34. Cao, K.L.; Zhang, X.L. An Improved Res-UNet Model for Tree Species Classification Using Airborne High-Resolution Images. *Remote Sens.* **2020**, *12*, 1128. [CrossRef]
35. Dong, Y.; Yu, K.; Hu, W. GC-UNet: An Improved UNet Model for Mangrove Segmentation Using Landsat8. In Proceedings of the 2021 3rd International Conference on Big Data Engineering, New York, NY, USA, 29–31 May 2021.
36. Hu, L.; Li, W.; Xu, B. Monitoring mangrove forest change in China from 1990 to 2015 using Landsat-derived spectral-temporal variability metrics. *Int. J. Appl. Earth Obs. Geoinf.* **2018**, *73*, 88–98. [CrossRef]

37. Li, L.; Liu, W.a. Dynamic changes of mangrove wetland landscape pattern in Lianzhou gulf of Guangxi and its causes. *J. For. Environ.* **2018**, *38*, 171–177.
38. Tian, Q.J.; Li, S.S. Mangrove Recognition and Extraction Using Multispectral Remote Sensing Data in Beibu Gulf. In Proceedings of the Conference on Multispectral, Hyperspectral, and Ultraspectral Remote Sensing Technology, Techniques and Applications VII, Honolulu, HI, USA, 24–26 September 2018.
39. Tao, Z.; Shucheng, Y.; Xiaomei, Y.; Shanshan, H. *Mangroves Map of China 2018 (MC2018) Derived from 2-Meter Resolution Satellite Observations and Field Data*; Science Data Bank: Beijing, China, 2020. [CrossRef]
40. Zhang, Y.; Mishra, R.K. A review and comparison of commercially available pan-sharpening techniques for high resolution satellite image fusion. In Proceedings of the IEEE International Geoscience and Remote Sensing Symposium (IGARSS), Munich, Germany, 22–27 July 2012; pp. 182–185.
41. You, N.; Dong, J.; Huang, J.; Du, G.; Zhang, G.; He, Y.; Yang, T.; Di, Y.; Xiao, X. The 10-m crop type maps in Northeast China during 2017–2019. *Sci. Data* **2021**, *8*, 41. [CrossRef] [PubMed]
42. Liao, B.; Zhang, Q. Area, Distribution and Species Composition of Mangroves in China. *Wetl. Sci.* **2014**, *12*, 435–440.
43. Zhang, Z.; Xu, N.; Li, Y.; Li, Y. Sub-continental-scale mapping of tidal wetland composition for East Asia: A novel algorithm integrating satellite tide-level and phenological features. *Remote Sens. Environ.* **2022**, *269*, 112799. [CrossRef]
44. Zhao, C.; Qin, C.-Z.; Teng, J. Mapping large-area tidal flats without the dependence on tidal elevations: A case study of Southern China. *ISPRS J. Photogramm. Remote Sens.* **2020**, *159*, 256–270. [CrossRef]
45. Wang, L.; Li, R.; Zhang, C.; Fang, S.; Duan, C.; Meng, X.; Atkinson, P.M. UNetFormer: A UNet-like transformer for efficient semantic segmentation of remote sensing urban scene imagery. *ISPRS J. Photogramm. Remote Sens.* **2022**, *190*, 196–214. [CrossRef]
46. Chaurasia, A.; Culurciello, E. LinkNet: Exploiting Encoder Representations for Efficient Semantic Segmentation. In Proceedings of the IEEE Visual Communications and Image Processing (VCIP), St. Petersburg, FL, USA, 10–13 December 2017.
47. He, K.; Zhang, X.; Ren, S.; Sun, J. Deep Residual Learning for Image Recognition. In Proceedings of the 2016 IEEE Conference on Computer Vision and Pattern Recognition (CVPR), Seattle, WA, USA, 27–30 June 2016; pp. 770–778.
48. Kingma, D.; Ba, J. Adam: A Method for Stochastic Optimization. *arXiv* **2014**, arXiv:1412.6980.
49. Jia, M.; Wang, Z.; Mao, D.; Ren, C.; Wang, C.; Wang, Y. Rapid, robust, and automated mapping of tidal flats in China using time series Sentinel-2 images and Google Earth Engine. *Remote Sens. Environ.* **2021**, *255*, 112285. [CrossRef]
50. Zhang, R.; Jia, M.; Wang, Z.; Zhou, Y.; Mao, D.; Ren, C.; Zhao, C.; Liu, X. Tracking annual dynamics of mangrove forests in mangrove National Nature Reserves of China based on time series Sentinel-2 imagery during 2016–2020. *Int. J. Appl. Earth Obs. Geoinf.* **2022**, *112*, 102918. [CrossRef]
51. Foody, G.M. Explaining the unsuitability of the kappa coefficient in the assessment and comparison of the accuracy of thematic maps obtained by image classification. *Remote Sens. Environ.* **2020**, *239*, 111630. [CrossRef]
52. Foody, G.M. Status of land cover classification accuracy assessment. *Remote Sens. Environ.* **2002**, *80*, 185–201. [CrossRef]
53. Tao, C.X.; Meng, Y.Z.; Li, J.J.; Yang, B.B.; Hu, F.M.; Li, Y.X.; Cui, C.L.; Zhang, W. MSNet: Multispectral semantic segmentation network for remote sensing images. *Giscience Remote Sens.* **2022**, *59*, 1177–1198. [CrossRef]
54. Maurya, K.; Mahajan, S.; Chaube, N. Remote sensing techniques: Mapping and monitoring of mangrove ecosystem—A review. *Complex Intell. Syst.* **2021**, *7*, 2797–2818. [CrossRef]
55. Zhao, C.; Qin, C.-Z. Identifying large-area mangrove distribution based on remote sensing: A binary classification approach considering subclasses of non-mangroves. *Int. J. Appl. Earth Obs. Geoinf.* **2022**, *108*, 102750. [CrossRef]
56. Giri, C.; Ochieng, E.; Tieszen, L.L.; Zhu, Z.; Singh, A.; Loveland, T.; Masek, J.; Duke, N. Status and distribution of mangrove forests of the world using earth observation satellite data. *Glob. Ecol. Biogeogr.* **2011**, *20*, 154–159. [CrossRef]
57. Peng, K.F.; Jiang, W.G.; Hou, P.; Wu, Z.F.; Ling, Z.Y.; Wang, X.Y.; Niu, Z.G.; Mao, D.H. Continental-scale wetland mapping: A novel algorithm for detailed wetland types classification based on time series Sentinel-1/2 images. *Ecol. Indic.* **2023**, *148*, 110113. [CrossRef]
58. Zhang, X.; Liu, L.Y.; Zhao, T.T.; Chen, X.D.; Lin, S.R.; Wang, J.Q.; Mi, J.; Liu, W.D. GWL_FCS30: A global 30 m wetland map with a fine classificationsystem using multi-sourced and time-series remote sensing imagery in 2020. *Earth Syst. Sci. Data* **2023**, *15*, 265–293. [CrossRef]
59. Gao, G.F.; Li, P.F.; Zhong, J.X.; Shen, Z.J.; Chen, J.; Li, Y.T.; Isabwe, A.; Zhu, X.Y.; Ding, Q.S.; Zhang, S.; et al. *Spartina alterniflora* invasion alters soil bacterial communities and enhances soil N_2O emissions by stimulating soil denitrification in mangrove wetland. *Sci. Total Environ.* **2019**, *653*, 231–240. [CrossRef] [PubMed]
60. Liu, M.Y.; Li, H.Y.; Li, L.; Man, W.D.; Jia, M.M.; Wang, Z.M.; Lu, C.Y. Monitoring the Invasion of *Spartina alterniflora* Using Multi-source High-resolution Imagery in the Zhangjiang Estuary, China. *Remote Sens.* **2017**, *9*, 539. [CrossRef]
61. Zheng, J.Y.; Wei, H.Y.; Chen, R.D.; Liu, J.M.; Wang, L.K.; Gu, W. Invasive Trends of *Spartina alterniflora* in the Southeastern Coast of China and Potential Distributional Impacts on Mangrove Forests. *Plants* **2023**, *12*, 1923. [CrossRef] [PubMed]
62. Zheng, X.J.; Javed, Z.; Liu, B.; Zhong, S.; Cheng, Z.; Rehman, A.; Du, D.L.; Li, J. Impact of Spartina alterniflora Invasion in Coastal Wetlands of China: Boon or Bane? *Biology* **2023**, *12*, 1057. [CrossRef]

remote sensing

MDPI

Article

Mangrove Forest Cover Change in the Conterminous United States from 1980–2020

Chandra Giri [1,*], Jordan Long [2] and Prapti Poudel [1]

1 Office of Research and Development, United States Environmental Protection Agency, 109 T.W. Alexander Drive, Research Triangle Park, Durham, NC 27709, USA; giri.prapti@epa.gov
2 Earth Resources Observation and Science (EROS) Center, 252nd Street, Sioux Falls, SD 47914, USA; jlong@usgs.gov
* Correspondence: giri.chandra@epa.gov; Tel.: +1-919-224-9817

Abstract: Mangrove forests in developed and developing countries are experiencing substantial transformations driven by natural and anthropogenic factors. This study focuses on the conterminous United States, including Florida, Texas, and Louisiana, where coastal development, urbanization, hydrological pattern alterations, global warming, sea level rise, and natural disasters such as hurricanes contribute to mangrove forest changes. Using time-series Landsat data and image-processing techniques in a cloud computing platform, we analyzed the dynamics of mangrove forests every five years from 1980 to 2020. Each thematic product was independently derived using a region of interest (ROI) suitable for local conditions. The analysis was performed using consistent data sources and a unified classification methodology. Our results revealed that the total mangrove area in the conterminous United States (CONUS) in 2020 was 266,179 ha. with 98.0% of the mangrove area in Florida, 0.6% in Louisiana, and 1.4% in Texas. Approximately 85% of the CONUS mangrove area was found between 24.5° and 26.0° latitude. Overall, mangrove forests in the CONUS increased by 13.5% from 1980 to 2020. However, the quinquennial variation in aerial coverage fluctuated substantially. The validation of 2020 using a statistical sample of reference data confirmed the high accuracy of 95%. Our results can aid policymakers and conservationists in developing targeted strategies for preserving the ecological and socio-economic value of mangrove forests in the conterminous United States. Additionally, all the datasets generated from this study have been released to the public.

Keywords: mangrove forests; remote sensing; image processing; Landsat; change analysis

1. Introduction

Mangrove forests in both developed and developing countries are undergoing significant transformations due to both natural and anthropogenic factors. In the conterminous United States encompassing Florida, Texas, and Louisiana, the forests are changing due to coastal development, urbanization, alterations in hydrological patterns, the effects of global warming, sea level rise, and the occurrence of natural disasters such as hurricanes [1–3].

The mangrove forests in various parts of the world contribute to ecosystem services to varying degrees, depending on the specific economic and ecological conditions present in each location [4,5]. In the continental United States, mangrove forests provide invaluable ecosystem goods and services that extend beyond their ecological significance, offering benefits in terms of coastal protection, biodiversity conservation, carbon sequestration, water filtration, fisheries support, and recreational opportunities [6–8].

In the face of diminishing forest resources and ecosystem goods and services, the remaining mangrove forests and their invaluable services are increasingly vulnerable to the compounding pressures of economic and biophysical factors [9–11]. Effective quantification of the rates, patterns, causes, and consequences of mangrove forest cover changes heavily relies on using time-series remote-sensing data. Such data play a crucial role in understanding and monitoring the dynamics of mangrove forests and their response to various drivers of change.

In this context, our research aimed to conduct a spatiotemporal analysis using Landsat satellite data spanning five-year intervals from 1980 to 2020. The goal was to assess both positive (afforestation) and negative (deforestation) changes in mangrove forest coverage within the contiguous United States.

2. Study Area, Data Basis, and Methods

2.1. Study Area

Mangrove forests of the conterminous United States (CONUS) are exclusively situated in the Gulf of Mexico states of Texas, Louisiana, and Florida (Figure 1). South Florida is home to the largest tract of continuous mangrove forests in the United States, of which 75% of the country's mangroves occur within the Everglades National Park alone [3]. These forests are predominantly composed of three mangrove species: red mangrove (*Rhizophora mangle*), black mangrove (*Avicennia germinans*), and white mangrove (*Laguncularia racemosa*). The mangrove forests in Texas and Louisiana mainly consist of black mangroves.

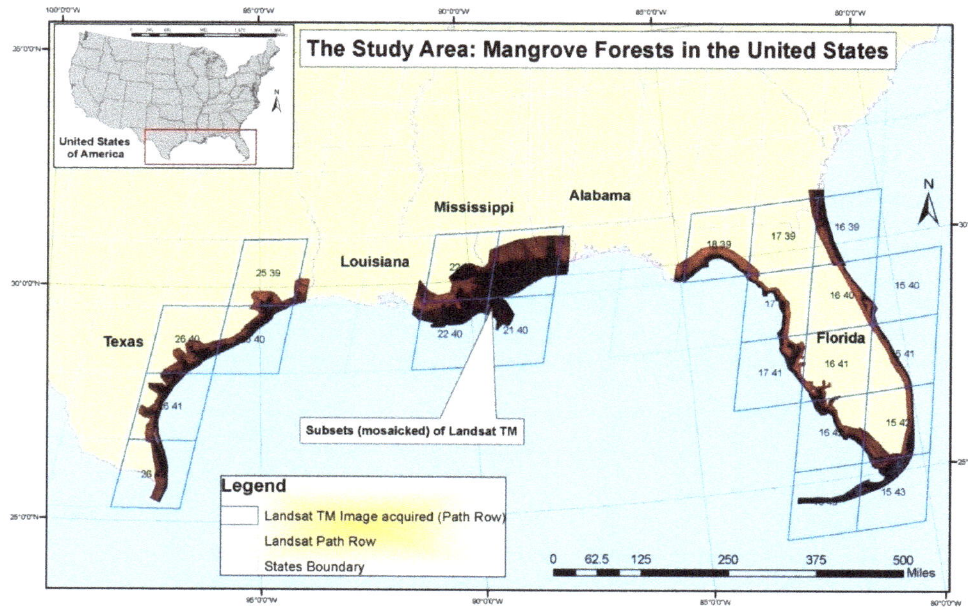

Figure 1. Approximate location of mangrove habitats in CONUS.

Mangrove forests in the United States are believed to be expanding towards the pole due to climate change [12]. Warmer winter air temperatures and less extreme sub-freezing winter air temperatures are likely causes of the expansion of black mangroves (*Avicennia germinans*) into salt marshes dominated by smooth cordgrass (*Spartina alterniflora*) [8,13,14]. However, long-term data dating back to the 1980s suggests the northernmost latitudinal limit of mangrove forests along the east and west coasts of Florida, in addition to Louisiana and Texas, has not systematically expanded toward the pole [15]. The forest is, however, expanding within the geographic limit [15]. Other factors such as fire, tropical storms, sea level rise, sedimentation/erosion, changing hydrology, and specific management practices are also believed to be contributing to the expansion in localized areas [16,17].

2.2. Data and Methods

The Landsat Multi Spectral Scanner (MSS), Thematic Mapper (TM), Enhanced Thematic Mapper Plus (ETM+), and Operational Land Imager (OLI) Tier 1 top-of-atmosphere (TOA) reflectance data were procured from the collections of the United States Geological Survey (USGS) and Google Earth Engine (GEE) archive. The collaborative Landsat program, initiated in the early 1970s by both the National Aeronautics and Space Administration (NASA) and USGS, represents the world's most extensive and uninterrupted repository of satellite-based Earth observation data. The complete data archive is openly accessible through USGS and is also made available on various other platforms including the Google Earth Engine. This study utilized data from the Landsat series corresponding to the years circa 1980, 1985, 1990, 1995, 2000, 2005, 2010, 2015, and 2020.

The Tier 1 Landsat data used in this study have undergone Level-1 Precision Terrain (L1TP) processing, which ensures well-defined radiometry and consistent inter-calibration across various Landsat sensors [18]. Additionally, Tier 1 scenes are meticulously georegistered and maintained in conformity with specified tolerances, ensuring a root mean square error (RMSE) of 12 meters or less. The specifications for the reflectance product and the constants applicable to all Landsat sensors including TM, ETM+, and OLI were comprehensively detailed and tabulated by Chander et al. (2009) [19].

It is important to note that all Tier 1 Landsat data are considered suitable for land-cover mapping and monitoring [19]. These data can be used to detect and estimate long-term dynamic land-cover change dating back to the 1970s [18,19].

We employed a cloud compositing technique to create mosaic images devoid of clouds for every five-year period from 1980 to 2020. This method utilizes a maximal value approach for each spectral band [20,21]. In situations where cloud-free data were not accessible for these specific five-year periods, Landsat data from one or two years prior to or after the target date were utilized. The increased availability of Landsat data over the years provided an opportunity to select data from the target date itself.

For the 1980 mosaic, we utilized Landsat-3 MSS images, which possess a spatial resolution of 60 m. To ensure compatibility with other timeframes, these images were resampled from 60 m to 30 m using the cubic convolution resampling technique. It is important to note that this resampling does not enhance the original spatial resolution of Landsat MSS data. In contrast, all other Landsat datasets consisted of TM or OLI images, boasting a spatial resolution of 30 meters. Since the launch of Landsat 4 in 1982, satellite data have been consistently captured at a uniform spatial resolution of 30 m per pixel, along with similar spectral bands. This uniformity facilitates a comprehensive multi-decadal analysis of both land cover and land use.

Training points were selected manually in the Landsat composite for each epoch by dividing the scene into ROI. ROI were selected based on the likelihood of mangrove presence or absence based on historical data. This method of ROI selection has been demonstrated to simplify image classification algorithms by reducing variability within the image [1,3].

The distinctive signature of mangrove forests within satellite data facilitated the identification of training points. The number of training points within each ROI differed, and this was an interactive procedure. Training samples were subject to modifications (additions or removals) based on a visual inspection of the classification outcomes. This adjustment process continued until a satisfactory classification outcome was achieved. In addition to the Landsat composite, previously published mangrove maps and very high-resolution satellite data of the area were utilized as points of reference.

Three major classes were selected for the classification: mangrove, non-mangrove, and water bodies. Brief descriptions of these classes are described in Table 1.

Table 1. Classification criteria.

Land Cover Types	Description
Mangrove	True mangroves that grow in brackish and saline water. In the United States, three main species are found: red, black, and white mangroves
Non-mangrove	All land use/land cover classes other than mangrove and water bodies including cropland, urban areas, barren land, shrubland, and grassland
Water bodies	Ocean water, brackish water, perennial river, streams and water reservoirs, and open water like lakes and ponds.

The classification was performed within each ROI due to unsatisfactory classification outcomes when attempting to classify expansive regions encompassing the entire image. As mentioned earlier, by employing ROIs, the heterogeneity within the Landsat image was mitigated, leading to improved classification results.

We used the JavaScript version of the code available from the GEE platform (https://code.earthengine.google.com/ for image classification, accessed on 16 June 2023). The GEE platform inherits several supervised classification algorithms that can be employed for image classification. The GEE and other cloud computing platforms are becoming standard for land-cover mapping and monitoring [22,23]. We used the pixel-based random forest (RF) classification algorithm. The RF is a tree-based classifier that includes K-decision trees [24]. It is a combination of multiple decision trees and is widely referred to as an ensemble learning technique. This classifier is a well-known and powerful classification model used to model complex non-linear processes with a lesser computational burden. The algorithm overcomes the problem of overfitting by constructing an ensemble of decision trees. Numerous studies have highlighted the accuracy and superior performance of RF classifiers in the realm of land-cover classification [23–26].

During image classification, all bands except the thermal infrared band were used. The thermal band was not used primarily due to the coarse spatial resolution (120 m).

After classification, all of the ROIs were mosaicked for Texas, Louisiana, and Florida in ERDAS Imagine, where these three were mosaicked to produce seamless coverage for the conterminous United States.

Mangrove forest cover change statistics were generated in each 0.5° latitudinal range on the east and west coasts of Florida except for the southernmost area where only one 0.5° latitudinal range was present. Texas and Louisiana were also delineated by these 0.5° latitudinal ranges and change statistics were calculated. (Figure 2). The goal of this analysis was to identify the distribution and dynamics of mangrove forests in each latitudinal zone.

Post-classification change detection was performed for 1980–2020, as well at every five-year interval from 1980 to 2020. An overlay method was used to compute from-to change detection. In this process, independently generated classifications at two time intervals are compared. The main advantage is to limit issues associated with radiometric calibration (i.e., solar illumination, atmospheric absorption and scattering, and sensor properties) in between intervals. However, the accuracy of the initial classification of each era dictated the accuracy of the post-classification comparison. In effect, the end accuracy approximates the compounding of the accuracies of every single classification.

Yet another benefit of post-classification change detection is its ability to reveal the characteristics and extent of alterations in land cover that have occurred over a specific period. The from-to-change analysis proved valuable in offering in-depth insights into the specific types of land-cover changes. In contrast, other researchers' have used spectral change analysis techniques [27–29].

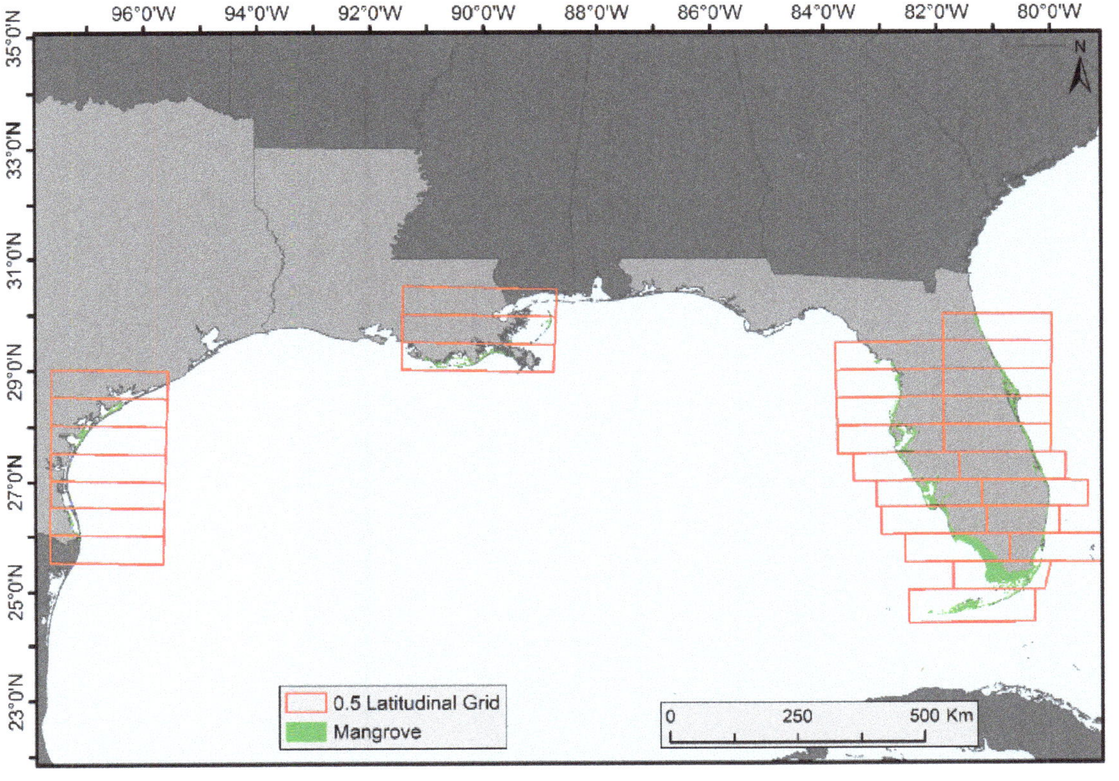

Figure 2. The 0.5° latitudinal zones in Florida, Louisiana, and Texas created for mangrove forest distribution and change analysis.

We generated a matrix reflecting shifts between different land cover classes. This was carried out at five-year intervals spanning from 1980 to 2020. The selection of a five-year time frame was driven by several factors: (i) the limited availability of cloud-free, high-quality images in the region, particularly during the early years of the Landsat archive, and (ii) the need to establish a consistent time series for the entire geographic area.

Specifically for the mangrove forests class, a comprehensive examination was conducted to ascertain the spatial extent and variations in area, including gains, losses, persisting regions, and transitions between different land-cover classes. Through cross-tabulation, an understanding of the fundamental change processes between mangrove and non-mangrove areas, as well as water bodies, was achieved.

We performed the validation in three steps: (i) comparison of our results visually with existing mangrove and land-cover maps, (ii) qualitative assessment, and (iii) quantitative validation. Published maps such as the National Oceanic and Atmospheric Administration (NOAA)'s Coastal Change Analysis Program (CCAP) and National Land Cover Dataset of the USGS were used for the first step. For the qualitative assessment, all of the classification maps were divided into a regularly spaced grid and checked visually with the goal to minimize gross errors during the image interpretation. For the quantitative validation of 2020, 100 stratified randomly sampled points were generated in ERDAS Imagine with a minimum of 25 points for each class. These points were compared with aerial photographs available in Microsoft Bing as reference data. Accuracy assessment measures (i.e., overall accuracy, producers' accuracy and users' accuracy) were computed from the classification-reference comparison.

3. Results and Discussion

We created a comprehensive database documenting the distribution and changes in mangrove coverage across the CONUS every five years spanning from 1980 to 2020. This endeavor involved utilizing uniform data sources, a standardized classification system, and a consistent methodology to carry out the analysis.

3.1. Mangrove Distribution

Figure 3 illustrates the geographical arrangement of mangrove forests within the contiguous United States (CONUS) during the year 2020. The overall expanse of mangrove coverage in the CONUS for 2020 encompassed 266,179 hectares, with the distribution being 98.0% in Florida, 0.6% in Louisiana, and 1.4% in Texas.

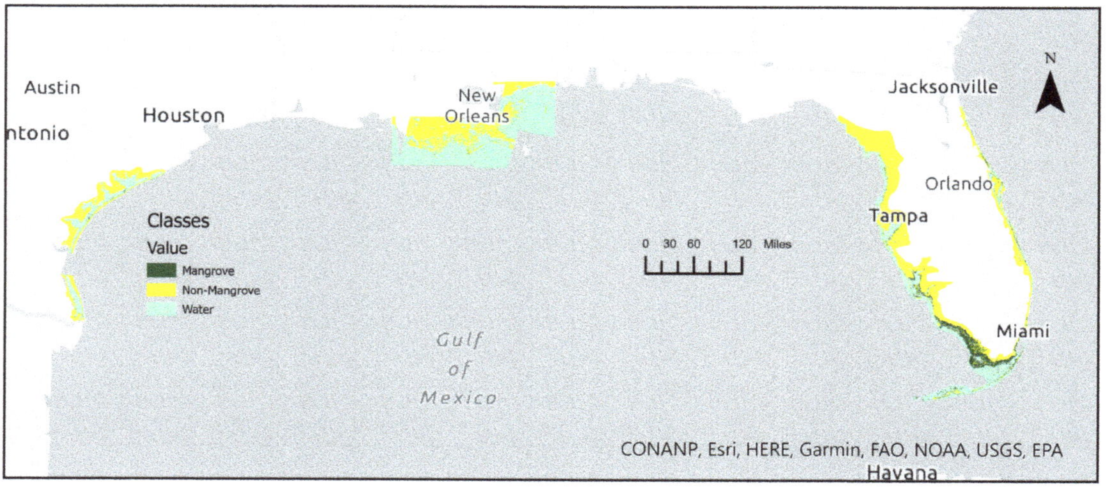

Figure 3. Spatial distribution of mangrove forests of the continental United States in 2020.

We also examined the distribution of mangrove forests along latitudinal gradients within the CONUS. In this analysis, statistics pertaining to changes in mangrove forest coverage were computed for each 0.5° latitudinal interval along the eastern and western coasts of Florida. However, it is worth noting that separation on the eastern coast was not feasible in the southernmost regions. Similar analyses were conducted for Texas and Louisiana, following the methodology outlined by Giri and Long (2016) [17]. Consistent with expectations, there was a general decline in the occurrence of mangroves as latitude increased from south to north. Notably, 85% of the total mangrove area in the CONUS was concentrated within the latitude range of 24.5° to 26.0°.

Alterations in the arrangement and prevalence of mangrove species, whether inside or beyond their traditional geographical boundaries, can yield significant repercussions for the array of ecological benefits and services they offer. In the CONUS, there is a prevailing notion that mangroves are advancing towards higher latitudes (northward) due to a reduction in the frequency and intensity of extreme cold events in recent years. Concurrently, the rise in sea levels frequently emerges as a contributing factor to the inland progression of mangroves on a local scale. Extensive documentation supports the idea that elevated temperatures expedite the expansion of mangroves and the accumulation of surface elevation in subtropical wetlands [30,31].

Employing four decades of satellite imagery and on-site observations focused on the CONUS, our findings indicate the following:

(i) The notion of a northward expansion of mangrove forests remains uncertain;

(ii) Simultaneously, a progressive expansion towards inland areas, within the established historical northern boundary, is presently underway.

Our findings illustrate that the northernmost latitudinal boundaries of mangroves along the east coast of Florida (81.317299°W, 29.945414°N) and the west coast of Florida (83.046396°W, 29.162046°N), as well as in Louisiana (88.860357°W, 30.038007°N) and Texas (96.410255°W, 28.428913°N), have not exhibited a consistent poleward expansion between 1980 and 2020 (as indicated in Table 2). Nevertheless, it is worth noting that the historical northern limit prior to the 1980s lacks comprehensive surveying or documentation, and the mapping of individual mangrove recruits was hindered by the lack of 30-meter resolution of Landsat satellite data.

Table 2. Geographic coordinates of the northernmost extent of mangrove distribution in eastern Florida, western Florida, Louisiana, and Texas.

State	Latitude (in Decimal Degrees)		Longitude (in Decimal Degrees)	
	1980	2020	1980	2020
Eastern Florida	29.86373	29.94541	−81.30328	−81.31730
Western Florida	29.16205	29.16232	−83.04640	−83.046480
Louisiana	30.03801	29.97985	−88.86036	−88.83519
Texas	28.42891	28.43685	−96.41026	−96.40120

3.2. Mangrove Change

In the aggregate, mangrove forests in the CONUS exhibited a notable growth of 13.5% spanning from 1980 to 2020. The rise in mangrove coverage between 1980 and 2015 accounted for 4.3%, implying that the alterations from 2015 to 2020 surpassed those of the preceding 35 years. However, there were substantial oscillations in quinquennial (i.e., "every five years") aerial expansion, with the lowest mangrove coverage recorded in 1990 (172,300 hectares) following a severe winter sub-freeze incident in late 1989, while the zenith was observed in 2020 (266,179 hectares) (depicted in Figure 4). The impact of winter freezes on mangroves can be profound. For example, during the winter freeze of 2017/2018, Louisiana experienced degradation in approximately 90% of its mangrove population [30].

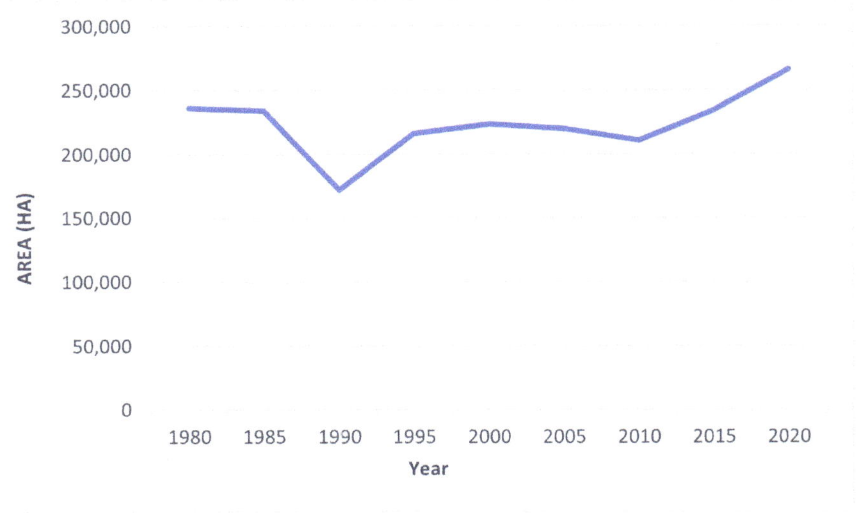

Figure 4. Mangrove forest areas every five years from 1980 to 2020 in the CONUS.

Figure 5 depicts the acreage of mangrove forest regions that remained unaltered, underwent deforestation, and experienced afforestation at five-year intervals spanning from 1980 to 2020. During each of these periods, more than 85% of the mangrove areas displayed constancy, except for the time frame between 1985 and 1990, where only 66% of the forests remained unchanged. Notably, the alteration patterns of both deforestation and afforestation areas varied across each era. Deforestation is characterized by the shift from mangrove to non-mangrove or from mangrove to water, while afforestation pertains to the transformation from non-mangrove to mangrove or from water to mangrove.

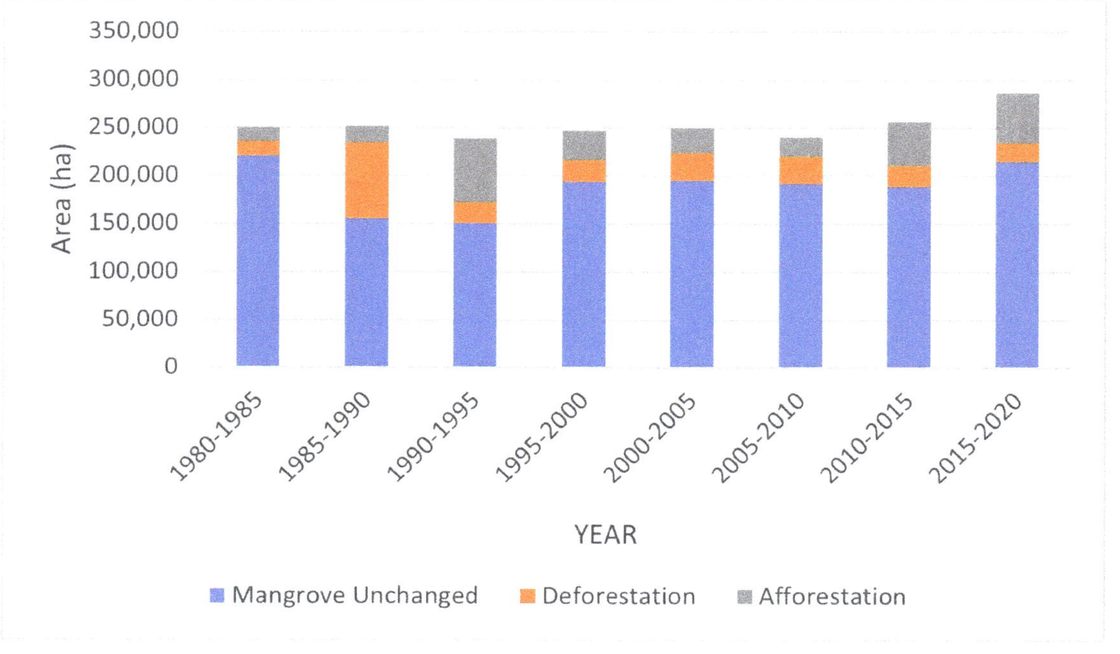

Figure 5. Comparison of mangrove areas unchanged, deforestation, and deforestation every five years from 1980–2020.

The observed changes, encompassing both deforestation and afforestation, exhibited irregularity along the latitudinal grid. Both mangrove expansion and reduction transpired across all latitudinal segments. Nevertheless, the alterations were more conspicuous in certain regions compared to others, as depicted in Figure 6.

3.3. Proximate Causes of Mangrove Change

Various factors, including but not limited to winter freeze, storms, land-use practices, hydrology, erosion, impoundment, sea level rise, and human plantation efforts, frequently contribute to alterations in mangrove forests (as illustrated in Figure 7). The influence of these factors is contingent upon the geographical areas in question (15), often with multiple factors concurrently shaping the observed changes.

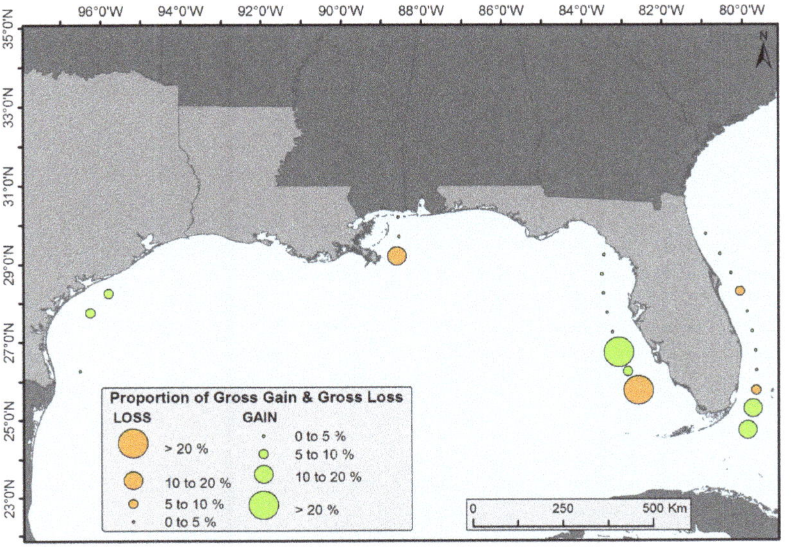

Figure 6. Proportion of total mangrove forest cover change per 0.5° latitudinal intervals during 1980 to 2020.

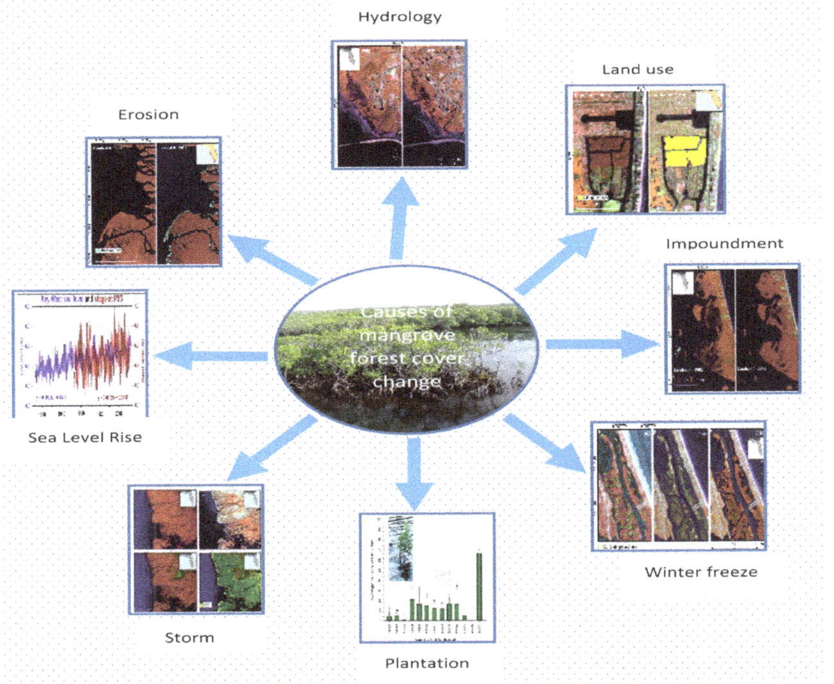

Figure 7. Proximate causes of mangrove forest cover change in the CONUS.

The role of winter freeze is pivotal in influencing the dynamics between mangrove and salt marsh ecosystems, particularly at the northern boundary of mangrove distribution in the continental United States. This phenomenon is notably prominent along the eastern coast of Florida and Louisiana, and to some extent, also on the western coast of Florida and Texas. Instances of significant winter freezes in 1983, 1989, and 2010 have led to the demise of mangrove forests, particularly those dominated by black mangroves (*Avicennia germinans* (L.)) in much of the northern boundaries. In these areas, the mangrove presence has been supplanted by salt marshes, primarily characterized by the dominance of smooth cordgrass (*Spartina alterniflora* Loisel.). However, following a period of mild winters spanning a decade or more, mangroves have once again reclaimed these regions. This intricate interaction between mangroves and Spartina grasses has conceivably been unfolding within the CONUS over an extended timeframe. Our findings, however, have specifically documented this phenomenon after the year 1980.

The impact caused by the significant winter freezes in 1983, 1989, and 2010 exhibited variability in terms of damage. Nonetheless, the affected mangrove areas managed to recuperate, surpassing 90% of their initial levels within a span of 5 to 10 years. Figure 8 illustrates this finding after the substantial damage caused during the winter freeze of 1983 in Cedar Keys, Florida.

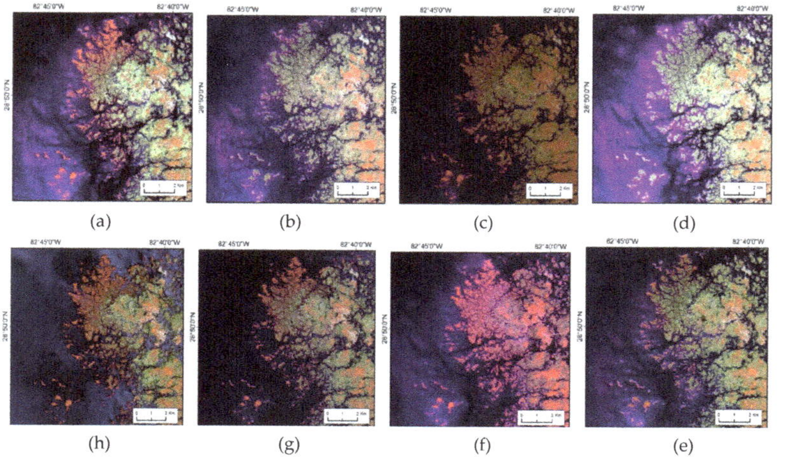

Figure 8. Mangrove damage due to severe winter freeze and subsequent recovery in Cedar Keys, Florida. The image is displaced in Landsat False Color Composite; dark red is mangrove forests, blue is water bodies, and other colors represent non-mangrove and non-water. (**a**) Landsat 1982, (**b**) Landsat 1984, (**c**) Landsat 1988, (**d**) Landsat 1990, (**e**) Landsat 1995, (**f**) Landsat 2008, (**g**) Landsat 2010, (**h**) Landsat 2014.

Mangrove forests within the CONUS are also frequently subjected to the impact of tropical storms and hurricanes. An illustrative instance is Hurricane Wilma in 2005, which inflicted significant harm upon a substantial portion of the mangrove forests along the Shark River within the Everglades National Park (Figure 9). The effects of tropical cyclones on mangrove forests span a spectrum from transient defoliation to extensive tree mortality. The passage of the hurricane of 2005 led to diverse outcomes: some trees lost their canopies, others endured breakage, uprooting, leaf detachment, and trunk fractures. However, despite the damage, mangrove regeneration managed to restore much of the impacted areas by 2010, except for certain coastal fringes, as indicated in Figure 9.

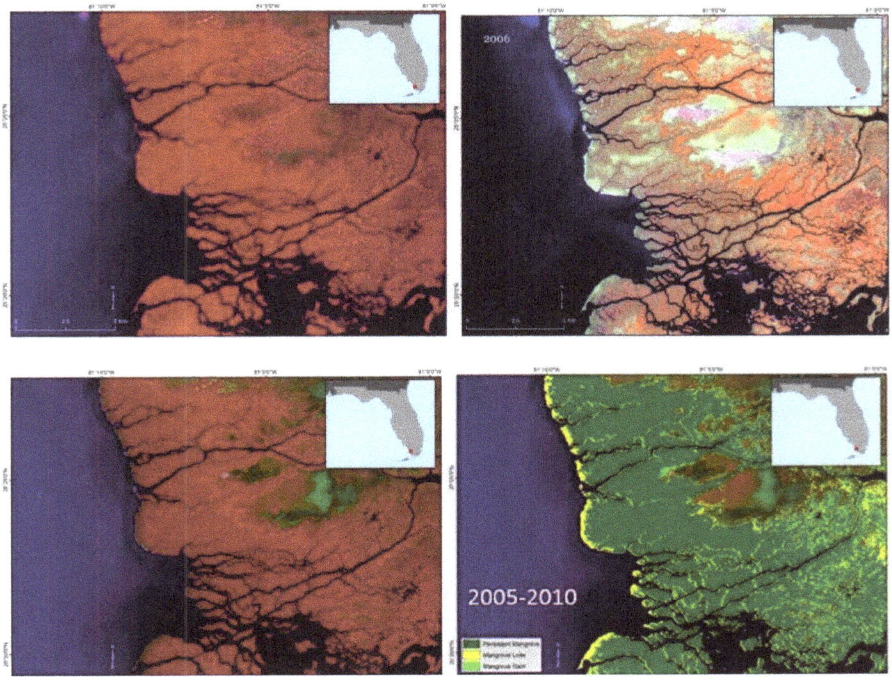

Figure 9. Mangrove damage just before hurricane Wilma in the winter of 2005 (**upper left**), extensive damage after Hurricane Wilma (**upper right**), and recovery in 2010 (**lower left**) as seen from the Landsat images. Landsat images are shown in False Color Composite, and dark red roughly corresponds to healthy mangrove vegetation. Much of the mangrove forest recovered except along the coastal fringes as seen in the change map from 2005 to 2010 (**lower right**).

Mangrove forests in the CONUS have also been impacted by other hurricanes, including Hurricane Andrew in 1992, Wilma in 2005, and Irma in 2017. However, the effects of the 1992 and 2017 hurricanes were not discernible in our study because much of the affected forest had already regenerated by the year of our observations, which were 1995 and 2020, respectively. This observation underscores the importance of conducting annual analyses of mangrove cover changes to effectively capture such dynamic shifts. Additionally, the damage caused by the 2005 hurricane is not evident in our study because it occurred later in the year and was not captured in the Landsat mosaic of 2005. These findings further highlight the potential need for future monthly monitoring of mangrove forests to comprehensively track their changes.

Other previous investigations have revealed that the subsequent recuperation from storm-induced damage varies, influenced by both the intensity of the storm and the underlying soil conditions [27–29]. Notably, certain storms deposit sediment rich in phosphorus, which in turn stimulates the growth of mangroves [29]. For instance, in the aftermath of Hurricane Irma, approximately 83% of the damaged mangrove forests exhibited recovery within the first year [30]. However, in some locations, the initial damage is compounded by delayed mortality, which can hinder mangrove recovery for extended periods following the storm. An illustrative case is the aftermath of Hurricane Irma, which incited one of the most substantial recorded instances of mangrove dieback in the region. Within the initial 15 months following Irma, a staggering 10,760 hectares of mangroves displayed signs of complete dieback, characterized by minimal or no regrowth [29–31].

Impoundments constructed during the period of 1940 to 1960 disrupted tidal circulation, leading to prolonged flooding, and resulting in a 76% decline in mangrove forests within the Indian River Lagoon (IRL) [25]. In response, rotational impoundment management (RIM) techniques were developed and introduced during the 1980s. These techniques enabled the natural cyclical flow of tidal waters into the impounded mangrove forests, on a rotational basis. This management intervention helped to recover the mangrove forests in this region. Similarly, during the 1950s and 1960s, predominantly along the central east coast of Florida, the common practice involved dredging estuarine sediments and depositing them onto coastal wetlands. This practice led to significant modifications in mangrove distribution. In some extensively ditched regions, up to 80% of mangrove forests were substituted with ditches and spoil piles. To counteract these impacts, the rehabilitation of coastal wetlands affected by dragline ditches was initiated in 1999. Since then, mangrove recovery has been observed in these areas.

Within the Ten Thousand Islands region, a notable occurrence of mangrove migration towards inland areas was observed. Thorough examination conducted by this study, along with research conducted by Krauss et al. (2011) [32], underscored that this transformation results from a combination of factors. These include the interplay of relative sea level rise, the establishment of waterways, localized precipitation patterns, and decreased freshwater discharge. The creation of waterways played a pivotal role in facilitating the dispersion of propagules necessary for the process of natural regeneration.

Likewise, a multifaceted and divergent pattern of mangrove alteration was evident in three neighboring sites within the Everglades [15]. Through their analysis of aerial photographs and images spanning from 1928 to 2004, Smith et al. (2013) [15] determined that fire contributed to the expansion of mangroves into salt marshes in one of the sites. However, in another site affected by fire, no discernible change was observed. In the third site, mangrove expansion was noted even in the absence of any notable fire incidents [15].

Significant transformations took place in Louisiana, particularly within the northernmost boundary spanning from 29° to 30.5° latitude. Giri et al. [17] demonstrated that a severe freeze occurring in late December of 1983 which led to a drastic reduction in Louisiana's mangrove coverage by approximately 90%. Subsequent harm from repeated sub-freezing temperatures in 1985, 1989, and 2010 impeded the recovery of the mangrove extent until the late 1990s and the following decades. Despite experiencing less frequent winter freeze events in the past ten years, Louisiana's mangrove coverage has not regained its historical extent from 1983. This is likely attributed to accelerated coastal retreat coupled with coastal subsidence, along with rapid sea level rise at a rate of 9.03 mm/year. Notably, Louisiana boasts the highest rates of relative sea level rise within the Gulf of Mexico.

Figure 10 illustrates the impact of a winter freeze on mangrove mortality and the subsequent hindered recovery caused by subsidence. By 2010, nearly all of Timbalier Island in southern Louisiana had become submerged, potentially attributed to rising sea levels.

In more recent times, there has been a noticeable emergence of red mangroves in previously unoccupied regions within Texas. Over the years, the expansion of mangrove forests in Texas has been propelled by relatively mild winters and potentially by rising sea levels. Within the timeframe of 1980 to 2015, the mangrove-covered area escalated by 2259 hectares, reflecting a remarkable 234% increase. An examination of changes indicated that the growth of mangroves predominantly occurred at the cost of salt marshes. A significant mangrove expansion was particularly notable between latitudes 27°30′ and 28°30′. In these specific zones, the expansion of mangroves primarily encroached upon existing marsh or upland areas. Within the last 15 years, new stands of mangroves emerged in the Corpus Christi Pass region.

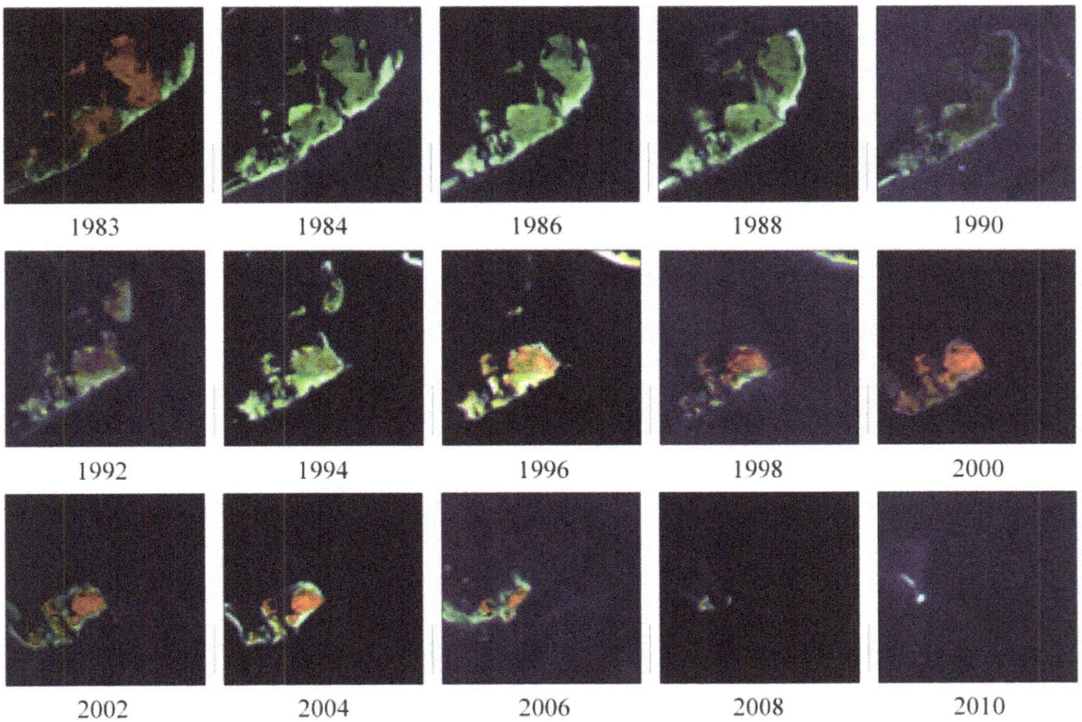

Figure 10. Extensive damage of mangrove forest due to the winter freeze of 1983 and subsequent recovery and near disappearance of the Timbalier Island, Louisiana.

3.4. Results Validation

While validating mangrove cover products using rigorous sampling methods and high-quality concurrent reference data is unquestionably preferable, resource constraints frequently hinder the attainment of fully comprehensive quantitative validation [33]. Nevertheless, we undertook a three-step validation process for our results: (i) visual comparison of our outcomes with existing mangrove maps, (ii) qualitative assessment, and (iii) quantitative validation.

In the initial method, we assessed the reliability of our database by comparing it with other available mangrove datasets covering both the CONUS and local areas. For the second approach, spanning all years, we partitioned the entire area into systematically distributed grids. Subsequently, we conducted visual inspections of each grid to detect and rectify any noticeable errors within the classified maps. Gross errors identified were corrected. These two processes not only contributed to a qualitative evaluation of the map but also enhanced the overall classification quality, while preserving consistency across the years.

During the assessment for the year 2020, we attained an overall accuracy of 95%. The producer accuracy and user accuracy for mangrove were 100% and 92%, respectively. The error matrix is provided in Table 3 below.

Table 3. Error matrix for 2020 classification.

Classified Data	Reference Data			Producers Accuracy	Users Accuracy
	Mangrove	Non-mangrove	Water		
Mangrove	23	1	1	100.00	92.00
Non-mangrove	0	35	1	92.11	97.22
Water	1	2	37	95.87	94.87
	Overall Accuracy = 95%				

4. Limitations

A significant drawback of the current study is the absence of high-quality Landsat satellite data for the intended target year. This limitation is especially pertinent in the case of earlier years, notably 1980 and 1985. This is because Landsat series, such as Landsat 4 and 5, lacked comprehensive global coverage for the desired target date. Consequently, the creation of a mosaic devoid of cloud cover necessitated the integration of data spanning multiple years. Additionally, a mechanical malfunction in the Scan-Line Corrector (SLC-Off) on the Landsat-7 satellite, causing a data loss ranging from 22% to 25% with each captured image, further compounded the difficulties in preparing a mosaic for the target date.

The availability of consistent, error-free, and cohesive composite images becomes vital for the consistent assessment and monitoring of mangrove forests over time. The five-year composites spanning from 1980 to 2020 demonstrate a remarkable seamlessness, unaffected by factors like atmospheric interference and sensor anomalies, such as the SLC-Off issue with Landsat-7, as well as gaps in pixel data. However, some of the changes were not captured on those mosaics.

Landsat satellite data with a spatial resolution of 30 meters did not facilitate the mapping of smaller mangrove patches. Mapping these smaller patches might require the utilization of very high-resolution satellite data, with spatial resolutions ranging from 1 to 5 m. It is important to note that larger and more uniform mangrove areas were mapped with greater accuracy compared to smaller, more complex, and diverse landscapes.

During the change analysis, we conducted a comparison of two classified thematic maps, which introduced errors into the change analysis process. This outcome reflects the fact that land-use/land-cover maps derived from remotely-sensed data inherently contain errors. The presence of cloud-free satellite data, pre-processing errors, and classification errors all contribute to limitations in post-classification change analysis.

5. Conclusions

We have compiled a comprehensive database detailing the distribution of mangroves across the entire CONUS at five-year intervals spanning from 1980 to 2020. This database was created utilizing Landsat satellite data with a spatial resolution of 30 m. Our findings unveiled that in the year 2020, the total extent of mangrove forests in the CONUS encompassed 266,179 hectares, with Florida accounting for 98.0% of the mangrove area, Louisiana contributing 0.6%, and Texas comprising 1.4%. The analysis of latitudinal distribution revealed a consistent trend of decreasing mangrove occurrence with the progression from south to north. Furthermore, our results indicated that approximately 85% of the total mangrove area within the CONUS was concentrated within the latitude range of 24.5° to 26.0° [17].

After conducting post-classification change analysis, it became evident that the overall mangrove area across the CONUS has experienced a 13.5% expansion from 1980 to 2020. Additionally, our findings indicate that the northernmost latitudinal boundaries of mangrove forests along the eastern and western coastlines of Florida, as well as in Louisiana and Texas, have not exhibited a systematic poleward expansion. While our analysis spanning

40 years has offered valuable insights into the contemporary drivers behind the range fluctuations, it is essential to acknowledge that this timeframe is relatively short. As a result, the historical dynamics of mangroves before the satellite era remain unknown.

The most notable increases in mangrove area from 1980 to the present were observed in the southern regions of Florida and Texas, well within their respective northern boundaries. The underlying causes of these changes were found to be location-specific rather than universal. A multitude of factors, including but not limited to sea level rise, the presence or absence of sub-freezing temperatures, alterations in land use, impoundment and dredging activities, shifts in hydrology, occurrences of fire, storms, sedimentation, erosion, and deliberate mangrove planting, collectively contribute to the observed changes. Particularly, the relatively milder winters and the reduced occurrence of sub-freezing temperatures in recent decades appear to facilitate local expansion.

Our findings underscore the intricate interplay of various forces influencing the northern limits of mangrove distribution. They underscore the necessity for sustained, long-term monitoring of this system's dynamics, especially as its significance grows in the context of adapting to rising sea levels and mitigating the impacts of heightened atmospheric CO_2. The compiled mangrove database could serve as a valuable tool for monitoring the potential impacts of climate change on mangrove forests in the future.

Author Contributions: Conceptualization C.G., methodology C.G. and J.L., software, validation and formal analysis J.L. and P.P., writing-original draft preparation and review and editing C.G. and P.P. All authors have read and agreed to the published version of the manuscript.

Funding: This research was funded by the U.S. Environmental Protection Agency through its Office of Research and Development (ORD) as a contribution to ORD's Safe and Sustainable Water Resources National Research Program. The APC was funded by the ORD.

Data Availability Statement: All the datasets generated from this study are freely available from www.epa.gov (Accessed on 13 June 2023).

Acknowledgments: We would like to extend our sincere thanks to Jeremy Schroeder who helped to analyze part of the 2020 Texas classification.

Conflicts of Interest: The author declares no conflict of interest.

References

1. Giri, C.; Ochieng, E.; Tieszen, L.L.; Zhu, Z.; Singh, A.; Loveland, T.; Masek, J.; Duke, N. Status and Distribution of Mangrove Forests of the World Using Earth Observation Satellite Data. *Glob. Ecol. Biogeogr.* **2011**, *20*, 154–159. [CrossRef]
2. Zhang, K.; Thapa, B.; Ross, M.; Gann, D. Remote Sensing of Seasonal Changes and Disturbances in Mangrove Forest: A Case Study from South Florida. *Ecosphere* **2016**, *7*, e01366. [CrossRef]
3. Giri, C. Frontiers in Global Mangrove Forest Monitoring. *Remote Sens.* **2023**, *15*, 3852. [CrossRef]
4. Vo, Q.T.; Künzer, C.; Vo, Q.M.; Moder, F.; Oppelt, N. Review of Valuation Methods for Mangrove Ecosystem Services. *Ecol. Indic.* **2012**, *23*, 431–446. [CrossRef]
5. Afonso, F.; Félix, P.M.; Chainho, P.; Heumüller, J.A.; de Lima, R.F.; Ribeiro, F.; Brito, A.C. Community Perceptions about Mangrove Ecosystem Services and Threats. *Reg. Stud. Mar. Sci.* **2022**, *49*, 102114. [CrossRef]
6. Bardou, R.; Osland, M.J.; Scyphers, S.; Shepard, C.; Aerni, K.E.; Alemu, I.J.B.; Crimian, R.; Day, R.H.; Enwright, N.M.; Feher, L.C. Rapidly Changing Range Limits in a Warming World: Critical Data Limitations and Knowledge Gaps for Advancing Understanding of Mangrove Range Dynamics in the Southeastern USA. *Estuaries Coasts* **2023**, *46*, 1123–1140. [CrossRef]
7. Parkinson, R.W.; Wdowinski, S. Accelerating Sea-Level Rise and the Fate of Mangrove Plant Communities in South Florida, USA. *Geomorphology* **2022**, *412*, 108329. [CrossRef]
8. Osland, M.J.; Hughes, A.R.; Armitage, A.R.; Scyphers, S.B.; Cebrian, J.; Swinea, S.H.; Shepard, C.C.; Allen, M.S.; Feher, L.C.; Nelson, J.A. The Impacts of Mangrove Range Expansion on Wetland Ecosystem Services in the Southeastern United States: Current Understanding, Knowledge Gaps, and Emerging Research Needs. *Glob. Chang. Biol.* **2022**, *28*, 3163–3187. [CrossRef]
9. Gilman, E.L.; Ellison, J.; Duke, N.C.; Field, C. Threats to Mangroves from Climate Change and Adaptation Options: A Review. *Aquat. Bot.* **2008**, *89*, 237–250. [CrossRef]
10. Romañach, S.S.; DeAngelis, D.L.; Koh, H.L.; Li, Y.; Teh, S.Y.; Barizan, R.S.R.; Zhai, L. Conservation and Restoration of Mangroves: Global Status, Perspectives, and Prognosis. *Ocean Coast. Manag.* **2018**, *154*, 72–82. [CrossRef]
11. Duke, N.C.; Meynecke, J.-O.; Dittmann, S.; Ellison, A.M.; Anger, K.; Berger, U.; Cannicci, S.; Diele, K.; Ewel, K.C.; Field, C.D. A World without Mangroves? *Science* **2007**, *317*, 41–42. [CrossRef] [PubMed]

12. Cavanaugh, K.C.; Kellner, J.R.; Forde, A.J.; Gruner, D.S.; Parker, J.D.; Rodriguez, W.; Feller, I.C. Poleward Expansion of Mangroves Is a Threshold Response to Decreased Frequency of Extreme Cold Events. *Proc. Natl. Acad. Sci. USA* **2014**, *111*, 723–727. [CrossRef]
13. Alongi, D.M. The Impact of Climate Change on Mangrove Forests. *Curr. Clim. Chang. Rep.* **2015**, *1*, 30–39. [CrossRef]
14. Godoy, M.D.; Lacerda, L.D.D. Mangroves Response to Climate Change: A Review of Recent Findings on Mangrove Extension and Distribution. *An. Acad. Bras. Ciênc.* **2015**, *87*, 651–667. [CrossRef] [PubMed]
15. Giri, C.; Long, J. Is the Geographic Range of Mangrove Forests in the Conterminous United States Really Expanding? *Sensors* **2016**, *16*, 2010. [CrossRef] [PubMed]
16. Smith, T.J.; Foster, A.M.; Tiling-Range, G.; Jones, J.W. Dynamics of Mangrove-Marsh Ecotones in Subtropical Coastal Wetlands: Fire, Sea-Level Rise, and Water Levels. *Fire Ecol.* **2013**, *9*, 66–77. [CrossRef]
17. Ansley, R.J.; Rivera-Monroy, V.H.; Griffis-Kyle, K.; Hoagland, B.; Emert, A.; Fagin, T.; Loss, S.R.; McCarthy, H.R.; Smith, N.G.; Waring, E.F. Assessing Impacts of Climate Change on Selected Foundation Species and Ecosystem Services in the South-Central USA. *Ecosphere* **2023**, *14*, e4412. [CrossRef]
18. Chander, G.; Markham, B.L.; Helder, D.L. Summary of Current Radiometric Calibration Coefficients for Landsat MSS, TM, ETM+, and EO-1 ALI Sensors. *Remote Sens. Environ.* **2009**, *113*, 893–903. [CrossRef]
19. Wulder, M.A.; White, J.C.; Loveland, T.R.; Woodcock, C.E.; Belward, A.S.; Cohen, W.B.; Fosnight, E.A.; Shaw, J.; Masek, J.G.; Roy, D.P. The Global Landsat Archive: Status, Consolidation, and Direction. *Remote Sens. Environ.* **2016**, *185*, 271–283. [CrossRef]
20. Roy, D.P.; Ju, J.; Kline, K.; Scaramuzza, P.L.; Kovalskyy, V.; Hansen, M.; Loveland, T.R.; Vermote, E.; Zhang, C. Web-Enabled Landsat Data (WELD): Landsat ETM+ Composited Mosaics of the Conterminous United States. *Remote Sens. Environ.* **2010**, *114*, 35–49. [CrossRef]
21. Qiu, S.; Zhu, Z.; Olofsson, P.; Woodcock, C.E.; Jin, S. Evaluation of Landsat Image Compositing Algorithms. *Remote Sens. Environ.* **2023**, *285*, 113375. [CrossRef]
22. Tamiminia, H.; Salehi, B.; Mahdianpari, M.; Quackenbush, L.; Adeli, S.; Brisco, B. Google Earth Engine for Geo-Big Data Applications: A Meta-Analysis and Systematic Review. *ISPRS J. Photogramm. Remote Sens.* **2020**, *164*, 152–170. [CrossRef]
23. Phan, T.N.; Kuch, V.; Lehnert, L.W. Land Cover Classification Using Google Earth Engine and Random Forest Classifier—The Role of Image Composition. *Remote Sens.* **2020**, *12*, 2411. [CrossRef]
24. Belgiu, M.; Drăguţ, L. Random Forest in Remote Sensing: A Review of Applications and Future Directions. *ISPRS J. Photogramm. Remote Sens.* **2016**, *114*, 24–31. [CrossRef]
25. Elmahdy, S.I.; Ali, T.A.; Mohamed, M.M.; Howari, F.M.; Abouleish, M.; Simonet, D. Spatiotemporal Mapping and Monitoring of Mangrove Forests Changes From 1990 to 2019 in the Northern Emirates, UAE Using Random Forest, Kernel Logistic Regression and Naive Bayes Tree Models. *Front. Environ. Sci.* **2020**, *8*, 102. [CrossRef]
26. Ghorbanian, A.; Zaghian, S.; Asiyabi, R.M.; Amani, M.; Mohammadzadeh, A.; Jamali, S. Mangrove Ecosystem Mapping Using Sentinel-1 and Sentinel-2 Satellite Images and Random Forest Algorithm in Google Earth Engine. *Remote Sens.* **2021**, *13*, 2565. [CrossRef]
27. Lagomasino, D.; Fatoyinbo, T.; Castañeda-Moya, E.; Cook, B.D.; Montesano, P.M.; Neigh, C.S.; Corp, L.A.; Ott, L.E.; Chavez, S.; Morton, D.C. Storm Surge and Ponding Explain Mangrove Dieback in Southwest Florida Following Hurricane Irma. *Nat. Commun.* **2021**, *12*, 4003. [CrossRef]
28. Xiong, L.; Lagomasino, D.; Charles, S.P.; Castañeda-Moya, E.; Cook, B.D.; Redwine, J.; Fatoyinbo, L. Quantifying Mangrove Canopy Regrowth and Recovery after Hurricane Irma with Large-Scale Repeat Airborne Lidar in the Florida Everglades. *Int. J. Appl. Earth Obs. Geoinf.* **2022**, *114*, 103031. [CrossRef]
29. Lagomasino, D.; Fatoyinbo, L.; Castaneda, E.; Cook, B.; Montesano, P.; Neigh, C.; Ott, L.; Chavez, S.; Morton, D. Storm Surge, Not Wind, Caused Mangrove Dieback in Southwest Florida Following Hurricane Irma. *Earth ArXiv* **2020**. *preprint*.
30. Coldren, G.A.; Langley, J.A.; Feller, I.C.; Chapman, S.K. Warming Accelerates Mangrove Expansion and Surface Elevation Gain in a Subtropical Wetland. *J. Ecol.* **2019**, *107*, 79–90. [CrossRef]
31. Cohen, M.C.L.; De Souza, A.V.; Liu, K.; Rodrigues, E.; Yao, Q.; Ryu, J.; Dietz, M.; Pessenda, L.C.R.; Rossetti, D. Effects of the 2017–2018 Winter Freeze on the Northern Limit of the American Mangroves, Mississippi River Delta Plain. *Geomorphology* **2021**, *394*, 107968. [CrossRef]
32. Krauss, K.W.; From, A.S.; Doyle, T.W.; Doyle, T.J.; Barry, M.J. Sea-Level Rise and Landscape Change Influence Mangrove Encroachment onto Marsh in the Ten Thousand Islands Region of Florida, USA. *J. Coast. Conserv.* **2011**, *15*, 629–638. [CrossRef]
33. Stehman, S.V.; Olofsson, P.; Woodcock, C.E.; Herold, M.; Friedl, M.A. A Global Land-Cover Validation Data Set, II: Augmenting a Stratified Sampling Design to Estimate Accuracy by Region and Land-Cover Class. *Int. J. Remote Sens.* **2012**, *33*, 6975–6993. [CrossRef]

remote sensing

Article

Uncovering Dynamics of Global Mangrove Gains and Losses

Valeria Contessa [1,*], Karen Dyson [2,3], Pedro Pablo Vivar Mulas [1], Adolfo Kindgard [1], Tianchi Liu [2], David Saah [4], Karis Tenneson [2] and Anssi Pekkarinen [1]

1 Food and Agriculture Organization of the United Nations, 00153 Rome, Italy; pedro.vivar@fao.org (P.P.V.M.); adolfo.kindgard@fao.org (A.K.); anssi.pekkarinen@fao.org (A.P.)
2 Spatial Informatics Group, Pleasanton, CA 94566, USA; kdyson@sig-gis.com (K.D.); tliu@sig-gis.com (T.L.); ktenneson@sig-gis.com (K.T.)
3 Dendrolytics, Seattle, WA 98105, USA
4 Geospatial Analysis Lab, University of San Francisco, San Francisco, CA 94117, USA; dssaah@usfca.edu
* Correspondence: valeria.contessa@fao.org; Tel.: +39-06-57053347

Abstract: Supporting successful global mangrove conservation and policy requires accurate identification of anthropogenic and biophysical drivers of mangrove extent, yet such studies are scarce. We apply a hybrid methodology, combining existing remote sensing mangrove maps with local expert knowledge of vegetation and land use dynamics. We conducted stratified random sampling in eight subregions, and local experts visually interpreted over 20,900 plots using high-resolution imagery in Collect Earth Online. Similar to previous estimates, we found 147,771 km^2 ($\pm$1.4%) of mangroves globally in 2020 and that rates of mangrove loss have decreased from 2000–2010 to 2010–2020, largely driven by South and Southeast Asia. Anthropogenic drivers of loss have shifted across subregions, with oil palm cultivation emerging in South and Southeast Asia and aquaculture in South America and Western and Central Africa, highlighting the need for ongoing monitoring and adaptable conservation efforts. Natural expansion outpaced natural retraction in both periods. This is the first global study uncovering land use drivers of mangrove decline and recovery, only made possible by collaboration with local experts. Key breakthroughs include successfully discerning spectrally similar anthropogenic from biophysical drivers, such as aquaculture from natural retraction, and creating data collection approaches that streamline visual interpretation efforts.

Keywords: mangrove; land use change; natural expansion; aquaculture; oil palm; agriculture; remote sensing; visual interpretation

1. Introduction

Identifying and quantifying the anthropogenic and biophysical drivers of global mangrove losses and gains at a global scale is essential for monitoring mangrove ecosystems and the services they provide [1]. Identifying these change drivers helps to inform mangrove conservation and carbon mitigation efforts [2–4]. This assessment can also inform resource allocation to regions where conservation may be more effective and can identify policies and programmes that can be actively implemented to improve conservation [5]. Identifying causes of mangrove loss can address continued threats necessary to move forward toward zero net loss in global mangroves [2].

With the advancement of automatic classification algorithms, freely available satellite imagery, and the ease of cloud computation, several global studies have created maps and estimates of mangrove areas and changes. These include the work of Giri et al. (2011) [6] based on Landsat data and the more recent Global Mangrove Watch analysis using ALOS-PALSAR [6–8]. While there is now an emerging consensus that the rate of mangrove deforestation is decreasing [2,8–10], estimates on the distribution and drivers of mangrove changes are very scarce and a robust methodology to reliably classify these is needed.

Two studies using remote sensing data have attempted to map drivers of global mangrove loss [2,11]. Goldberg et al. (2020) [2] used machine learning and post-analysis classification trees relying on open data and noted the limitations of this approach, including difficulty discerning different drivers from satellite data alone and error propagation from other datasets [2]. Thomas et al. (2017) [11] used manual interpretation of radar mosaic composite images to detect changes in mangroves and used contextual information and additional imagery sources to classify them into predefined driver categories. They found their classification based on human interpretation was time-consuming and potentially subjective [11].

Understanding mangrove dynamics and supporting global mangrove conservation and policy needs requires an approach that identifies drivers at a fine thematic level and collects enough samples to accurately assess anthropogenic and biophysical drivers of global mangrove losses and gains at a global. Our study attempts to overcome previous thematic limitations by involving local experts in the interpretation of samples, providing them clear guidelines on how to separate the different classes, and relying on their knowledge of the local vegetation ecology and land use dynamics.

To improve the precision of change estimates, we used stratified random sampling and collected data from over 20,900 plots using visual interpretation of medium- and high-resolution imagery in cloud-based Collect Earth Online [12]. This sample-based approach allowed us to gain efficiencies in visual interpretation and collect high thematic resolution data. Our approach covers the years 2000–2020 and complements existing knowledge by providing additional information on specific mangrove change drivers (oil palm, aquaculture, rice cultivation, etc.).

2. Materials and Methods

2.1. Study Area

We examined the distribution of mangroves at the global scale and examined regional patterns in mangrove extent and change [13–15] (Figure 1). The global extent of mangroves was considered as there is a lack of consistent, reliable and ground-truthed data on the area of mangroves and changes over time. FAO was in a unique position to fill this gap, given existing partnerships with local field experts from countries around the world.

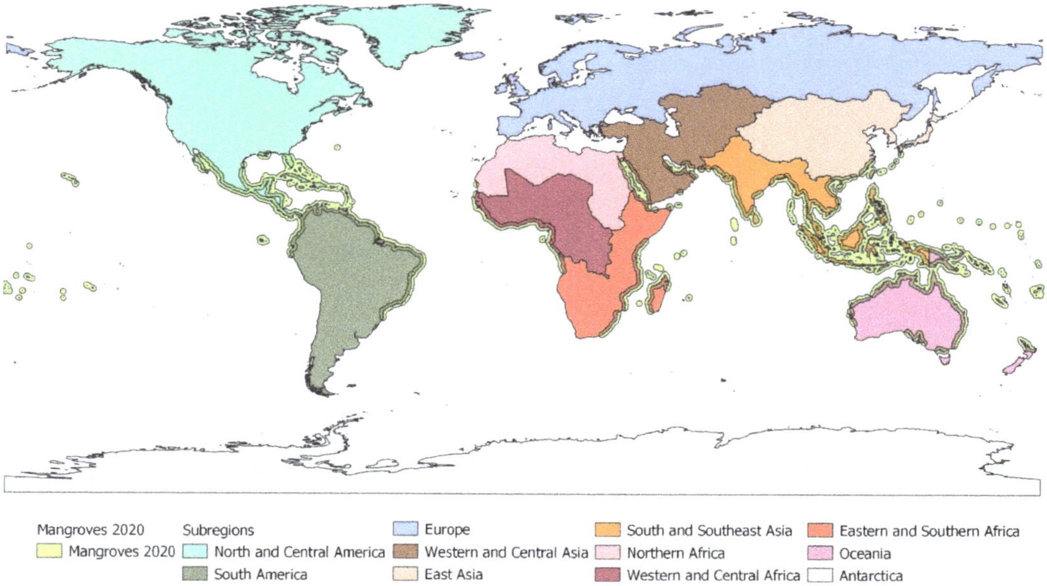

Figure 1. The global extent of mangroves. Mangrove extent from GMW v3 [8]. Subregions based on [15].

Mangroves (plant families) may be divided into three groups according to their features and morphological adaptations: major elements (strict or true mangroves, which are exclusively found in the mangrove habitat), minor elements and mangrove associates [13,14]. To define our study area, we considered the presence of major and minor mangrove species as defining criteria for mangrove vegetation (see Table S1). We excluded mangrove associates from our definition of the mangrove ecosystem as they are not limited to the mangrove environment and do not feature morphological specialisation and/or physiological mechanisms developed in true mangrove species.

2.2. Sampling Design

Our hybrid methodology used existing remote-sensing mangrove maps for the stratification and sample allocation and local interpreters for the data collection (Figure 2). To define the strata for our sampling design, we created three key maps: a Mangrove Vegetation Index (MVI) map, a mangrove presence map and a mangrove change map. The MVI layers were produced using cloud-free Landsat 7 and Landsat 8 Tier 1 images following [16], and the band arithmetic equation MVI = (NIR − Green)/(SWIR1 − Green). Higher values of MVI in a pixel indicate a higher likelihood of mangroves being found in the pixel.

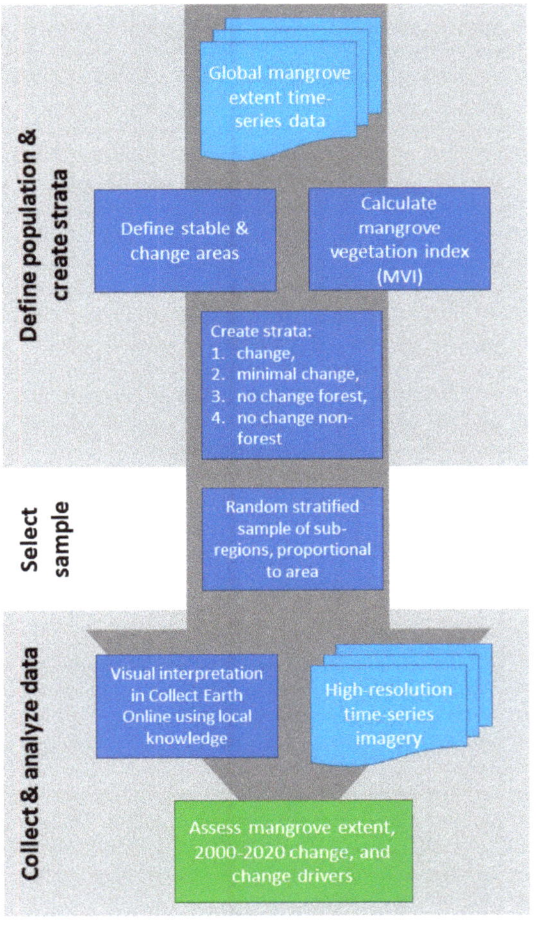

Figure 2. Overview of the hybrid methodology used to assess mangroves.

For the mangrove presence and change maps, the global mangrove mapping products used were the Global Mangrove Forest Distribution, v1 2000 [6], the Joint Research Center of the European Commission (JRC) dataset on forest cover change in tropical moist forests (TMF) 2000–2019 [17] and the Global Mangrove Watch v2 product [7] (Table S2).

First, a 600 m buffer around the Global Mangrove Watch v2 (GMW) layer was created to capture all mangrove areas off the coast. Within this buffered area, pixels from the JRC mangrove dataset and the Global Mangrove Forest Distribution v1 2000 dataset were selected. From the JRC mangrove dataset's classes, we used the undisturbed mangroves, mangrove deforestation (2000–2019), mangrove degradation (2000–2019) and mangrove degraded or regrown before 2000 classes. From the Global Mangrove Forest Distribution v1 2000 dataset, we used the mangrove and non-mangrove pixel information for the year 2000.

This combined dataset was used to generate the two maps. The mangrove presence (stable mangrove) layer was generated using only pixels where all datasets agreed on the presence of mangroves. Pixels with disagreements between the datasets were considered the mangrove change areas. For both maps, we applied a 120 m buffer around mangroves and changed pixels to ensure we captured all mangroves and changes.

Once we defined our strata, we created our sampling frame. As the basis, we used a tessellation of the Earth's surface into equal area hexagons (39.62 hectares each), originating from a discrete global grid of equally sized hexagons (shared with the FRA 2020 RSS; [18]). Each hexagon contains a 1-hectare square centroid to collect more detailed information on land use, land use change and related drivers. All hexagons that overlapped, even partially, with the mangrove presence and mangrove change areas were selected as part of the sampling frame.

The final stratification of hexagons into stable and change classes was performed using the MVI layers for the years 2000 and 2020, along with the mangrove presence and change maps. If more than 40% of the hexagon area was covered by pixels with change, it was assigned to the "change" stratum. If between 5–40% of the hexagon area was change pixels, and there was a maximum of 30% of stable mangroves, it was assigned to the "minimal change" stratum. If the hexagon area included less than 5% change, less than 10% stable mangrove, and the MVI indices for 2000 and 2020 were less than 4.5, it was assigned to the "no change, no forest" stratum. Finally, hexagons meeting the following criteria were assigned to the "no change forest" stratum: less than 40% of the hexagon area covered by change pixels and more than 10% covered by stable pixels, or between 5–40% coverage with change pixels, but more than 30% stable pixels, or less than 5% change coverage, more than 10% stable mangrove coverage and the 2000 MVI is greater than 4.5, or less than 5% change, less than 10% stable mangrove and the 2020 MVI is higher than 4.5.

We used stratified random sampling to select samples from each stratum (Table S2). A total of 20,900 samples were collected, with more samples collected from strata with larger areas and from the mangrove change strata [18]. All the computations to compile the maps were performed on the cloud-based platform Google Earth Engine (GEE) [19], and the random stratified sample selection was performed in Excel [20].

2.3. Data Collection and Validation

Local mangrove experts classified 20,954 samples. Experts were drawn from the FAO-FRA Remote Sensing network. When local experts were not available, experts from FAO performed the classification. Visual interpretation of the samples was completed using Collect Earth Online (CEO), a free and open-source cloud platform for viewing and interpreting high-resolution satellite imagery [12].

The assessment of mangroves in 2020 was completed based on Sentinel-2 cloud-free composite mosaics to ensure the consistency of the results globally. Land use changes were assessed using Landsat mosaics from 2000, 2010, 2018 and 2020. Very high-resolution images, freely available from Google Earth and Bing Maps, were used as auxiliary data to facilitate the understanding and classification of the samples.

Each sample had two levels of analysis: discrete classes at the centroid level (1 ha-square) and quantitative estimation of each class at the hexagon level (39.6 ha), see Figure 3. The main land use classes used in this mangrove assessment are structured according to the terms and definitions used in [21] (Appendix A). The mangrove-specific classes for land use, land-use change and deforestation drivers are based on a review of the relevant literature [22–24]

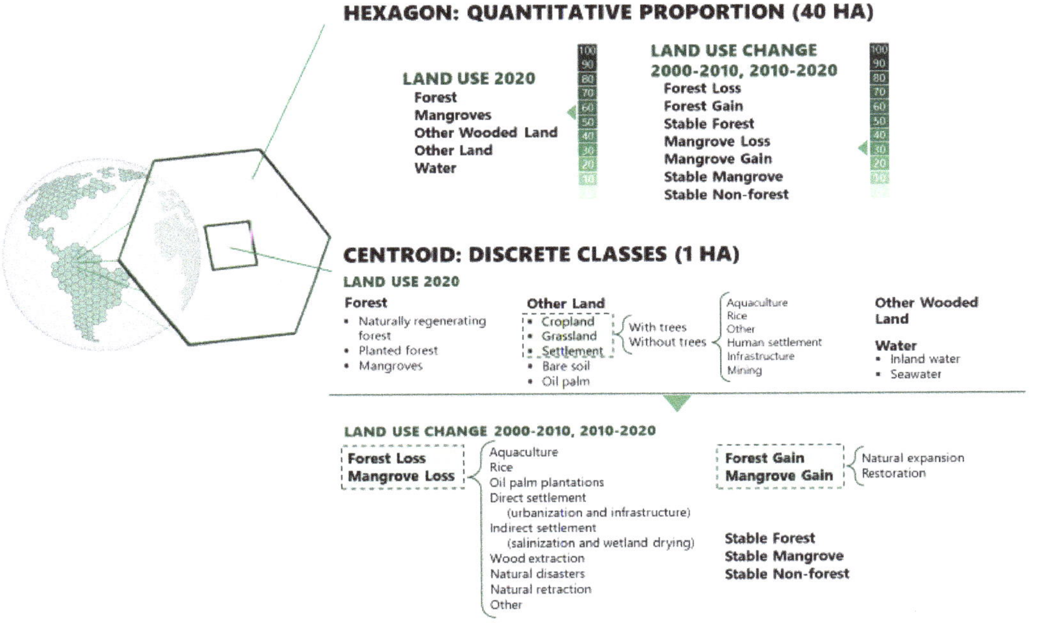

Figure 3. Data were collected for each plot. Modified from [18].

2.4. Data Analysis

Three time periods were collected to facilitate other analyses (2000–2010, 2010–2018 and 2018–2020). For this paper, to facilitate comparison with other research, we combined the raw data at the plot level prior to any downstream analyses to examine changes only between 2000–2010, 2010–2020 and 2000–2020. As a result, our numbers will slightly differ from those in the FAO FRA Mangrove report [25].

All land use and land use change variables were first put in terms of percentages at the unit of analysis scale (hexagon or centroid) prior to being scaled up to the subregional level (Table S3). For data derived from the hexagons, where data were collected as a percentage of the hexagon area, the mean percentage of each variable across sample plots was calculated for each stratum (e.g., % of Mangrove Loss). For data derived from the centroids, the percentage of the centroid with that land use or land use change was calculated. Then, for both hexagons and centroids, the area for the stratum was calculated by multiplying the percentage value for each land use or land use change variable by the total area of the plots in the stratum. These areas were then summed within each subregion. Variances were calculated based on these estimators using the formulas found in Table S3.

Area data estimated at the hexagon level were more precise than the centroid level. Thus, estimates of drivers collected at the centroid level were calibrated by multiplying the area of the specific change by the ratio of the area of the total change in hexagons to centroids.

3. Results

3.1. Mangrove Extent in 2020

This study found that the total area of mangroves globally in 2020 was 147,771 km^2 (±1.4%). The largest share was in South and Southeast Asia (44%), with smaller areas found in South America (14.5%), Western and Central Africa (14.2%), North and Central America (12.5%) and Oceania (9.9%; Table 1).

Table 1. Area and 95% confidence interval of mangroves in 2020 globally and broken down by subregion.

Subregion	Mangrove Area (km^2)	± (km^2)	± (%)
North and Central America	18,455	821	4.4%
South America	21,400	727	3.4%
Western and Central Africa	20,926	795	3.8%
Eastern and Southern Africa	7264	493	6.8%
Western and Central Asia	209	95	45.6%
East Asia	150	52	34.7%
South and Southeast Asia	64,755	1313	2.0%
Oceania	14,611	703	4.8%
World Total	147,771	2075	1.4%

3.2. Mangrove Change 2000–2020

Overall, we observed a global trend, which shows a decrease in the net rate of mangrove deforestation over the last 20 years. While global net change remains negative, the rate of decline decreases from 1815 km^2 lost between 2000–2010 to 1067 km^2 lost between 2010–2020.

However, these global trends mask important differences between subregions (Figure 4). In South and Southeast Asia, mangrove losses have almost halved between the first and second decades. This is globally impactful, as most of the world's mangroves by area are located in this region. North and Central America and Eastern and Southern Africa also saw improvements, moving from net losses to net gains in mangrove areas. Conversely, South America and Oceania all saw increased rates of mangrove loss. Western and Central Asia, Western and Central Africa and East Asia showed small or no difference in the rate of change.

Disaggregating net change to look at losses and gains separately, around half of the total 6314 km^2 mangroves lost have been offset by the establishment of 3475 km^2 of new mangroves. Global losses decrease by 25% between the first and the second decade of the study period. Global gains show a smaller decrease of 8.3%.

Examining subregional changes, we can identify trends behind differences in net change (Figure 5). South and Southeast Asia saw declines in the rate of both mangrove gain and mangrove loss, with a greater magnitude of change in the rate of mangrove loss leading to a net decrease in the rate of mangrove loss. North and Central America and Eastern and Southern Africa both saw decreased rates of mangrove loss and increased rates of mangrove gains, leading to a net increase in mangrove gain. Conversely, South America and Western and Central Africa both saw increased rates of mangrove loss and decreased rates of mangrove gain, leading to a net increase in mangrove loss between the two decades. However, Oceania's observed net increase in mangrove loss can be attributed to the decreased rate of mangrove gain instead of any change in the rate of mangrove loss.

3.3. Drivers of Mangrove Loss and Gain

Globally, we found that the most prominent anthropogenic driver of mangrove loss is conversion to aquaculture (Figures 6 and 7). When comparing the first decade of this study with the second, the contribution of aquaculture, direct settlement and rice cultivation notably decreased, while the contribution of indirect settlement, oil palm cultivation and other drivers notably increased. The "Other" category includes other forms of agriculture and conversion to Grasslands for livestock grazing. Breaking down the sources of anthropogenic mangrove gain by subregion shows that there are some large contributions from artificial restoration in South and Southeast Asia and Eastern and Southern Africa.

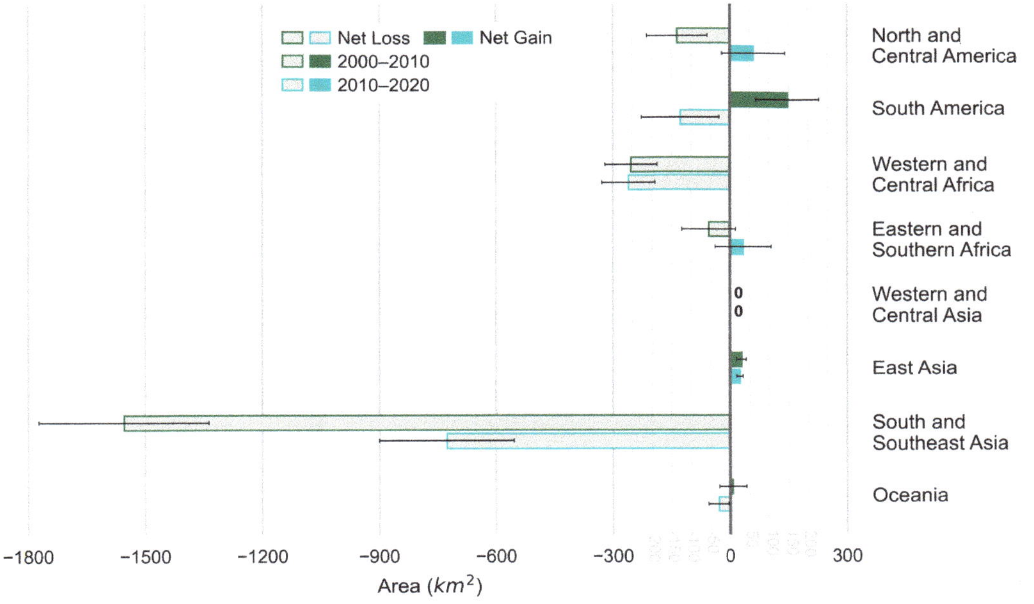

Figure 4. Net change in mangrove area by decade and subregion. Error bars represent standard errors.

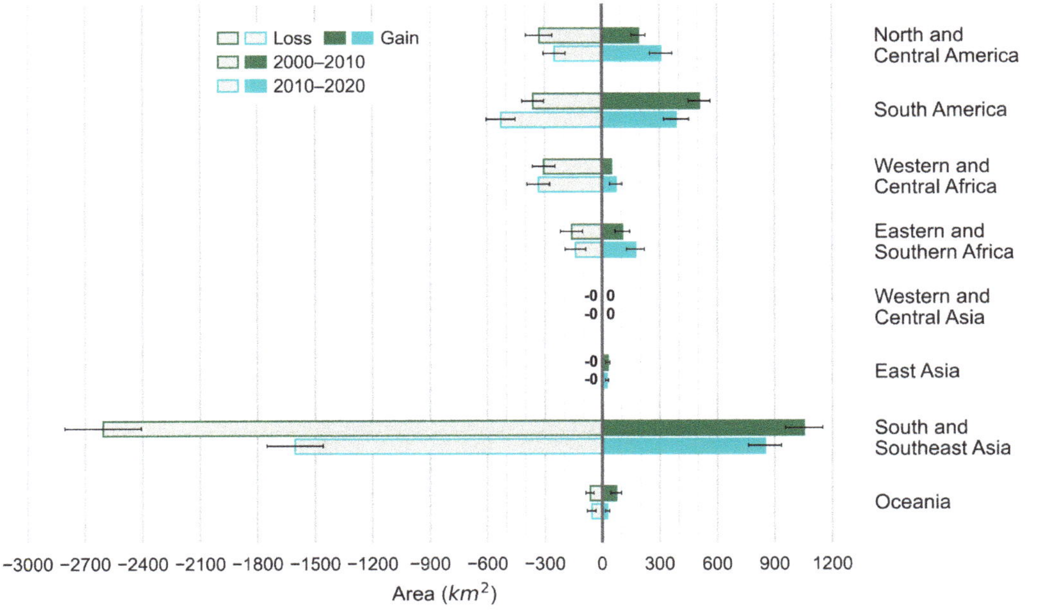

Figure 5. Mangrove gain and loss by decade and subregion. Error bars represent standard errors.

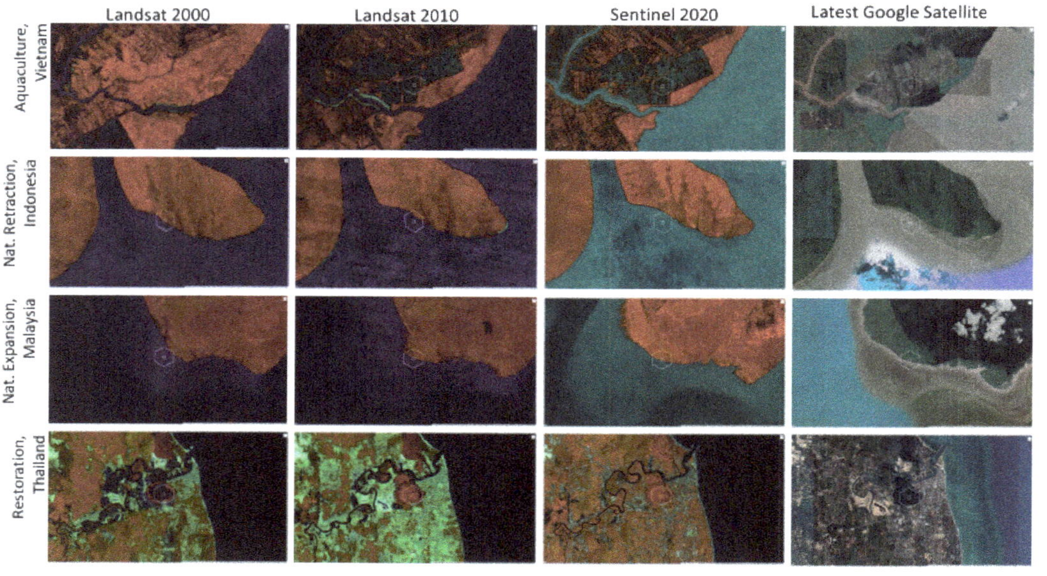

Figure 6. Examples of anthropogenic and biophysical drivers of mangrove loss and gain. Latest Google Satellite image refers to the latest very high-resolution image available in Google Earth for the location.

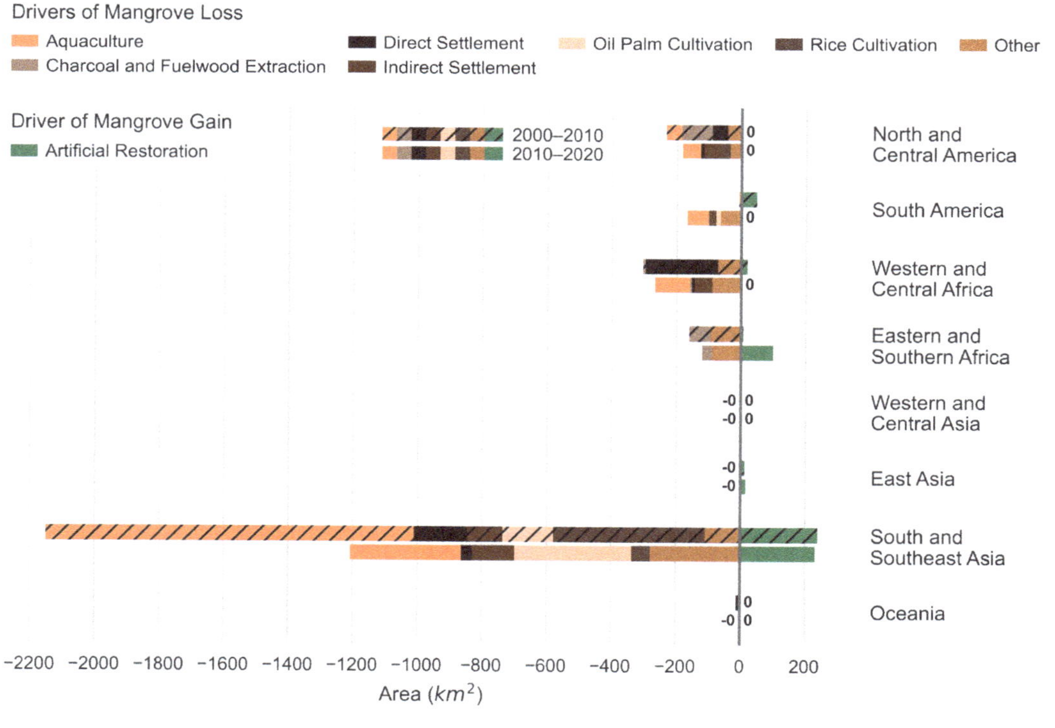

Figure 7. Anthropogenic drivers of mangrove loss and gain by decade and subregion.

Globally, natural retraction is the largest biophysical driver of mangrove loss, and natural expansion is the largest driver of mangrove gain in both decades (Figures 6 and 8). While the net global mangrove change was negative over the study period, we found that areas of natural expansion far surpassed areas lost to natural retraction and natural disasters. The natural expansion is 158% of the area of natural losses between 2000 and 2010 and 170% of natural losses between 2010 and 2020. Across both decades, this represents an area of expansion of 2968 km^2 compared with an area of retraction of 1812 km^2.

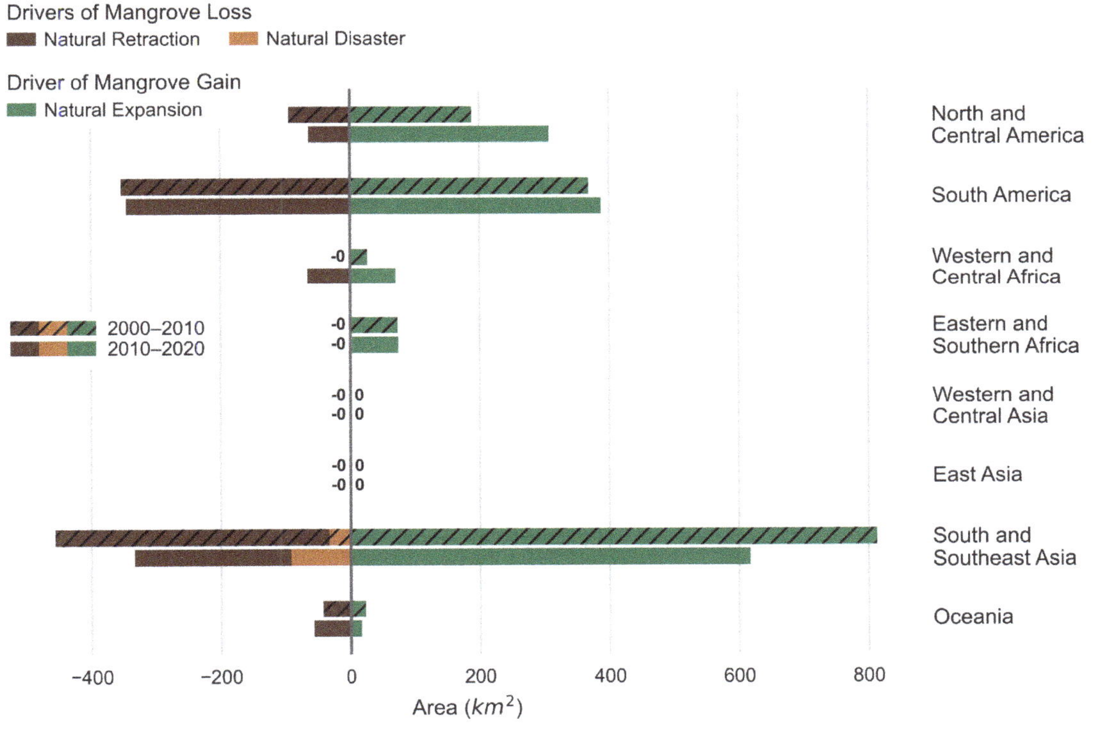

Figure 8. Biophysical drivers of mangrove loss and gain by decade and subregion.

We found differences in the area estimates calculated from the hexagons and the centroids. However, the uncertainty of estimates from centroids is larger than that from the hexagons, which supported our use of the calibration procedure (Supplementary Materials).

4. Discussion

4.1. Estimates of Mangrove Extent

At the global level, there have been several attempts at mapping mangrove extent and mangrove area change [6,8,10,17,23,26]. Our estimate of 147,771 km^2 (±1.4%) of mangroves globally in 2020 is in line with other available estimates and is particularly close to the GMW mangrove cover estimate [8] (Table 2).

Similarly, our finding that global mangrove deforestation has slowed in the past 20 years agrees with previous research [8,10]. However, slightly more of the global mangrove losses were offset by the establishment of new mangrove areas (Table 3). Confidence intervals varied considerably between studies (e.g., compare this study with [8,10]). Confidence intervals were not calculated for other studies due to methodological limitations [2,23].

Table 2. Mangrove area estimates by FAO Subregion according to different global mangrove maps. Areas for the JRC and ESA products calculated using a pixel-counting approach for the digital product. Only studies with estimates for 2020 are included in the table.

Subregion	JRC (2021, km^2)	ESA (2020, km^2)	GMW 3.0 (2020, km^2)	This Paper (2020, km^2)
East Asia	2	247	228	150
Eastern and Southern Africa	4189	9155	7917	7264
North and Central America	14,432	27,961	22,827	18,455
Oceania	11,900	20,920	16,518	14,611
South America	19,538	23,966	20,378	21,400
South and Southeast Asia	52,504	74,009	57,772	64,755
Western and Central Africa	19,849	25,004	21,428	20,926
Western and Central Asia	0	379	285	209
Total	122,414	181,639	147,352	147,771

Table 3. Estimates of total and annual mangrove loss and gain from multiple global studies. Confidence intervals included where studies calculated them. Due to differences in study length, annual loss and gain estimates may be the most accurate for comparison. Due to differences in how geographical subdivisions were made, we compared global estimates. NA indicates that the estimate was not calculated in the study. Adapted from [8].

Study	Period	Loss in Period (km^2)			Annual Loss (km^2)	Gain (km^2)			Annual Gain (km^2)
		Est.	Lower CI	Upper CI		Est.	Lower CI	Upper CI	
Hamilton and Casey 2016 [23]	2000–2012	1646	NA	NA	−137	NA	NA	NA	NA
Goldberg et al. 2020 [2]	2000–2016	−3363	NA	NA	−210	NA	NA	NA	NA
Murray et al. 2022 [10]	1999–2019	−5561	−6827	−3326	−278	1828	932	2960	91
GMW v3.0 [8]	1996–2020	−9348	−15,825	−5568	−390	4130	2238	7012	171
FRA Mangrove RSS (This Study)	2000–2020	−6314	−6923	−5706	−316	3475	3093	3857	174

4.2. Anthropogenic Drivers of Mangrove Loss and Gain

Previous studies have used remote sensing imagery and sophisticated machine learning approaches to identify and categorise different types of change in mangrove forests [2,11]. However, these approaches have limitations, including the similarity in spectral signatures that different land use classes present and reliance on existing datasets which can propagate error. This is the only global study of land use change leveraging local experts instead of models.

Using visual interpretation of satellite imagery and local expertise allowed us to use the surrounding landscape context and location-specific knowledge to identify multiple difficult-to-classify plots correctly. For example, during data collection, we found instances where some mangrove wood was harvested in a sustainable way, and it was still classified as stable mangrove, whereas the same sample could have been classified as mangrove loss using only satellite imagery. Further, while some anthropogenic and biophysical drivers have similar spectral signatures, we were able to distinguish between these, allowing us to distinguish between these.

Despite these different approaches, our results agreed with the earlier findings that the majority of total global mangrove loss was concentrated in the South and Southeast Asia region [2,11,23], and the key drivers were anthropogenic [2,11,22]. Broadly, we found that 72% of losses between 2000 and 2020—75% in the first decade and 68% in the second decade—were the result of human impacts. This was higher than Goldberg et al.'s (2020) [2] estimate, which found that 62% of mangrove loss was due to anthropogenic drivers using remote sensing only [2]. This difference could be attributed to differences in the types of anthropogenic losses considered, as our drivers included more granularity, e.g., aquaculture, oil palm and rice vs. commodities, but could also be due to the driver classification method or methodological approach. The Goldberg et al. (2020) [2] study used a decision-tree model based on open-source datasets to assign Landsat pixels to each driver;

however, due to spectral signature similarities between anthropogenic and biophysical land uses/land covers, they could have failed to correctly assign all the losses [2].

The largest of the anthropogenic drivers is conversion for the production of agricultural commodities, including aquaculture, oil palm and rice. We found that these accounted for 44.5% of the global mangrove losses between 2000 and 2020; Goldberg et al. (2020) [2] found that these same commodities accounted for 47 per cent of mangrove losses between 2000 and 2016 [2]. Overall, we anticipate that while the rate of loss has slowed over the past 20 years, agriculture will remain the predominant anthropogenic driver of mangrove losses globally.

However, this global pattern hides some important regional differences. For example, declines in the rate of loss due to aquaculture are driven by reductions in South and Southeast Asia. In this region, the overall rate of net loss declined in the past 10 years due to growing awareness of the importance of mangroves for climate mitigation and adaptation, as well as for biodiversity, fisheries and livelihoods [27]. Other positive factors in this region include efforts to restore mangroves by governments and communities, improved regulation of the use and conversion of mangroves and increasing the area of mangroves under protection [28]. Yet, while conversion for aquaculture declined, conversion to oil palm in South and Southeast Asia sharply increased between the first and second decade of this study, as anticipated by Richards and Friess (2016) [22]. In comparison, in North and Central America, South America and Western and Central Africa, mangrove conversion to aquaculture sharply increased, indicating that better management of this driver will be needed to ensure a suitable combination of conservation and maintenance of local livelihoods [24,29].

In Western and Central Africa, the subregion with the highest rate of mangrove loss in 2010–2020, drivers of mangrove deforestation have shifted considerably in the last two decades from direct settlement to indirect settlement and agriculture. In the coming years, mangroves could continue to be threatened in this subregion by conversion to aquaculture and other forms of agriculture, natural retraction and indirect settlement, as large areas of mangroves (69%) remain unprotected [1]. Overall, anthropogenic pressure is still the most relevant through shrimp farming in South and Southeast Asia, while drivers in Africa seem closely connected to increasing demographic pressure in the region [30,31].

4.3. Biophysical Drivers of Mangrove Loss and Gain

Natural retraction due to shoreline erosion and extreme weather events was an important driver of mangrove loss observed in all sub-regions, in line with other research [2,11]. Yet, natural expansion was also an important driver of mangrove gain and mangrove retreat and regrowth were commonly observed simultaneously (see also [11]).

Moreover, the area of natural expansion was much greater than natural retraction. Given the impacts of global climate change on mangroves, including sea level rise; increases in atmospheric CO_2; rise in temperature; changes in rainfall; and a predicted increase in the frequency and severity of extreme weather, we might expect that natural retraction would outpace natural expansion [32]. Other researchers have suggested that climate change will impact different regions in different ways, likely increasing mangrove growth in areas with increased precipitation and warming and decreasing it in areas with sea level rise, increased cyclonic activity or where precipitation decreases [33–36]. Yet, where the effect sizes have been estimated, the negative impacts are much greater [35].

However, our results found the opposite trend, highlighting the difficulty in predicting the influence of global climate change on mangrove communities, given the complex interplay between local biophysical conditions and the consequences of global warming [37]. Other possible explanations include that humans are changing places where mangroves can naturally retract, e.g., by sediment removal, and the resilience of mangroves in responding to environmental changes and rapidly colonising suitable habitats when propagules are available (e.g., [38]).

4.4. Methodological Strengths, Tradeoffs and Lessons Learned

This hybrid methodology fills a gap in the literature and provides improved resolution for both anthropogenic and biophysical drivers of mangrove loss and gain while matching existing estimates of global mangrove areas and rates of change. The two broad types of approaches for evaluating mangroves—existing approaches using machine learning and our hybrid approach—each have different strengths and weaknesses and roles to play in mangrove conservation policy.

Many recent studies have leveraged advancements in automatic classification algorithms, freely available satellite imagery and the ease of cloud computation to produce maps of regional and global mangrove extent using automatic and semi-automatic classification, e.g., [10,17]. These approaches allow for wall-to-wall mapping and manually collected verification data, area estimates and confidence intervals.

Yet, distinguishing land use and between specific drivers requires additional datasets, and there are limitations imposed both by the sensor characteristics of the imagery used, persistent cloud cover and machine learning methods [2,8,39]. Further, area estimates obtained with these methods may have omission and commission errors introduced using additional data layers [2]. Many maps do not provide error analyses or confidence intervals to users, making them less useful for decision-making [40].

Our approach allows for improved resolution of drivers using a network of people knowledgeable about the local environment. The approach is robust, and local experts were able to classify samples even with high cloud cover and differentiate between different land use and land use change dynamics, including those that cannot be discriminated between using remote sensing-only approaches, e.g., [2]. Using Collect Earth Online (CEO) allowed for rapid data collection as the cloud platform is easy to use and teach to remote users in a short period of time. The use of the CEO platform allowed data collection to proceed quickly with an easy-to-use interface and our simple questionnaire that could be remotely taught in a half-day webinar. The approach is technically straightforward, requiring only basic remote sensing knowledge for the data collection and standard area estimation techniques for obtaining uncertainty with confidence intervals around land use change drivers, e.g., [41]. While no maps are produced, this approach is flexible, and it is possible to increase sampling density where additional granularity is required, such as at the country scale. In addition, the approach is repeatable, and organisations can replicate this assessment periodically, using the same network of local interpreters to maintain consistency.

This work involved 48 interpreters from all around the world, recruited from a previously established global network of remote sensing experts [18]. Organising this data collection effort was challenged by the ongoing COVID-19 pandemic in 2020 and 2021, as in-person workshops could not be held. While contact had already been established with most of the interpreters during past in-person workshops, it has sometimes been difficult to follow up with them and obtain quick results due to the remote-working modality. In addition, conversations about different land use dynamics occurring in different parts of the world emerged from the remote workshops. These discussions made it clear that local knowledge of the landscape is essential. They also reinforced the opinion that at least part of a data collection process of this type should be live, to better facilitate these emergent discussions and achieve better results.

During this data collection effort, we also learned some lessons regarding the sample plots. First, the interpretation of the hexagons using 10% area intervals was very successful. These were visually assessed by the interpreters, often without the use of a ruler. This approach was much faster than having many sample points within the plot (e.g., 100 evenly spaced grid points) and was just as accurate. In fact, after 5–6 days of data collection, the interpreters were very accurate, and an internal check found that these visual estimates were very close to estimates obtained by precisely measuring using the ruler tool in Google Earth. To ensure and improve accuracy, we used exercises ahead of data collection to become accustomed to estimating the hexagon's area visually.

Additionally, combining the easy classification of the centroids with the more precise area estimates from the hexagons was a good approach. This approach requires benchmarking the centroid estimates of driver loss and gain with the hexagon estimates of loss and gain. The approach was successful at engaging local experts, as classifying driver loss and gain across the entire hexagon is too large of an effort. However, only using the centroids loses precision, something we have also found with using small plots of half a hectare [42].

Employing robust QAQC protocols was also important. To reduce bias in our area estimates as much as possible, we used the same definition for mangrove, mangrove change and mangrove drivers globally and employed quality control after the data collection. In addition, due to resource constraints and COVID-19 travel limitations, it was more difficult to engage interpreters for some areas of the world who were highly knowledgeable about all different types of mangrove ecosystems, requiring robust QAQC protocols. This shows how essential it is to involve local interpreters to obtain correct land use classifications.

From a policy standpoint, FAO is especially interested in reporting on land use and land use change in forestry, instead of land cover, to support countries in their assessment and monitoring of forest resources. Understanding forest resources and their changes is key to national and international environmental and developmental policy processes and is required by many international commitments, including the Paris Agreement of the United Nations Framework Convention on Climate Change (UNFCCC), the Kunming-Montreal Global Biodiversity Framework of the Convention on Biological Diversity (CBD), the United Nations Convention to Combat Desertification (UNCCD), the United Nations Strategic Plan for Forests 2030 and the Sustainable Development Goals of the Agenda 2030 (SDGs; [43]).

5. Conclusions

Using remote sensing and the knowledge of local specialists, we devised a reliable and repeatable technique. This approach was successfully used to collect data on mangrove areas in 2020, mangrove area changes between 2000 and 2020 and, most importantly, both anthropogenic and biophysical drivers of mangrove change for these periods. This is the first global study that provides information on land use based on direct interpretation of satellite imagery and not land cover for those variables, a result which was only possible with the involvement of local experts.

In line with previous estimates, our results found that globally there were 147,771 km^2 ($\pm$1.4%) of mangroves in 2020 and that the rate of mangrove loss has decreased between 2000–2010 and 2010 to 2020, largely driven by decreasing rates of loss in South and Southeast Asia. The contribution of different anthropogenic and biophysical drivers of loss and gain during these periods has shifted considerably. For anthropogenic drivers, aquaculture is an emerging concern in South America and Western and Central Africa, and oil palm cultivation is an emerging concern in South and Southeast Asia. For biophysical drivers, natural expansion continues to be larger than natural retraction and natural disasters combined.

Areas of future research to be considered start with deepening the analysis of the drivers. This includes collecting data on mangrove degradation, a parameter difficult to map using remote sensing but which could be addressed with sample-based interpretation by human interpreters. Our land use change drivers had relatively broad confidence intervals, so increasing the number of samples would reduce the confidence intervals for the drivers' estimates.

Another opportunity that our sampling-based methodology offers is the intensification at a national or sub-national level to assess changes and trends in mangroves and obtain better mangrove statistics for a particular country's national inventory [44]. FAO is already collaborating with some countries, such as Indonesia and Cuba, to use this methodology to possibly integrate the new data in their own national forest inventories. The resulting area estimates can be used to report to international conventions and processes.

Finally, there is a need for a robust global drivers database for mangrove deforestation [45] that is periodically updated. Our future assessments will focus on collecting data about drivers of loss and gain using this hybrid methodology.

Supplementary Materials: The following supporting information can be downloaded at: https://www.mdpi.com/article/10.3390/rs15153872/s1, Table S1: Mangrove species; Table S2: Strata names and sizes; Table S3: Key equations used; Supplementary Data: Project data.

Author Contributions: Conceptualisation, V.C., K.D., A.K. and K.T.; Data curation, V.C., K.D. and P.P.V.M.; Formal analysis, V.C., K.D., P.P.V.M. and A.K.; Funding acquisition, D.S., K.T. and A.P.; Investigation, V.C. and P.P.V.M.; Methodology, P.P.V.M. and A.K.; Project administration, A.K.; Supervision, D.S., K.T. and A.P.; Validation, P.P.V.M. and A.K.; Visualization, K.D. and T.L.; Writing—original draft, V.C., K.D., P.P.V.M. and K.T.; Writing—review and editing, V.C., K.D., P.P.V.M., A.K., T.L., D.S., K.T. and A.P. All authors have read and agreed to the published version of the manuscript.

Funding: This study was implemented with financial support from the European Union, the Government of Norway and the Government of Finland. The writing support was funded by SERVIR. The views expressed in this publication are those of the author(s) and do not necessarily reflect the views or policies of the Food and Agriculture Organization of the United Nations.

Data Availability Statement: The data presented in this study are available as a supplement.

Acknowledgments: This study is the result of a collective effort by the FAO Forestry Division, FAO member countries, institutional and resource partners and many individuals. We thank the 48 experts from 26 countries who participated in the data collection. They include FRA national correspondents, their alternates and collaborators. The methodology was developed by FAO with initial support from Radhika Bhargava. The authors would like to thank Kenichi Shono (FAO), who contributed to the study conceptualisation, Erica Lupi (FAO), who contributed to coordinating and carrying out the data collection, and Enikoe Bihari (SIG-GIS) for her comments on data analysis. The authors are grateful to all countries and territories, institutions and individuals who made this study possible.

Conflicts of Interest: The authors declare no conflict of interest. The funders had no role in the design of the study; in the collection, analyses or interpretation of data; in the writing of the manuscript; or in the decision to publish the results.

Appendix A. Classification Scheme

Appendix A.1. Terms and Definitions

The following terms and definitions for the mangrove assessment are structured according to the FRA 2020 Terms and Definitions.

Some of the definitions and terms used in the land use classification categories that are explained below are based on the definitions of the 2006 IPCC (Intergovernmental Panel on Climate Change) Guidelines for National Greenhouse Gas Inventories, in which volume 4 provides guidance for preparing annual Greenhouse Gas Inventories in the Agriculture, Forestry and Other Land Use (AFOLU) sector.

- Global Forest Resources Assessment 2020 Terms and Definitions Document: http://www.fao.org/3/I8661EN/i8661en.pdf (accessed on 30 July 2023).
- 2006 IPCC Guidelines for National Greenhouse Gas Inventories, Volume 4 Agriculture, Forestry and Other Land Use (Chapter 3 Consistent Representation of lands, 3.2. Land Use Categories): https://www.ipcc-nggip.iges.or.jp/public/2006gl/pdf/4_Volume4/V4_03_Ch3_Representation.pdf (accessed on 30 July 2023).

Appendix A.2. Definitions for Centroid and Hexagon Current Land Use 2020

See Figure A1.

Figure A1. Centroid and hexagon current land use 2020.

Level 1

Forest:

Land spanning more than 0.5 hectares with trees higher than 5 m and a canopy cover of more than 10 per cent or trees able to reach these thresholds in situ.

It does not include land that is predominantly under agricultural or urban land use.

Forest is determined both by the presence of trees and the absence of other predominant land uses.

The trees should be able to reach a minimum height of 5 m in situ.

The Forest definition also includes forest roads, firebreaks and other small open areas inside the forest.

It also includes Windbreaks, Shelterbelts and Corridors of trees, but those should provide forest ecosystem services and not be used for livestock breeding or for crops under the trees.

It also includes abandoned shifting cultivation land with the regeneration of trees that have, or are expected to reach, a canopy cover of 10 per cent and a height of 5 m.

Please refer to the FRA 2020 Terms and Definitions document for other explanatory notes.

Other Wooded Land:

Land not classified as "Forest", spanning more than 0.5 hectares, with trees higher than 5 m and a canopy cover of 5–10 per cent, or trees able to reach these thresholds in situ, or with a combined cover of shrubs, bushes and trees above 10 per cent.

It does not include land that is predominantly under agricultural or urban land use.

Other Land:

All land that is not classified as "Forest" or "Other wooded land".

Explanatory notes:

Includes agricultural lands, meadows and pastures, built-up areas, barren lands, lands under permanent ice, etc.

Includes all areas considered as "Other land with tree cover".

Level 2

Under Forest:

Naturally regenerated forest:

Forests are predominantly composed of trees established through natural regeneration.
Explanatory notes:

1. Includes forests for which it is not possible to distinguish whether planted or naturally regenerated.
2. Includes forests with a mix of naturally regenerated native tree species and planted/seeded trees, and where the naturally regenerated trees are expected to constitute the major part of the growing stock at stand maturity.
3. Includes coppice from trees originally established through natural regeneration.
4. It includes naturally regenerated trees of the introduced species.

Planted Forest:

Forests are predominantly composed of trees established through planting and/or deliberate seeding.

Explanatory notes:

1. In this context, it predominantly means that the planted/seeded trees are expected to constitute more than 50 per cent of the growing stock at maturity.
2. Includes coppice from trees that were originally planted or seeded.

And

A planted forest is intensively managed and meets all the following criteria at planting and stands maturity: one or two species, even age class and regular spacing.

Explanatory notes:

1. Specifically includes short rotation plantation for wood, fibre and energy.
2. Specifically excludes forests planted for protection or ecosystem restoration.
3. Specifically excludes forest established through planting or seeding which at stand maturity resembles or will resemble naturally regenerating forest.

Mangroves:

Forest predominantly composed of true mangroves (listed in Table S1) established through natural regeneration or through planting and/or deliberate seeding.

Under Other Land:

Cropland

This category includes arable and tillable land, rice fields and agroforestry systems.

Cropland includes all annual and perennial crops as well as temporary fallow land (i.e., land set at rest for one or several years before being cultivated again).

Annual crops include cereals, oilseeds, vegetables, root crops and forages. Perennial crops include trees and shrubs, in combination with herbaceous crops (e.g., agroforestry) or as orchards, vineyards and plantations such as cocoa, coffee, tea, oil palm, coconut and bananas.

Grassland

This category includes all pasture lands, all-natural Grasslands and also agricultural and silvo-pastoral systems.

The term Grassland for our RSS is closely linked to livestock breeding, independent of whether the tree cover is high in situ.

If the use is to raise livestock, it must be categorised as Grassland regardless of whether there is a high density of trees, bushes or a mixture of shrubs with trees, such as the two pictures on the side.

Settlement

This category includes all developed land, including transportation infrastructure and human settlements of any size.

It includes trees in urban settings such as in parks and gardens. It also includes mining areas, which are not considered bare soil because it is soil bared by human activities.

Bare Soil

This category includes all-bare soil for natural site conditions: rocks, sand (beaches or desert) and snow-covered mountain tops.

Oil palm

This category includes all oil palm (*Elaeis* sp.) plantations for commercial agriculture in the production of palm oil.

Level 3

Other Land with Tree Cover (subcategory for Cropland, Grassland and Settlement):

Land classified as "other land", spanning more than 0.5 hectares with a canopy cover of more than 10 per cent of trees able to reach a height of 5 m at maturity.

Explanatory notes:

1. Land use is the key criterion for distinguishing between forest and other land with tree cover.
2. Specifically includes palms (coconut, dates, etc.), tree orchards (fruit, nuts, olive, etc.), agroforestry and trees in urban settings.
3. Includes groups of trees and scattered trees (e.g., trees outside the forest) in agricultural landscapes, parks, gardens and around buildings, provided that area, height and canopy cover criteria are met.
4. It includes tree stands in agricultural production systems, such as fruit tree plantations/orchards. In these cases, the height threshold can be lower than 5 m.
5. Includes agroforestry systems when crops are grown under tree cover and tree plantations established mainly for purposes other than wood.

6. Excludes scattered trees with a canopy cover of less than 10 per cent, small groups of trees covering less than 0.5 hectares and tree lines less than 20 m wide (the latter are included under Forest).

Level 4

Aquaculture:

Aquaculture or farming in water is the aquatic equivalent of agriculture or farming on land. Agriculture, defined broadly, includes farming both animals (animal husbandry) and plants (agronomy, horticulture and forestry in part). Similarly, aquaculture covers the farming of both animals (including crustaceans, finfish and molluscs) and plants (including seaweeds and freshwater macrophytes). While agriculture is predominantly based on the use of freshwater, aquaculture occurs in both inland (freshwater) and coastal (brackish water, seawater) areas.

FAO (1988) introduced a clear definition of aquaculture:

Aquaculture is the farming of aquatic organisms, including fish, molluscs, crustaceans and aquatic plants. Farming implies some form of intervention in the rearing process to enhance production, such as regular stocking, feeding, protection from predators, etc. Farming also implies individual or corporate ownership of the stock being cultivated. For statistical purposes, aquatic organisms that are harvested by an individual or corporate body which has owned them throughout their rearing period contribute to aquaculture, while aquatic organisms, which are exploitable by the public as a common property resource, with or without appropriate licences, are the harvest of fisheries. http://www.fao.org/3/x6941e/x6941e04.htm (accessed on 30 July 2023)

Rice fields:

Any cultivation of rice (*Oryza sativa* sp.), submerged or not.

It includes rice-cum-fish cultures, which are the mixed rice cultivation and harvesting of fish in the same ponds.

Natural Mangrove Grasslands:

Natural grasses that grow in mangrove habitats, which are mangrove associate species (not true mangroves).

Human Settlement:

Any human settlement of any size.

Infrastructure:

It is any transportation infrastructure, such as railways and highways.

Mining:

It is any area designated for the extraction of valuable minerals or other geological materials from the Earth.

Appendix A.3. Definitions for Centroid and Hexagon Changes 2000–2010, 2010–2018 and 2018–2020

See Figure A2.

Figure A2. Centroid and hexagon changes.

Level 1

Forest Loss or Mangrove Loss

This category indicates any land use change from Forest Use to any other Non-Forest Use (Other Land Use, water or—in very rare cases—OWL).

Forest Gain or Mangrove Gain

This category indicates any land use change from Non-Forest Use to Forest Use. It can be a new forest plantation established on a previously barren area or natural forest expansion.

Stable N. R. Forest

This category refers to areas in which forest land use remains during the study period. It includes temporarily unstocked areas because of forest management or natural causes.

It includes planted forests.

Stable Mangrove

This category refers to areas in which the mangrove vegetation and forest land use remains during the study period.

Stable Non-Forest

This category refers to areas in which the OWL or Other Land Use (Cropland, settlement, Grassland, etc.) remains during the study period.

Level 2

Loss to Aquaculture

Change of land use from forest to any type of aquaculture.

Loss to Rice fields

Change of land use from forest to any type of rice field.

Loss to Oil Palm plantations

Change of land use from forest to any type of oil palm plantation.

Loss to Direct Settlement (Urbanisation and infrastructure)

Forest loss to urbanisation and other types of infrastructures, such as roads or mining activities.

Loss to Indirect settlement (salinisation, wetland drying)

Forest loss because of pedological, microclimatic and hydrologic changes in the area indirectly generated by human actions (e.g., the construction of a dam upstream).

Loss to Charcoal and fuel wood extraction

Forest loss because of any type of wood extraction. It includes wood extraction for timber, fuel and charcoal production.

Wood extraction for fuelwood and charcoal can be a gradual process, starting with the loss of a few trees at a time, which, however, if continuous, will lead to ecosystem degradation. Only when the ecosystem is degraded, and the trees cannot reach the forest threshold anymore, with visible forest loss in the imagery, will it be classified as forest loss.

Loss to Natural disasters

In case of particularly severe natural disasters such as floods, storm surges, tsunamis or landslides, the pedological, microclimatic and hydrologic conditions of the area can irrevocably change and not allow the growth of mangrove vegetation anymore.

Loss to Natural retraction

Regarding mangroves, natural changes or movements in riverbeds and sediments input or sea level rise may lead to the local extinction of the mangrove ecosystem.

Loss to Others

Any other type of land use change from forest to non-forest, which is not included in the previous categories.

Gain—Natural expansion

Regarding mangroves, natural changes or movements in riverbeds and sediment input may lead to the local colonisation of the new areas by mangrove vegetation.

Gain—Restoration

Change of land use from non-forest to forest because of direct human action.

It includes reforestation and afforestation projects, both through direct planting/seeding and through hydrological restoration and control of disturbances that result in the natural regeneration of mangroves. It also includes protected areas ensured by new regulations, where the forest is naturally regrowing because the laws have banned human disturbance.

References

1. Leal, M.; Spalding, M.D. (Eds.) *The State of the World's Mangroves*; Global Mangrove Alliance: Washington, DC, USA, 2022.
2. Goldberg, L.; Lagomasino, D.; Thomas, N.; Fatoyinbo, T. Global declines in human-driven mangrove loss. *Glob. Change Biol.* **2020**, *26*, 5844–5855. [CrossRef] [PubMed]
3. Taillardat, P.; Friess, D.A.; Lupascu, M. Mangrove blue carbon strategies for climate change mitigation are most effective at the national scale. *Biol Lett.* **2018**, *14*, 20180251. [CrossRef] [PubMed]
4. Romañach, S.; Deangelis, D.; Koh, H.; Li, Y.; Teh, S.; Sulaiman, R.; Lu, Z. Conservation and restoration of mangroves: Global status, perspectives, and prognosis. *Ocean. Coast. Manag.* **2018**, *154*, 72–82. [CrossRef]
5. Hagger, V.; Worthington, T.A.; Lovelock, C.E.; Adame, M.F.; Amano, T.; Brown, M.B.; Friess, D.A.; Landis, E.; Mumby, P.J.; Morrison, T.H.; et al. Drivers of global mangrove loss and gain in social-ecological systems. *Nat. Commun.* **2022**, *13*, 6373. [CrossRef]
6. Giri, C.E.; Ochieng, L.; Tieszen, L.; Zhu, Z.; Singh, A.; Loveland, T.; Masek, J.; Duke, N. Status and distribution of mangrove forests of the world using earth observation satellite data. *Glob. Ecol. Biogeogr.* **2011**, *20*, 154–159. [CrossRef]
7. Bunting, P.; Rosenqvist, A.; Lucas, R.; Rebelo, L.-M.; Hilarides, L.; Thomas, N.; Hardy, A.; Itoh, T.; Shimada, M.; Finlayson, M. The Global Mangrove Watch—A New 2010 Global Baseline of Mangrove Extent. *Remote Sens.* **2018**, *10*, 1669. [CrossRef]
8. Bunting, P.; Rosenqvist, A.; Hilarides, L.; Lucas, R.M.; Thomas, N.; Tadono, T.; Worthington, T.A.; Spaldingg, M.; Murray, N.J.; Rebelo, L.M. Global mangrove extent change 1996–2020: Global Mangrove Watch version 3.0. *Remote Sens.* **2022**, *14*, 3657. [CrossRef]
9. Friess, D.A.; Rogers, K.; Lovelock, C.E.; Krauss, K.W.; Hamilton, S.E.; Lee, S.Y.; Lucas, R.; Primavera, J.; Rajkaran, A.; Shi, S. The state of the world's mangrove forests: Past, present, and future. *Annu. Rev. Environ. Resour.* **2019**, *44*, 89–115. [CrossRef]
10. Murray, N.; Worthington, T.A.; Bunting, P.; Duce, S.; Hagger, V.; Lovelock, C.E.; Lucas, R.; Saunders, M.I.; Sheaves, M.; Spalding, M.; et al. High-resolution mapping of losses and gains of Earth's tidal wetlands. *Science* **2022**, *376*, 744–749. [CrossRef]
11. Thomas, N.; Lucas, R.; Bunting, P.; Hardy, A.; Rosenqvist, A.; Simard, M. Distribution and drivers of global mangrove forest change, 1996–2010. *PLoS ONE* **2017**, *12*, e0179302. [CrossRef]
12. Saah, D.; Johnson, G.; Ashmall, B.; Tondapu, G.; Tenneson, K.; Patterson, M.; Poortinga, A.; Markert, K.; Quyen, N.H.; San Aung, K.; et al. Collect Earth: An online tool for systematic reference data collection in land cover and use applications. *Environ. Model. Softw.* **2019**, *118*, 166–171. [CrossRef]
13. FAO. *The World's Mangroves 1980–2005*; FAO Forestry Paper 153; FAO: Rome, Italy, 2007.
14. Tomlinson, P.B. *The Botany of Mangroves*; Cambridge University Press: Cambridge, UK, 1986.
15. FAO. *Global Forest Resources Assessment 2020*; FAO: Rome, Italy, 2020.
16. Baloloy, A.B.; Blanco, A.C.; Rhommel, R.; Ana, C.S.; Nadaoka, K. Development and application of a new mangrove vegetation index (MVI) for rapid and accurate mangrove mapping. *ISPRS J. Photogramm. Remote Sens.* **2020**, *166*, 95–117. [CrossRef]
17. Vancutsem, C.; Achard, F.; Pekel, J.-F.; Vieilledent, G.; Carboni, S.; Simonetti, D.; Gallego, J.; Aragão, L.E.O.C.; Nasi, R. Long-term (1990–2019) monitoring of forest cover changes in the humid tropics. *Sci. Adv.* **2021**, *7*, eabe1603. [CrossRef]
18. FAO. *FRA 2020 Remote Sensing Survey*; FAO Forestry Paper No. 186; FAO: Rome, Italy, 2022. [CrossRef]
19. Gorelick, N.; Hancher, M.; Dixon, M.; Ilyushchenko, S.; Thau, D.; Moore, R. Google Earth Engine: Planetary-scale geospatial analysis for everyone. *Remote Sens. Environ.* **2017**, *202*, 18–27. [CrossRef]
20. Microsoft Corporation. Microsoft Excel. Available online: https://office.microsoft.com/excel (accessed on 30 July 2023).
21. FAO. *Terms and Definitions-FRA 2020*; Forest Resources Assessment Working Paper 188; FAO: Rome, Italy, 2018. Available online: www.fao.org/3/I8661EN/i8661en.pdf (accessed on 1 August 2023).
22. Richards, D.R.; Friess, D.A. Rates and drivers of mangrove deforestation in Southeast Asia, 2000–2012. *Proc. Natl. Acad. Sci. USA* **2016**, *113*, 344–349. [CrossRef] [PubMed]
23. Hamilton, S.E.; Casey, D. Creation of a high spatio-temporal resolution global database of continuous mangrove forest cover for the 21st century (CGMFC-21). *Glob. Ecol. Biogeogr.* **2016**, *25*, 729–738. [CrossRef]
24. Feka, N.Z.; Ajonina, G.N. Drivers causing decline of mangrove in West-Central Africa: A review. *Int. J. Biodivers. Sci. Ecosyst. Serv. Manag.* **2011**, *7*, 217–230. [CrossRef]
25. FAO. *The World's Mangroves 2000–2020*; Forest Resources Assessment Working Paper; FAO: Rome, Italy, 2023; *in press*.

26. Zanaga, D.; Van De Kerchove, R.; De Keersmaecker, W.; Souverijns, N.; Brockmann, C.; Quast, R.; Wevers, J.; Grosu, A.; Paccini, A.; Vergnaud, S.; et al. ESA WorldCover 10 m 2020 v100. 2021. Available online: https://zenodo.org/record/5571936 (accessed on 1 August 2023).
27. Gerona-Daga, M.E.B.; Salmo, S.G., III. A systematic review of mangrove restoration studies in Southeast Asia: Challenges and opportunities for the United Nation's Decade on Ecosystem Restoration. *Front. Mar. Sci.* **2022**, *9*, 987737. [CrossRef]
28. Spalding, M.; Kainuma, M.; Collins, L. *World Atlas of Mangroves*; Earthscan: London, UK; Washington, DC, USA, 2010.
29. de Lacerda, L.D.; Borges, R.; Ferreira, A.C. Neotropical mangroves: Conservation and sustainable use in a scenario of global climate change. *Aquatic Conserv Mar Freshw Ecosyst.* **2019**, *29*, 1347–1364. [CrossRef]
30. Duguma, L.; Bah, A.; Muthee, K.; Carsan, S.; McMullin, S.; Minang, P. Drivers and Threats Affecting Mangrove Forest Dynamics in Ghana and The Gambia. In *Women Shellfishers and Food Security Project*; WSFS2022_01_CRC; World Agroforestry (ICRAF): Nairobi, Kenya; Coastal Resources Center, Graduate School of Oceanography, University of Rhode Island: Narragansett, RI, USA, 2022; 53p.
31. Lontsi, F.R.Z.; Tchawa, P.; Mbaha, J.P. Mapping and Botanical Study of Pressures Causing Mangrove Dynamics of Tiko (Southwest Cameroon). *Open Access Libr. J.* **2023**, *10*, 1–19. [CrossRef]
32. CMEP. *Caribbean Marine Climate Change Report Card 2017*; Buckley, P., Townhill, B., Trotz, U., Nichols, K., Murray, P.A., Samuels, C.C., Gordon, A., Eds.; Commonwealth Marine Economies Programme (CMEP): London, UK, 2017; 12p.
33. Alongi, D.M. The impact of climate change on mangrove forests. *Curr. Clim. Chang. Rep.* **2015**, *1*, 30–39. [CrossRef]
34. Coldren, G.A.; Langley, J.A.; Feller, I.C.; Chapman, S.K. Warming accelerates mangrove expansion and surface elevation gain in a subtropical wetland. *J. Ecol.* **2019**, *107*, 79–90. [CrossRef]
35. Friess, D.A.; Adame, M.F.; Adams, J.B.; Lovelock, C.E. Mangrove forests under climate change in a 2 C world. *Wiley Interdiscip. Rev. Clim. Chang.* **2022**, *13*, 792. [CrossRef]
36. Arnaud, M.; Baird, A.J.; Morris, P.J.; Dang, T.H.; Nguyen, T.T. Sensitivity of mangrove soil organic matter decay to warming and sea level change. *Glob. Change Biol.* **2020**, *26*, 1899–1907. [CrossRef]
37. Bindiya, E.S.; Sreekanth, P.M.; Bhat, S.G. Conservation and Management of Mangrove Ecosystem in Diverse Perspectives. In *Conservation and Sustainable Utilization of Bioresources*; Springer Nature: Singapore, 2023; pp. 323–352.
38. Fickert, T. To plant or not to plant, that is the question: Reforestation vs. natural regeneration of hurricane-disturbed mangrove forests in Guanaja (Honduras). *Forests* **2020**, *11*, 1068. [CrossRef]
39. Keenan, R.J.; Reams, G.A.; Achard, F.; de Freitas, J.V.; Grainger, A.; Lindquist, E. Dynamics of global forest area: Results from the FAO Global Forest Resources Assessment 2015. *For. Ecol. Manag.* **2015**, *352*, 9–20. [CrossRef]
40. Olofsson, P.; Foody, G.M.; Stehman, S.V.; Woodcock, C.E. Making better use of accuracy data in land change studies: Estimating accuracy and area and quantifying uncertainty using stratified estimation. *Remote Sens. Environ.* **2013**, *129*, 122–131. [CrossRef]
41. Olofsson, P.; Foody, G.M.; Herold, M.; Stehman, S.V.; Woodcock, C.E.; Wulder, M.A. Good practices for estimating area and assessing accuracy of land change. *Remote Sens. Environ.* **2014**, *148*, 42–57. [CrossRef]
42. Kindgard, A. (FAO, Rome, Italy). Personal Communication, 2023.
43. FAO. Regional Technical Platform on Green Agriculture. Available online: https://www.fao.org/platforms/green-agriculture/areas-of-work/natural-resources-biodiversity-green-production/forestry/forest-and-land-use-cover-assessment-and-monitoring/en (accessed on 31 May 2023).
44. Sandker, M.; Carrillo, O.; Leng, C.; Lee, D.; d'Annunzio, R.; Fox, J. The importance of high–quality data for REDD+ monitoring and reporting. *Forests* **2021**, *12*, 99. [CrossRef]
45. Bhowmik, A.K.; Padmanaban, R.; Cabral, P.; Romeiras, M.M. Global Mangrove Deforestation and Its Interacting Social-Ecological Drivers: A Systematic Review and Synthesis. *Sustainability* **2022**, *14*, 4433. [CrossRef]

remote sensing

MDPI

Article

Extent, Severity, and Temporal Patterns of Damage to Cuba's Ecosystems following Hurricane Irma: MODIS and Sentinel-2 Hurricane Disturbance Vegetation Anomaly (HDVA)

Hannah C. Turner [1,2], Gillian L. Galford [2,3,*], Norgis Hernandez Lopez [4], Armando Falcón Méndez [4], Daily Yanetsy Borroto-Escuela [4], Idania Hernández Ramos [4] and Patricia González-Díaz [3,5]

1 Stantec, Portland, ME 04101, USA
2 Rubenstein School of Environment and Natural Resources, University of Vermont, Burlington, VT 05405, USA
3 Gund Institute for Environment, University of Vermont, Burlington, VT 05405, USA
4 Caguanes National Park, Ministerio de Ciencia, Tecnología y Medio Ambiente, Yaguajay 62100, Cuba
5 Centro de Investigaciones Marinas, Universidad de La Habana, La Habana 11300, Cuba
* Correspondence: gillian.galford@uvm.edu; Tel.: +1-802-656-2920

Abstract: Mangrove forests provide a range of ecosystem services but may be increasingly threatened by climate change in the North Atlantic due to high-intensity storms. Hurricane Irma (Category 5) hit the northern coast of Cuba in September 2017, causing widespread damage to mangroves; losses have not yet been extensively documented due to financial and logistical constraints for local scientists. Our team estimated Irma's impacts on Cuban ecosystems in a coastal and upland study area spanning over 1.7 million ha. We developed a multi-resolution time series "vegetation anomaly" approach, where post-disturbance observations in photosynthetically active vegetation (Enhanced Vegetation Index, EVI) were normalized to the reference period (dry season mean over a historical time series). The Hurricane Disturbance Vegetation Anomaly (HDVA) was used to estimate the extent, severity, and temporal patterns of ecological changes with Sentinel-2 and MODIS data and used vicarious validation with microsatellite interpretation (Planet). HDVA values were classed to convey qualitative labels useful for local scientists: (1) Catastrophic, (2) Severe, (3) Moderate, (4) Mild, and (5) No Loss. Sentinel-2 had a limited reference period (2015–2017) compared to MODIS (2000–2017), yet the HDVA patterns were similar. Mangrove and wetlands (>265,000 ha) sustained widespread damages, with a staggering 78% showing damage, largely severe to catastrophic (0–0.81 HDVA; >207,000 ha). The damaged area is 24 times greater than impacts from Irma as documented elsewhere. Caguanes National Park (>8400 ha, excluding marine zones) experienced concentrated, severe mangrove and wetland damages (nearly 4000 ha). The phenological declines from Irma's impacts took up to 17 months to fully actualize, a much longer period than previously suggested. In contrast, dry forests saw rapid green flushes post-hurricane. With the increase of high-intensity storm events and other threats to ecosystems, the HDVA methods outlined here can be used to assess intense to low-level damages.

Keywords: mangroves; multitemporal remote sensing; deforestation; climate change; Sentinel-2; Cuba

1. Introduction

Mangroves and their ecosystems provide many benefits, or ecosystem services. These include a wide variety of ecosystem services to coastal communities, such as producing timber, supplying food as important fisheries habitat, and growing plants used as traditional medicine [1–3]. Coastal communities are often buffered from storm damages by mangroves, which can dissipate wave energy, absorb sediment, and buffer storm surges [4,5]. Mangroves are vital for carbon storage and sequestration, as they are among the most productive ecosystems in the world, containing, on average, 1023 Mg of carbon per hectare [6–8]. Globally, it is estimated that 30–50% of the world's mangroves have been lost within the last 50 years [8].

Mangrove declines and deforestation are caused by human activities such as increased coastal urban development, agriculture, mariculture, and tourism [7,9]. Mangrove decline is further affected by climatic changes (e.g., sea level rise [10]) and tropical cyclones, with storms causing the greatest mortality [11]. Intact mangroves are increasingly threatened by tropical cyclones ("hurricanes" in the North Atlantic and North Pacific; "typhoons" in the Northwest Pacific) due to climate change, which could cause widespread mortality [8] and eliminate vital ecosystem services.

Mangroves are low- and mid-latitude ecosystems; thus, hurricanes and tropical storms are common disturbances [12,13] and have been prevalent in these communities for centuries [14]. The structure (shallow root system) and position (coastal, low lying) of mangroves makes them highly vulnerable to high-energy storms [12,13]. During high-intensity storm events, mangrove trees can be uprooted, severely defoliated, and suffer damage to their bark and limbs [15–17]. Hurricanes can even cause sudden mass tree mortality from these direct impacts, as well as delayed mortality [18] from indirect impacts such as flooding and siltation [17]. Extra sediment from storm surges and standing water from heavy rainfall can suffocate mangroves through inhibition of gas exchange [16,17].

Climate change, and a warmer, wetter atmosphere in particular, is increasing the frequency of high-intensity storms [19]. With these stronger storms comes greater risk of direct destruction of mangroves (e.g., physical damage) as well indirect damage (e.g., increased rainfall leading to flooding mortality), which may be the most devastating impacts of future storms [20]. Future hurricane events may pose more of a threat to the remaining mangrove ecosystems than gradual climate changes [15].

Studies of mangroves over space and time, including their productivity, have used remote sensing techniques for over half a century [21]. Recently, remote sensing has been used to assess the impact of hurricane events on mangrove populations (e.g., [22–25]), primarily utilizing LiDAR [26], Landsat [22,24,27], and MODIS data [3,23,25]. MODIS-derived analysis has become the first choice for mapping on a country level, as it provides a long data record (2000–present) for understanding changes over time and has high temporal frequency [25]. Although MODIS is useful at a regional scale, these data (250–500 m pixels) do not match the scale of information needed by local communities or decision makers for assessment of damage. Landsat products (30 m pixels) provide a finer spatial resolution but may lack the continuity of coverage over time or relevant overpass frequency for punctuating events like hurricanes. New sensors with higher spatial resolution and temporal cover have the potential to address the need for damage detection and mapping relevant at local scales, such as the European Space Agency's recently launched Sentinel-2 MultiSpectral Instrument (2015–present, 10 m resolution). Sentinel-2 has rapidly been adopted for spatiotemporal analysis of mangroves and hurricane damage [28–30]. New high-resolution imagery (i.e., Planet) aids in the interpretation and rapid evaluation of remote mangrove complexes that have limited field-based access [31].

The 2017 North Atlantic hurricane season exemplified increased storm intensities, tying the record for the most storms reaching hurricane strength [19]. We focused on Hurricane Irma, a Category 5 storm that made landfall on the northern coast of Cuba, where there were large coastal mangrove complexes [32]. Irma was the first Category 5 to make landfall in 93 years within the Cuban archipelago. At landfall, Irma was ~550 km in diameter and was over the area from 8 September 2017 to early 10 September 2017 [33]. The region experienced coastal flooding, storm surges, and high winds [33]. As the storm tracked along the coast and keys, winds were sustained at 157 km/h (85 kt), with gusts up to 226 km/h (122 kt) [33]. Additional damage was caused by heavy rainfall across the archipelago [33]. The eye of Irma passed directly over the Buena Vista Reserve (UNESCO Biosphere Reserve, Ramsar Wetland of International Importance), which includes Caguanes National Park (CNP). Due to limited resources, CNP managers were unable to fully assess the extent and severity of damage to the Park's ecosystems and requested our assistance (personal comm. N.H. López, March 2018).

The ecological damage of Hurricane Irma was extensive, as witnessed first-hand by this team, but has not previously been adequately documented in the literature in terms of the spatial extent, severity, and temporal patterns. Here, we quantify the short-term (years) damages of a high-intensity storm to a mangrove ecosystem while testing new methods to provide rapid evaluation and locally relevant scale information. We apply our new method, Hurricane Disturbance Vegetation Anomaly (HDVA), which utilizes historical phenology represented by the Enhanced Vegetation Index (EVI) to normalize the magnitude of hurricane disturbances on land cover, or the so-called anomaly (HDVA ≥ 1 when there is no damage and is less than one when damages occur). Land covers such as forests or mangroves have some interannual variations (EVI $\pm$5–10%); however, large variations may suggest more extreme impacts to the ecosystems, including mortality. We compared analyses with Sentinel-2 and MODIS-derived damage detection (e.g., [25]). We addressed the following questions: (1) What is the potential for detection of mangrove damage and damage severity with Sentinel-2 data? (2) How do the analysis process and information vary when MODIS and Sentinel-2 data are compared? (3) What is the extent and duration of ecological disturbance of Cuban mangroves from Hurricane Irma, particularly in Caguanes National Park?

2. Materials and Methods

We developed and tested a pixel-based change detection method that compared post-storm vegetation cover with the mean in the historical data record prior to the storm. We also analyzed several years of post-hurricane phenology to assess recovery. Vegetation cover was assessed using the Enhanced Vegetation Index (EVI) during the dry-season, as this period is relatively (1) cloud-free, (2) stable in phenology, and (3) consistent year to year. We assessed a new metric, the Hurricane Damage Vegetation Anomaly, as the change in EVI post-storm relative to the historical observations. MODIS provides a long period of historical observations for assessing forest disturbance [25], whereas Sentinel-2 had only a 2-year period of record prior to Hurricane Irma. All mangroves, as well as other land covers and land uses, were previously identified for Cuba [32], which we used within our study area (Figure 1).

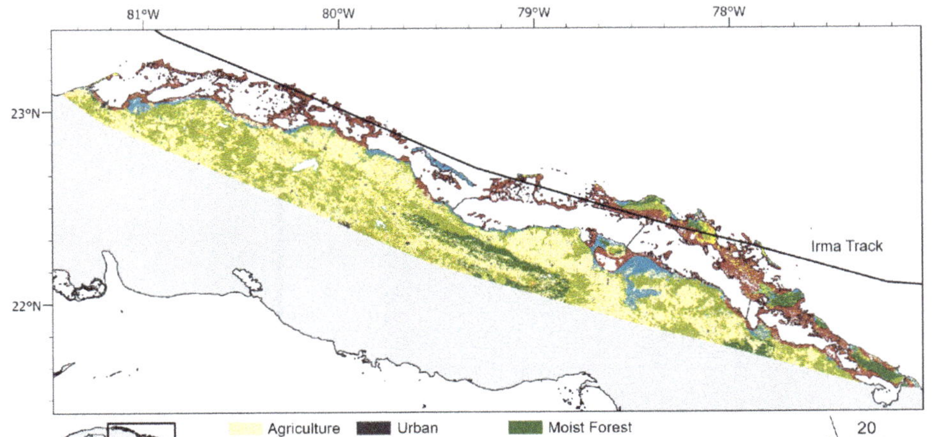

Figure 1. Land cover map of Cuba modified with permission from [32]. Here we focused on damage to mangroves (red), wetlands (aqua), and moist and dry forest (dark and light green). The track of the eye of Hurricane Irma is shown to parallel a long swath of coastline from east to west.

2.1. Study Area and Context

Hurricane Irma hit the northern coast of Cuba as a Category 5 Storm on 9 September 2017 (Figure 2; [33]). The eye of the hurricane swept along the Cuban keys, and as it did, it weakened to a Category 4 storm after a few hours, then down to a Category 2 before it headed north on 10 September [33]. As it paralleled the coast for ~400 km, the eye of the storm was ~30 km from the mainland, and the storm was nearly 550 km wide, measured perpendicular to the storm track at its greatest intensity (Figure 2, A to A′), spanning many ecologically important areas (Figure 3). Severe damage to vegetation on the Cuban keys and coastline, specifically the mangrove populations, was clear upon visual inspection of high-resolution images (Planet) before and after the hurricane. Most of the damage was within 50 km of the eye of the storm; thus, we defined the study area with a 60 km buffer from the eye of the storm. The total study area (land area, excluding ocean) was 1,700,390 ha (ArcGIS calculation of study area polygon), although MODIS and Sentinel-2 estimates varied due to their spatial resolution (MODIS: 1,757,981 ha; Sentinel-2: 1,710,779 ha). The study area is comprised of many land uses and land covers: agriculture, dry forests, mangroves and wetlands, moist forest, and a minor presence of xeric salt flats, barren, and urban areas [32]. Mangrove and wetlands comprised over 260,000 ha (MODIS: 288,931 ha; Sentinel-2: 264,258 ha).

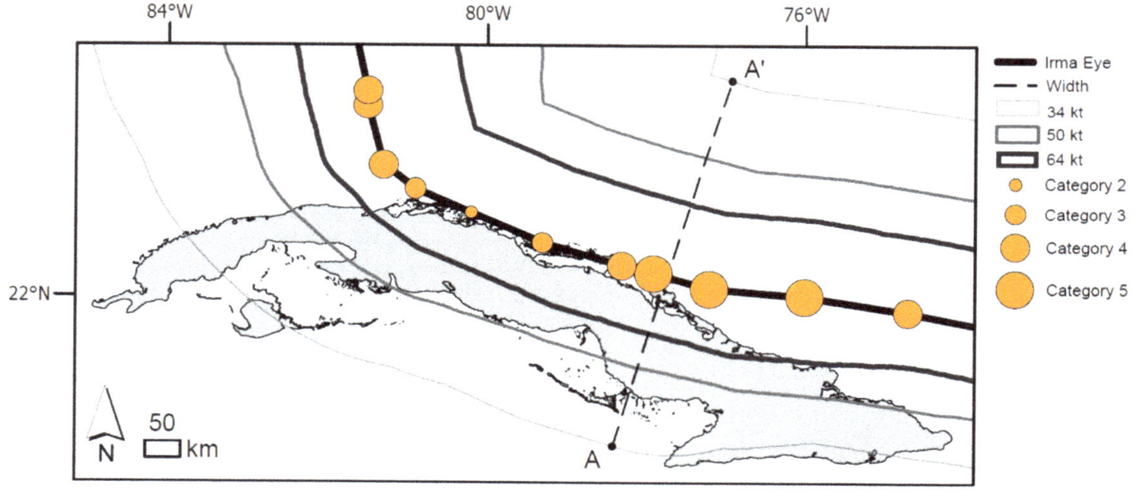

Figure 2. Track of Hurricane Irma (NOAA National Hurricane Center GIS Archive—Tropical Cyclone Best Track). Distance between A and A′ is ~550 km.

Previous storm paths that were <200 km from the study area include Ike (2008, Category 1 at closest proximity to the study area), Ernesto (2006, Tropical Storm), Rita (2005, Tropical Storm), Dennis (2005; storm track more than ~200 km to the NW but a Category 4 storm), Michelle (2001; entered NW corner of study area at Category 4, de-intensifying to Category 1), and Georges (1998; Category 1 with a similar path as Irma). Note that Georges was prior to the period of record for the MODIS data used here but might have affected vegetation in the early 2000s. The authors recall Hurricane Kate passing through the region in 1985 (Category 2); however, although the eye of the storm passed through the south end of CNP, the ecological damage was not extensive. One author's grandfather spoke of the devastation of the mangroves from the 1933 storm, which is noted as Category 3 in our study region, passing 60 km north of CNP (Major Hurricane 1933234N13335 in the IBTrACS database [34]).

Caguanes National Park (CNP) experienced intensive ecological damage from Hurricane Irma. Our team visited CNP in March 2018. We used this as a case study area to train ourselves on visual image interpretation for the broader study area due to our familiarity with the disturbances from Hurricane Irma as well as requests from the CNP staff to assist in assessing the damages. Due to limited resources and infrastructure, a complete census of the ecological damage could not be conducted in CNP. We developed remote sensing methods to assess hurricane damage at 10 m resolution, a scale relevant to CNP managers and stakeholders. CNP is an area of high importance to Cuba. In addition to being classified as a National Park, it is also a Protected Area of Managed Resources, a Ramsar site (Wetland of International Importance), and a UNESCO-MAB Biosphere Reserve (a site that aims to merge social systems based on research, tourism, education, and settlement with the protection and conservation of species and ecosystems [35]). Figure 3 shows the multiple protected areas encompassing CNP and the study area along the northern coast where the hurricane hit.

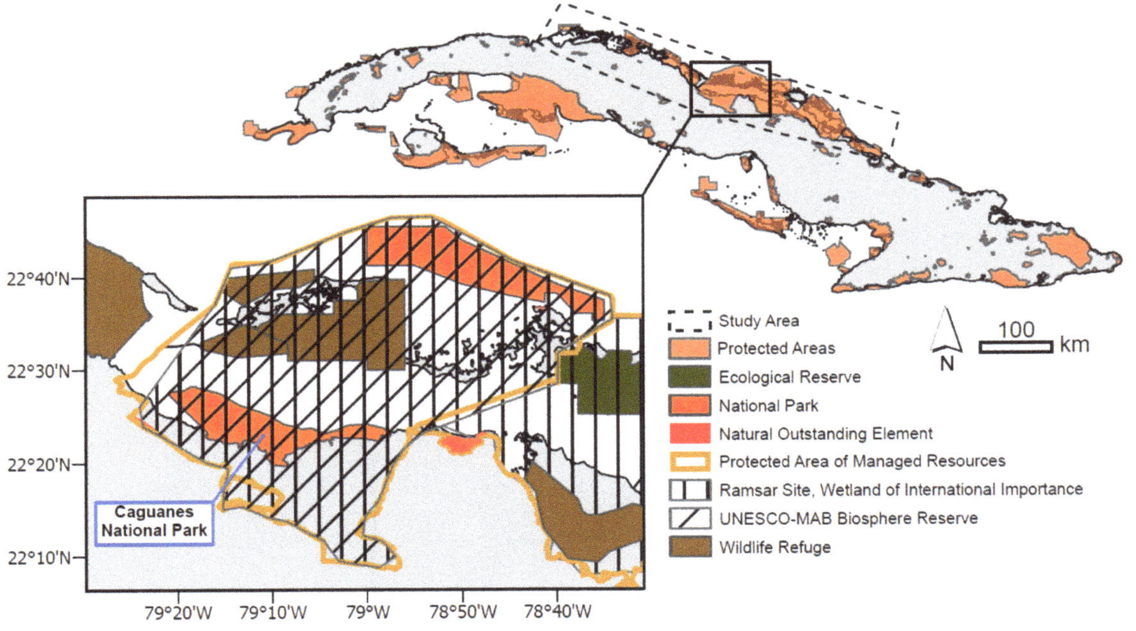

Figure 3. Protected areas in Cuba and Caguanes National Park (CNP) [inset] [36], with the broader study area (60 km buffer from eye of Hurricane Irma) shown within dashed line.

These protected areas exist to safeguard the natural elements of this region (Figure 3). A feature of the study site is a large area of relatively undisturbed mangroves, at least, prior to Hurricane Irma. Mangroves are brackish ecosystems with four species (roughly, from coastline proceeding inland): (1) *Rhizohora mangle I* (L.) (red mangrove), closest to the ocean, with roots perpetually inundated with salt water; (2) *Avicennia germinans* (L.) (black mangrove); (3) *Laguncularia racemose* (L.) C.F. Gaertn (white mangrove), found at higher elevations; and (4) *Conocarpus erectus* (L.) (buttonwood) [32]. While the focus of this study was mangroves, we also considered Cuban moist and dry forests. Small farms in the region include grasslands for pasture grazing and small-scale croplands, although historically the region had extensive, large-scale sugarcane production involving large water diversions and canals. The upland areas in the study area are largely karstic topography. These upland characteristics can have a strong effect on coastal freshwater hydrology, such as the delivery of agricultural pollutants, factory biproducts, and freshwater flow paths. The semi-deciduous dry forests predominate on limestone geology with little soil substrate, losing their leaves during the dry season.

The nested design of study sites and data sets included analysis of high-resolution images for CNP (See Section 2.2) and temporal, moderate-resolution images (MODIS and Sentinel; see Section 2.3; study area shown in Figure 3). The data scenes and tiles used in this study are described in Table 1.

Table 1. Remote sensing data sets described by tile or scene ID and time period.

Data Set	Tile or Scene ID	Date	Years
Planet	1749423_2017-05-27_RE2_3A 1749422_2017-05-27_RE2_3A 20180424_183410_0f02_3B 20190302_152754_1044_3B 20200419_160223_95_106e_3B 20210417_150620_27_2429_3B 20220428_154341_49_2403_3B	27 May 2017 24 April 2018 2 March 2019 19 April 2020 17 April 2021 28 April 2022	2017–2022
Sentinel-2	17QMF,17QNF, 17QNE, 17QPF, 17QPE, 17QQE, 17QQF, 18QTK	All observations during the dry season (1 November to 21 March)	2015–2020
MODIS	v6h10	All observations during the dry season (1 November to 21 March)	2000–2022

2.2. High-Resolution Images

We obtained Planet microsatellite images (Planetscope and Rapid Eye; <5 m pixels) for April and May 2017, as this timeframe allowed for cloud-free views of CNP prior to Hurricane Irma. After Hurricane Irma, cloud-free microsatellite images were found for March through May 2018 for CNP, as well as subsequently in 2019, 2020, 2021, and 2022. Planet provided these images through their Education and Research Program for limited, non-commercial access to the Planetscope Imagery and Rapid Eye Archive. These images were used to interpret potential changes in land cover to inform analysis and interpretation of MODIS and Sentinel-2 (Table 1). Additionally, monthly pixel-based reflectance composites of Sentinel-2 were used to visually assess the duration of freshwater inundation in CNP, changes in green vegetation, and recovery trajectories over the first 8 months post-hurricane. Monthly pixel-based composites were needed to ensure a cloud-free, or nearly cloud-free, look at CNP.

2.3. Data Processing of Spatiotemporal Data

Studies in tropical areas are hampered by cloud cover or cloud contamination at the time of satellite overpass (e.g., midday), and our study area is no exception. We focused on the dry season (1 November to 21 March) to minimize clouds and atmospheric moisture as well as reduce missing data due to persistent cloud cover of the rainy season. This period has relatively stable phenology and strong contrast in vegetation indices between land covers (e.g., forest, grassland).

To create pixel-based image composites, we built a workflow in the Google Earth Engine Code Editor API (GEE) [37] based on our previous pixel-based compositing methods for input to land use classification [32]. The GEE allows access to an extensive public data catalogue and to cloud-based computing power [37]. Conceptually, the same approach was used for MODIS and Sentinel-2 image processing, where our algorithm selects the best quality observations by pixel during the dry season for a given year to create a seasonal mean EVI. All good observations were retained and then were transformed as described below. The resulting imagery was the input for MODIS and Sentinel-2 analyses described below.

We utilize the Enhanced Vegetation Index (EVI, Table 2) to represent vegetation. The EVI includes the blue band to remove residual atmospheric contamination, such as from smoke or thin clouds [38], and has been shown to have a greater dynamic range and sensitivity than other vegetation indices [38,39]. We utilized the Enhanced Vegetation Index (EVI) from the MODIS Terra Vegetation Indices 16-Day Global data set (250 m, MOD13Q1 v. 006, [38]) (Table 2). The MODIS composite algorithm selects the best available pixel from all acquisitions (2–3 day repeat overpass) during the 16-day period, filtering for low clouds, low view angle, and greatest EVI value. After selecting dry season observations from MOD13Q1 EVI, each pixel's annual dry season mean EVI was calculated for each year of the historical period (2000–2017) and for the post-hurricane period (2017–2018). These EVI metrics were input to the damage detection algorithms discussed below.

The Sentinel-2 (10 m, Copernicus Sentinel-2 MSI: MultiSpectral Instrument, Level-1C processed by the European Space Agency) data archive begins in 2015, a substantially shorter historical reference period (2015–2017) with regard to Hurricane Irma. We utilized the Sentinel-2 Quality Assurance information to mask cloud and cirrus cloud contamination (QA 60). If no cloud-free pixels existed, the best observation was selected based on quality assurance. After selecting good observations, Sentinel-2 EVI was derived following García-Llamas et al. [40] for the annual dry season means (2017, 2018) and the historical dry season mean (2015–2017).

Table 2. Enhanced Vegetation Index (EVI) formulas by sensor, where ρ is atmospherically corrected or partially atmospherically corrected surface reflectance [41]; EVI metrics (disturbance, reference) calculated after quality assurance filtering.

	MODIS	Sentinel-2
EVI	$2.5\left[\frac{\rho5-\rho4}{\rho5+6\rho4-7.5\rho2+1}\right]$	$2.5\left[\frac{\rho8-\rho4}{\rho8+6\rho4-7.5\rho2+1}\right]$
EVI Disturbance	Dry Season Mean, 2017–2018	Dry Season Mean, 2017–2018
EVI Reference Period	Dry Season Mean, 2000–2017	Dry Season Mean, 2015–2017

EVI can naturally vary over time due to disturbances such as storms. To determine whether the EVI observations after Hurricane Irma were significantly different than the reference period, a matched pairs t-test was run on a sample of 500 random points to determine whether there was a significant difference between the mean EVI value of a pixel for the reference period and the EVI value of the same pixel for the disturbance year. The test was run on both MODIS and Sentinel-2 outputs and run for the mangrove and wetlands as well as the dry forest land cover types. Points that only had a value in either the reference period or disturbance year and not in both (a few seen mainly around the northern islands, coastal pixels) were removed. The rasters of the mean EVI of the reference period and disturbance period were generated in Google Earth Engine using the methods described above. The points were generated in ArcGIS Pro, and all t-tests were run using JMP 17. Additionally, we utilized a z-score metric to evaluate whether the MODIS EVI disturbance period was significantly different from the MODIS EVI interannual variance during the reference period.

2.4. Hurricane Disturbance Vegetation Anomoly (HDVA)

We created the Hurricane Disturbance Vegetation Anomaly (HDVA) to assess the damage to vegetation cover (HDVA; Equation (1), Table 2). Conceptually, the HDVA is a comparison of EVI after the storm (EVI disturbance mean; 1 November 2017 through 21 May 2018 for Hurricane Irma) relative to "normal" (EVI reference period mean; mean based on available historic data). The historical dry season mean was used as the reference period (Table 2), and the disturbance mean was the dry season following Hurricane Irma (2017–2018). MODIS and Sentinel-2 have very different reference period lengths, so the utility of a reference period could be compared. For MODIS, the reference period included 2000 through 2017; Sentinel-2 was limited to two dry seasons (2015 through 2017) prior to

Hurricane Irma. An advantage to using a reference period not immediately adjacent to the hurricane (e.g., days or weeks after landfall) is that it allows for analysis of potential persistent disturbances (e.g., slow mortality), rather than sudden changes that might rebound quickly (e.g., defoliation), persistent standing water (lowering EVI), or green flush with new leaves. Further, mangrove mortality can be delayed due to slow suffocation from flooding and might be better captured months after the storm [16,18].

HDVA represents the relative vegetation abundance assessed by EVI as a fraction compared with the historical reference period. The HDVA formulation is calculated by pixel, as given in Equation (1):

$$\overline{(EVI_{disturbance})} / \overline{\left(EVI_{reference}\right)} \quad (1)$$

where $\overline{EVI}$ is the mean EVI value by pixel over the disturbance or preference period (Table 2).

These HDVA data were restricted from 0 to 1 for MODIS and 0.001 to 1 for Sentinel-2 (due to outliers around zero in water features) for damage severity analyses; HDVA values of one indicated no loss and values greater than one signified a gain in comparison to the reference period. An HDVA value near zero corresponded to nearly a 100% decline in EVI (Table 2). We displayed HDVA values as percentages for easy interpretation and communication, where a value of 100% means no discernable damages. As HDVA approaches zero, the greater the disturbances to vegetation cover or more serve damage from the storm; at zero, HDVA indicates all photosynthetically active vegetation cover is absent compared with the reference period.

2.5. Determining Severity

To communicate the relative severity of disturbances, we transformed HDVA data by quartiles. After restricting the data to less <1, HDVA data was exported for statistical categorization by quartiles, where each quartile represents 25% of the data set. Statistical analysis was run in JMP Pro (14.2.0) to determine quartiles defining five severity categories: (1) Catastrophic (0–25%), (2) Severe (25–50%), (3) Moderate (50–75%), (4) Mild (75–100%), (5) No Loss (>100%). Once the statistical classes (severity levels) were defined, the classes were mapped back to pixels in GEE, creating a severity map. The results were mapped to show the final assessment of severity along the northern coast). We tested the concurrence of MODIS and Sentinel-2 using a confusion matrix and Kappa statistic. We created 10,000 random points and removed points with no Sentinel-2 data (as the MODIS footprint was larger), for a total of 9144 points. We used Sentinel-2 as the "ground truth" data and MODIS as the test data.

2.6. Damages by Land Cover

To determine how different land use types were affected by the hurricane, the severity maps from HDVA were analyzed with a contemporary land cover map generated from Landsat data, representing 2015 [32] (Figure 1). Natural land cover classes of interest were mangrove, wetlands, moist forest, and Cuban dry forest. The area of damage, and the severity of that damage, were tabulated by land cover and severity level. The land cover map had some errors where wetlands were labeled but where we could identify mangroves in Planet data; we combined wetland and mangrove land cover types to be a single class. Pixel counts were converted to area based on their native size in an equal area projection.

2.7. Reference Time Series and Images

To gather a sense of how variable the MODIS-derived reference period was, we looked at EVI values from 2000–2022 for the entire study area as well as at 10 points in and near CNP in both mangrove and dry forest classes. Representative points were randomly selected within areas that were well-distinguished and defined in terms of land cover types based on the user's visual interpretation of Planet and Sentinel-2 data and the authors' familiarity of the area. These points were selected to avoid mixed pixel effects. We also utilized cloud-free Planet scenes to train our interpretation of post-hurricane recovery time series.

3. Results

3.1. Image Interpretation

High-resolution imagery obtained from Planet (3 to 5 m) in May 2017 and April 2018 gave a clear view of CNP during the same time of year (end of dry season) (Figure 4). False-color infrared displays of CNP show as bright red (high photosynthetic activity) in the mangrove and wetland complex in 2017. In 2018, the central portion of the mangroves and wetlands remains devoid of photosynthetically active vegetation, with some potential recovering in the western region (Figure 4).

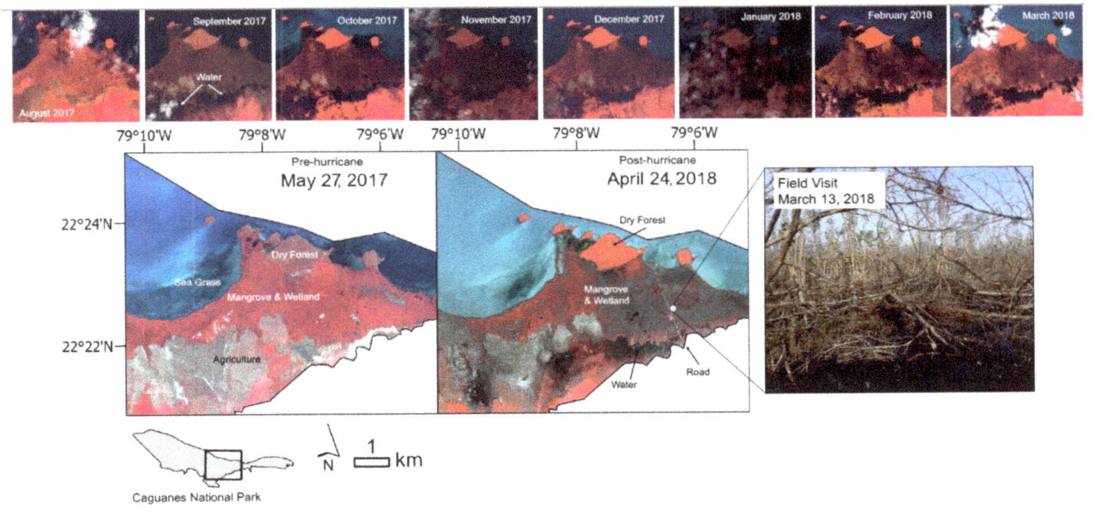

Figure 4. Pre- and post-hurricane imagery for CNP (false-color infrared). Single scene images from Planet (center panel) for 27 May 2017 (5 m) and 24 April 2018 (3 m) show changes in vegetation cover of mangroves and wetlands, as well as increased water in agriculture and wetlands. Monthly Sentinel-2 composites (top panels) show sustained freshwater inundation (black) well after the storm.

Sentinel-2 pixel-based monthly composites displayed in false-color infrared show large changes in photosynthetically active vegetation after Hurricane Irma, most notably in the mangroves and wetlands of CNP. Some recovery of vegetation in the Western portion of the mangroves can be seen in the October and December images, and quite clearly by March 2018, yet the majority of the mangrove and wetland vegetation remains severely changed after the Hurricane. This time series illustrates the duration of freshwater inundation. While there is typically standing brackish water in these mangroves, areas of open standing water are evident through the entire Sentinel-2 time series shown here (Figure 4).

3.2. Data Processing

Figure 5A,B shows the best single scene images from Sentinel-2 before and after the storm; while the damage can clearly be seen, there are clouds blocking large areas where additional info would have been missed (standing water in southern portion of CNP).

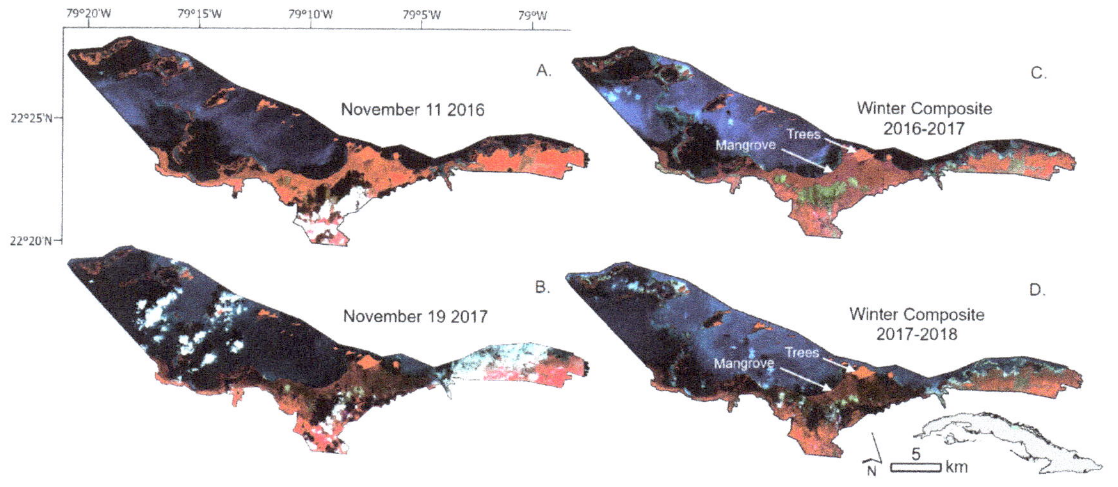

Figure 5. Sentinel-2 imagery shown in false-color infrared (RGB Bands 8, 4, 3) for single dates (**A**,**B**) and pixel-based seasonal composites (**C**,**D**). (**A**) Sentinel-2 11 November 2016 (pre-hurricane), (**B**) 19 November 2017 (post-hurricane), (**C**) 1 November 2016–23 March 2017 (pre-hurricane), (**D**) 1 November 2017–23 March 2018 (post-hurricane). The same stretch and gamma were used in all displays.

Our pixel-based dry season composites show changes in vegetation cover without large areas of cloud contamination (Figure 5C,D), although there are still cloud-contaminated pixels. The composite images key into specific areas of CNP affected by Hurricane Irma that may have been missed with single scene imagery.

We found that the EVI post-hurricane was significantly different than the reference period. Results for the matched pairs t-test showed the mangrove and wetland land cover type had a significantly lower dry season mean EVI in the disturbance year than the reference period (MODIS: t(453) = −17.68, $p < 0.001$; Sentinel-2: t(477) = −18.29, $p < 0.0001$). In dry forests, the disturbance year dry season mean EVI was significantly greater than the reference period (MODIS: t(499) = 12.09, $p < 0.0001$; Sentinel-2 t(499) = 8.50, $p < 0.001$). Considering all land cover types, both sensors agreed that the disturbance year mean EVI was significantly greater than the reference period, although at a lower significance with Sentinel-2 (MODIS: t(486) = 6.96; $p < 0.0001$, Sentinel-2: t(491) = 2.94, $p = 0.0017$).

3.3. HDVA Results

The HDVA values were categorized based on statistical quartiles, so they had slightly different break points for MODIS and Sentinel-2 HDVA values (Table 3). Overall, the distribution of HDVA observations were similar between the two sensors. MODIS had a slightly lower range for catastrophic and severe losses; that is, more pixels were detected as having HDVA values in the order of 0.8 or less (Figure 6). Statistically, the Kappa of 0.27 (Table 4) is a fair agreement; however, the confusion matrix shows MODIS underestimated severe damage and overestimated mild damage compared with Sentinel-2.

Table 3. MODIS and Sentinel-2 severity HDVA ranges.

	Area (ha)	Catastrophic	Severe	Moderate	Mild	No Loss	Mean HDVA
MODIS HDVA		0–0.595	0.596–0.805	0.806–0.905	0.906–0.98	>0.98	-
Study area (ha)	1,757,981	120,869	137,175	116,813	146,350	1,236,775	1.05
mangroves and wetlands	288,931	30%	27%	13%	8%	22%	0.77
dry forests	520,275	2%	3%	5%	9%	81%	1.1
moist forests	121,394	3%	8%	8%	12%	69%	1.04
Sentinel-2 HDVA		0.001–0.645	0.646–0.805	0.806–0.905	0.906–0.98	>0.98	-
Study area (ha)	1,710,779	177,885	152,845	161,797	159,273	1,058,980	1.08
mangrove and wetlands	264,258	41%	19%	11%	7%	22%	0.77
dry forests	517,995	4%	6%	9%	10%	71%	1.13
moist forests	120,242	8%	10%	13%	13%	56%	1.01

Table 4. Confusion matrix using Sentinel-2 HDVA classes as "ground truth" and MODIS HDVA classes as classified.

		Sentinel-2 HDVA						
		Catastrophic	Severe	Moderate	Mild	No Loss	Total	User's Accuracy
MODIS HDVA	Catastrophic	1141	154	51	15	124	1485	0.768
	Severe	785	510	207	103	216	1821	0.280
	Moderate	268	492	352	227	578	1917	0.184
	Mild	115	284	292	288	974	1953	0.147
	No Loss	26	94	156	186	1506	1968	0.765
	Total	2335	1534	1058	819	3398	9144	
	Producer's accuracy	0.489	0.332	0.333	0.352	0.443		0.415
	Kappa							0.271

Spatially, maps show the distribution of damages estimated by the HDVA have distinct patterns (Figure 7). Generally, the spatial patterns with MODIS and Sentinel-2 are similar with catastrophic damages in many coastal areas. A majority of damage occurred along the northern coast and Cuban keys (Figure 8).

Zooming into CNP, clear damage to mangroves is visible, and standing water can also be seen towards the south of the park (Figures 4 and 9).

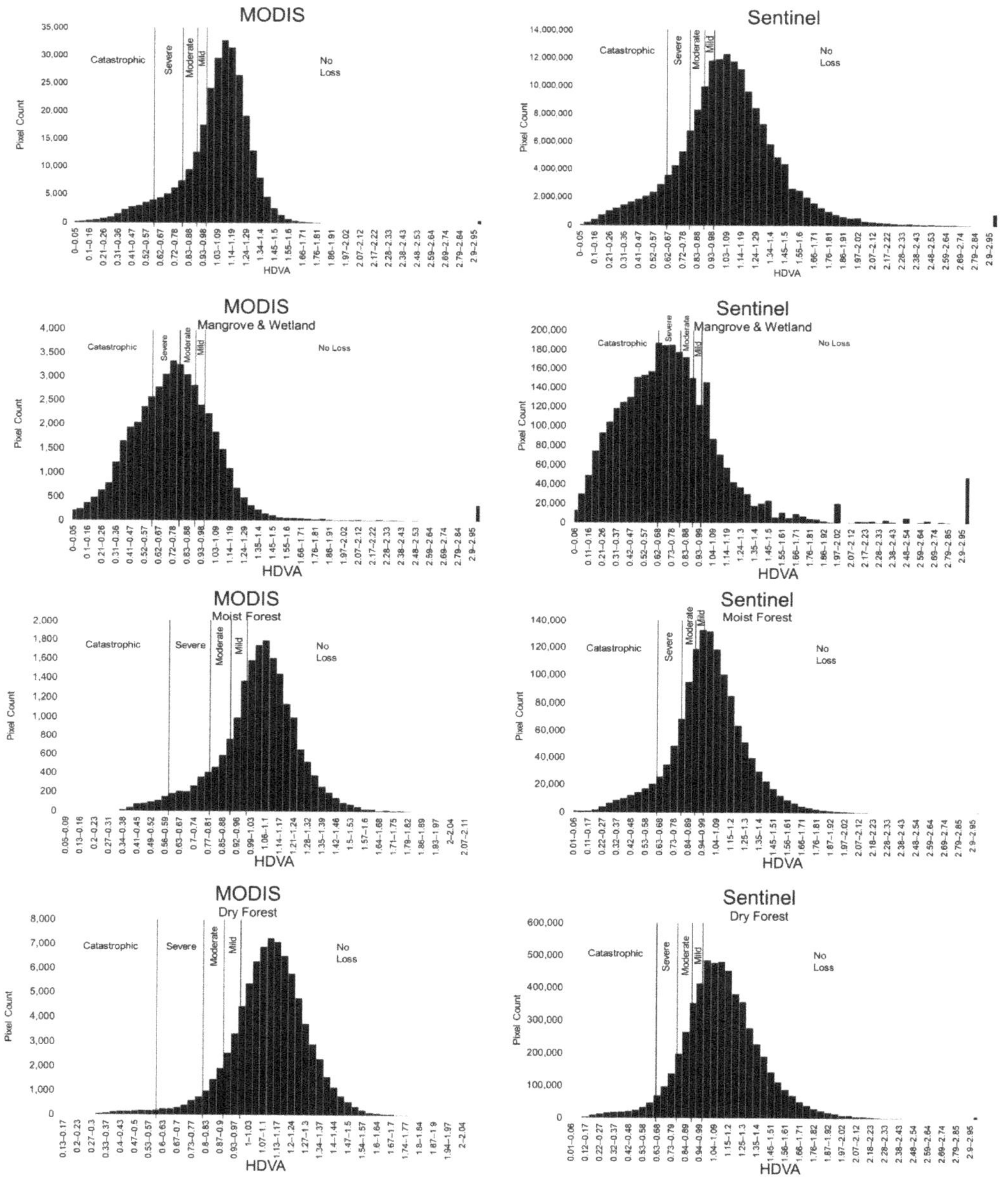

Figure 6. Distribution of all pixels in the study area with severity categories overlaid. Severity classes were determined based on these distributions restricted to <1. Mean values of the entire distribution can been seen in Table 3.

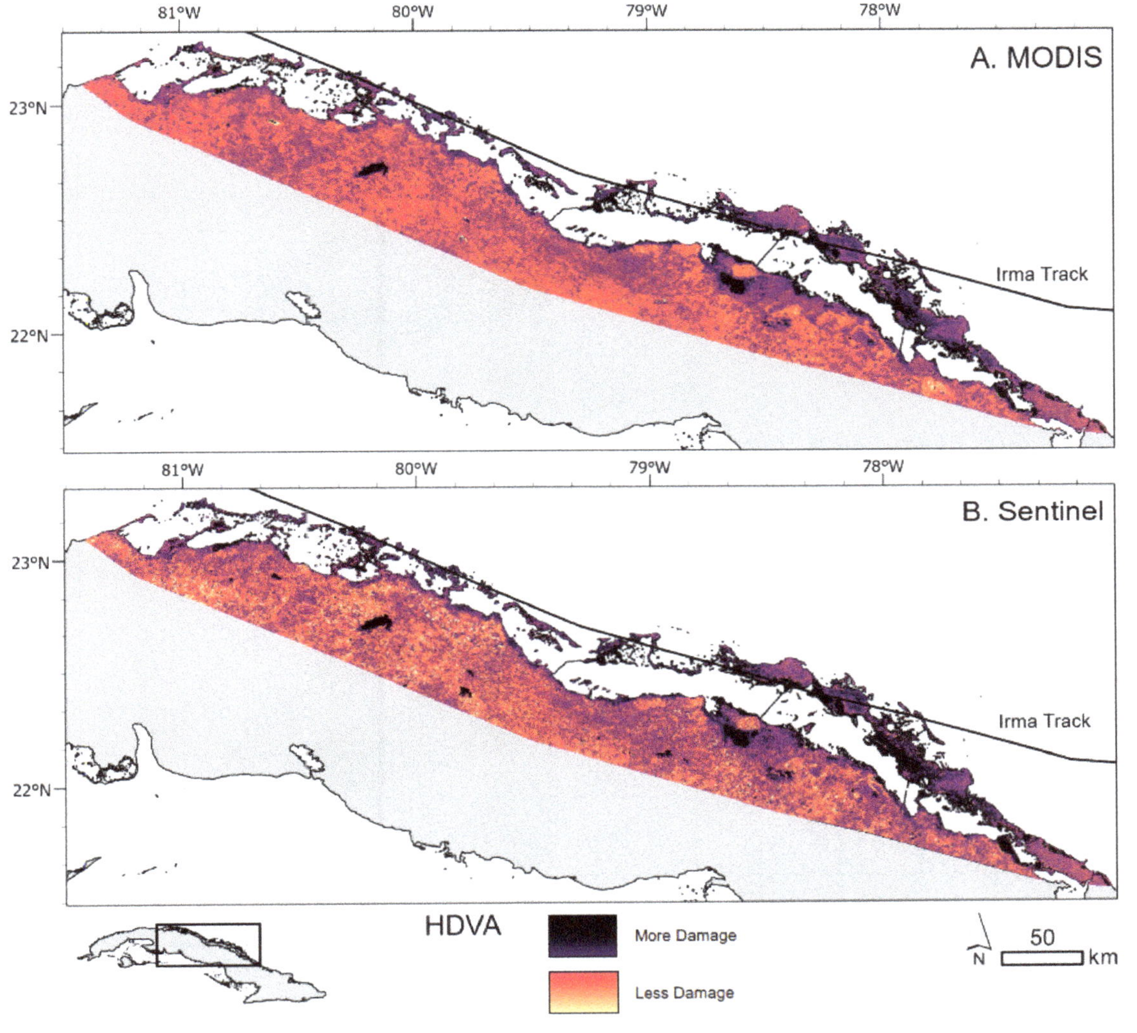

Figure 7. MODIS (**A**) and Sentinel-2 (**B**) HDVA maps illustrating spatial patterns in vegetation damage from Hurricane Irma. Darker colors indicate greater damages in vegetation in the year after the storm.

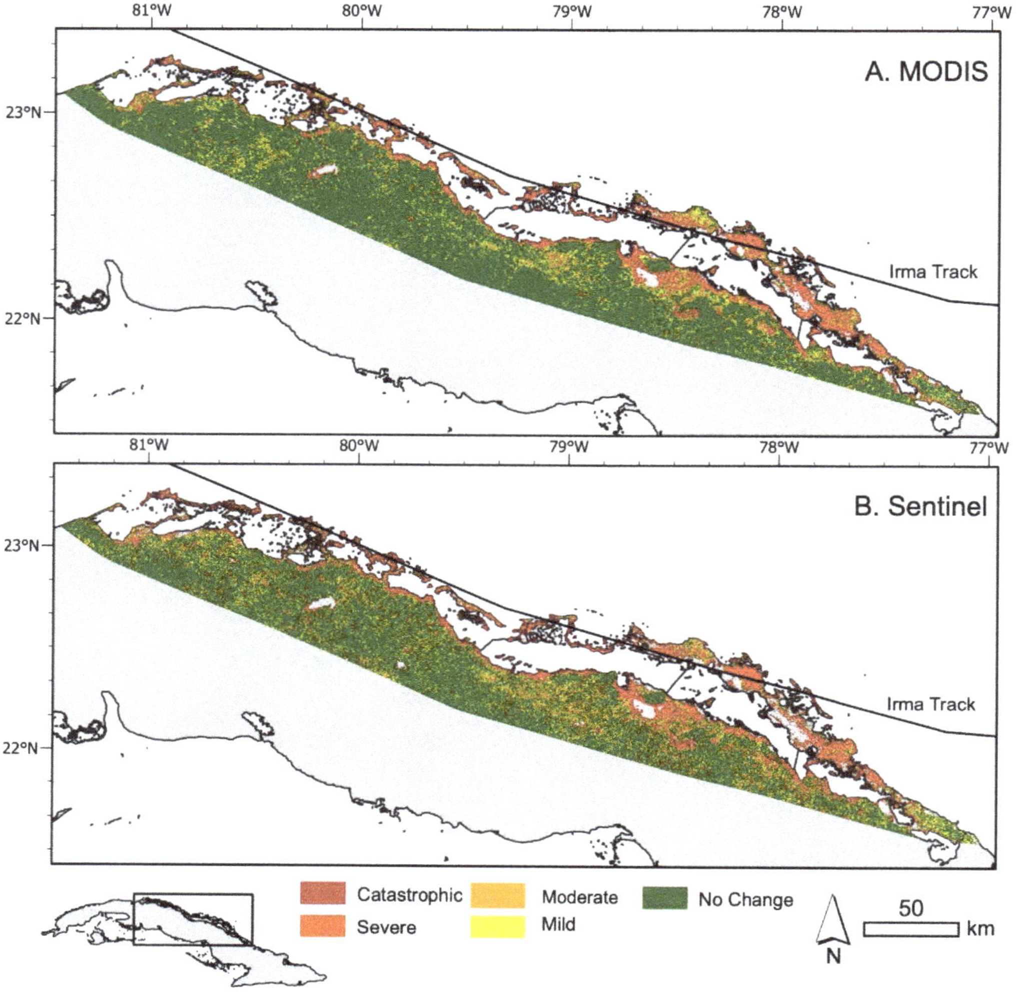

Figure 8. MODIS (**A**) and Sentinel-2 (**B**) severity maps from HDVA data classes.

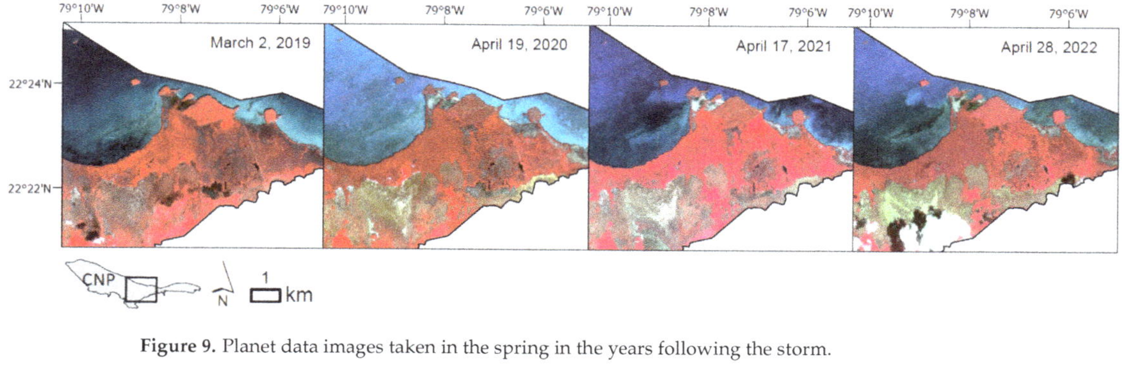

Figure 9. Planet data images taken in the spring in the years following the storm.

3.4. Reference Variability and following Years

The MODIS EVI time series (Figure 10) for mangrove and dry forest in and around CNP was relatively stable historically, even through prior hurricanes. The full study area mean shows little variation other than a dip seen in both land cover types in the 2005–2006 range. During the historical period, for the 20 points in and near CNP, the average EVIs were relatively stable (0.46 average $\pm$ 0.02 standard deviation in mangrove, 0.54 average $\pm$ 0.02 standard deviation in dry forest). Both land covers had substantial declines in EVI for the two winters, declining an average of 41% (mangrove) and 37% (dry forest) by winter 2019 relative to the historical period. Since winter 2019, EVI values have begun to recover, to 9% higher than the historical EVI mean in mangrove and to −17% of the historical EVI mean in dry forest (Figure 10). The mean dry season MODIS EVI following Hurricane Irma (2017–2018) was significantly different than the reference period's mean dry season MODIS EVI within the mangrove and wetland land cover type (z-score = −4.03, $p < 0.01$). No significant difference was found in the mean dry season MODIS EVI of 2017–2018 in comparison to the reference period when considering just the dry forest area or the study area (all pixels).

HDVA values were higher in the northeast of the Cuban archipelago. Within the land use classes that included mangroves and wetlands, the severity of damage was extremely high (Figure 11). MODIS detected 225,363 ha (78%) of the mangrove and wetland class were damaged by the storm, of which 86,231 ha (30%) of mangroves and wetlands were catastrophically damaged and 78,650 ha (27%) were severely damaged (Table 3).

Sentinel-2 analysis detected a slightly lower disturbance of mangroves and wetlands (207,113 ha or 78%), with 109,431 ha (41%) catastrophically damaged and 51,278 ha (19%) severely damaged. While MODIS detected a larger area of disturbance within mangroves, Sentinel-2 picked up more area classified as catastrophically damaged.

Both sensors indicated that the damage to dry forest was mostly moderate or mild (Figure 11). More damage was picked up in dry forest with Sentinel-2, estimating around 50,000 ha more damage in comparison to MODIS.

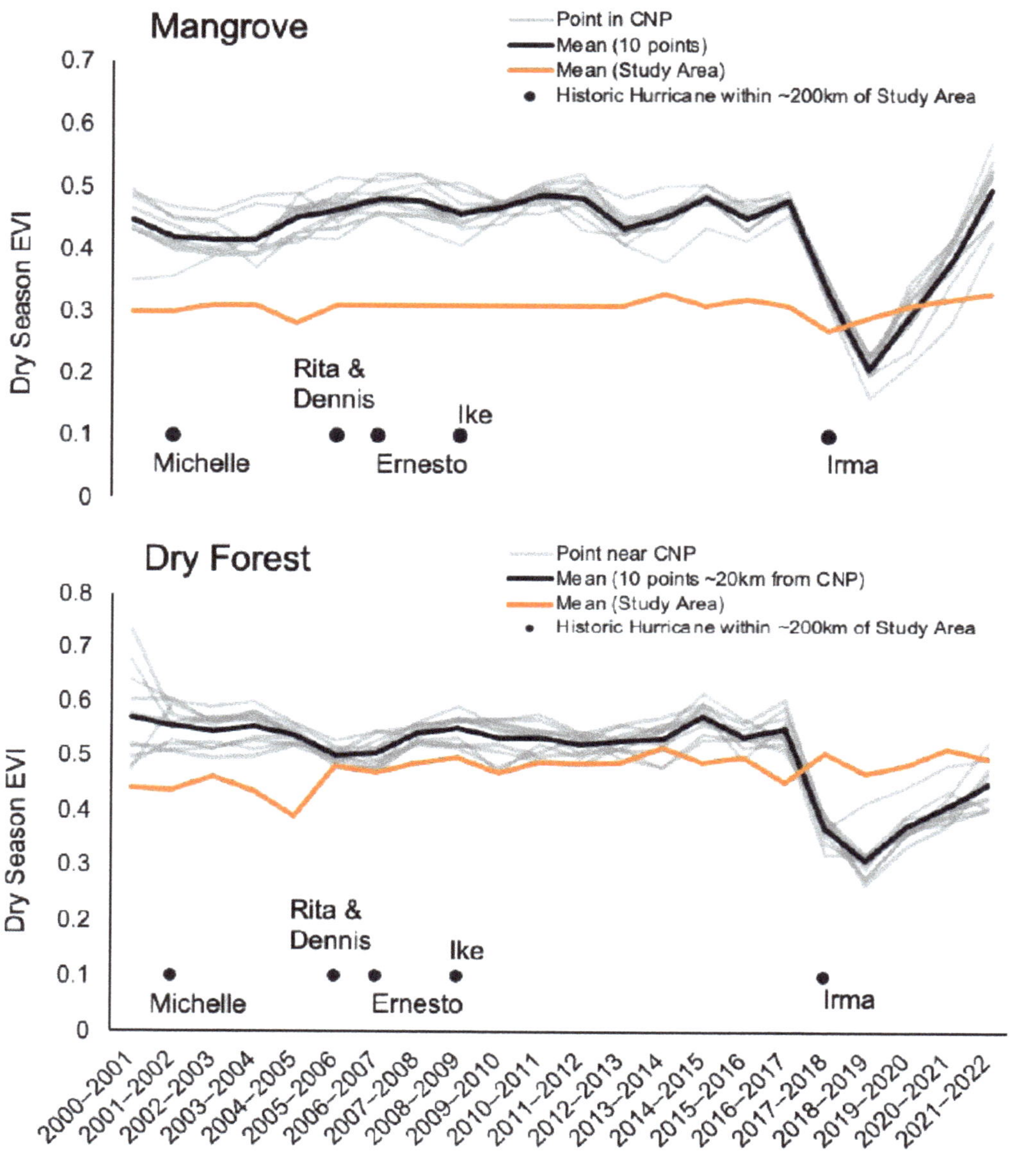

Figure 10. MODIS dry season EVI values representing 10 points in mangrove (CNP) and dry forest classes (~20 km east of CNP). Overall mean EVI of the entire study area overlaid (orange). Historic hurricanes within 200 km of the study area noted.

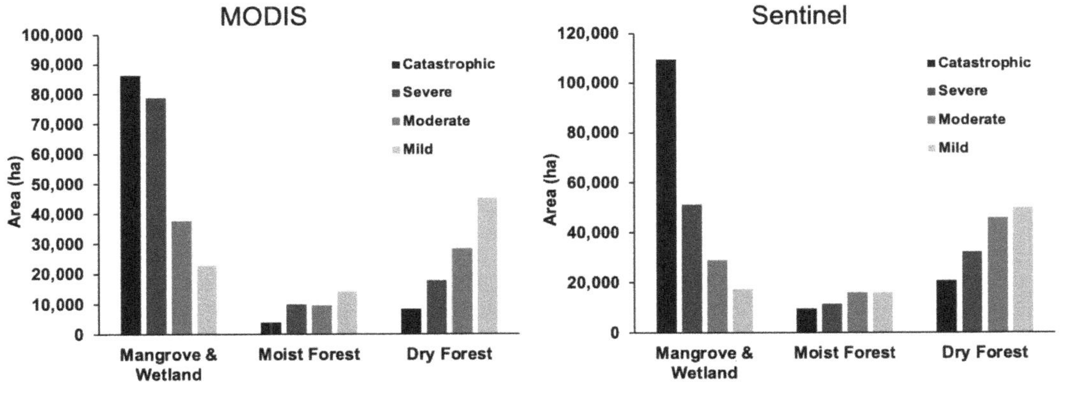

Figure 11. Total damaged area (ha) by severity classification within three land use types: mangrove and wetland, moist forest, and dry forest.

4. Discussion

We found there to be a great amount of catastrophic and severe ecological damage, specifically to mangrove and wetland land covers, along the northern coast of the Cuban archipelago following Hurricane Irma. This confirms the field-based observations of our team in and around CNP.

4.1. Ecological Impacts

Based on our 2018 field visit and regular reports from authors based at CNP, stands of mangroves with an HDVA categorized at catastrophic had widespread mortality, whereas those areas categorized on the lower end of catastrophic and severe were extremely damaged. Defoliation and organic decomposition in the catastrophic zones were so intense that the transparency of coastal waters plummeted to zero for more than a year due to tannins.

Concentrated zones of catastrophic mangrove damage coincide with stands of black mangrove, which occupy the low-lying saline mud flats and tidal lagoons, as the species is very salt tolerant. Red mangroves were also catastrophically affected in CNP, although to a lesser extent than black mangroves. The black mangrove species is vulnerable to prolonged submersion of its pneumatophores, which are the site of oxygen exchange for the root system in low-oxygen sediments. Extending vertically ~20 cm above the sediment, the black mangroves have pneumatophores that exchange oxygen when above water at low tide but are also vulnerable to suffocation if submerged for long periods of time. In this study, we see prolonged pooling of water in the tidal lagoons (Figures 4, 5 and 9)—the same areas that are dominated by black mangrove and experienced catastrophic damage from Hurricane Irma. Complete defoliation of mangroves was combined with the pooling of water.

Pooling of water after a hurricane is known to deplete oxygen in soil and water, which can exacerbate damage, particularly through submersion of black mangroves' pneumatophores. Catastrophic damage related to prolonged (months) water inundation could be related to anaerobic changes in soils, including reduced soil oxygen, lower redox potential (affecting denitrification and reduction of iron, manganese, and sulfate), and increased concentration of phytotoxins [42]. Typically, inundation after a hurricane increases salinity in soils and waters, but we find the opposite in this case. Lagomasino et al. [43] found similar pooling and black mangrove damage from Hurricane Irma in the mangroves of Florida, USA, which they speculate is due to prolonged salt exposure and physiological stresses from prolonged inundation. Unlike in Florida, the pooling of water in this study was not due to human-made structures restricting water flow but rather the large volume of water delivered by the storm (tidal surge, rainfall) plus the long inundation of freshwater that flowed slowly from the initial rainfall through the upland karst bedrock. Authors based in CNP measured coastal lagoon salinity as approaching 0 PSU for more than a year following Hurricane Irma due to freshwater inundation.

4.2. Changes in Land Cover Phenology

Recent remote sensing literature has highlighted the green leaf vegetation phenology of mangrove and coastal ecosystems, including seasonal and long-term trends (e.g., [44]). Using EVI and NDVI time series to represent green leaf phenology, it is known that phenology is correlated to annual precipitation and solar radiation for red mangrove and white mangrove and min/max temperatures and monthly precipitation for seasonal trends in black mangrove [44]. Other changes in remotely sensed phenology have been documented due to land cover change (e.g., deforestation), rainfall, storm events, and fragmentation [25,44–46]. Disturbance to mangroves from fragmentation in Malaysia led to some increases in MODIS Gross Primary Productivity [46], whereas cyclones in Tanzania produced mixed results—some increases and some decreases in MODIS EVI [45] in the order of ± 0.05 to ± 0.22 EVI units. While there was a significant decrease in EVI within mangroves and wetlands, a significant increase in EVI was detected across the entire study area as well as within the dry forest land cover type (see Section 3.2, Data Processing). Increases (HDVA > 1) can be explained by ecological responses such as increased light availability and new plant growth with high cellulose content relative to non-photosynthetic vegetation. Additionally, there were several years of below-normal rainfall or drought prior to Hurricane Irma that could have affected a post-hurricane green-up.

Ground-truth campaigns in other locations have shown a high degree of correlation between remotely sensed indices of vegetation and observed live vegetation cover [47,48]. In this study area, it was not possible to conduct ground-based surveys of vegetation cover due to accessibility and resource limitations, making the remote sensed mapping results highly useful to park managers who could not reach all areas. To address this data-limited situation, we considered the "typical" MODIS EVI observations in the reference period from 20 locations in and near CNP from 2000–2017. Mangroves showed low variability in dry season EVI during the reference period ($\pm 5\%$ from the mean), with the greatest decrease being 13% change in EVI compared with the mean in the 2001–2002 dry season, which could be related to Hurricane Michelle (Category 2 on 5 November 2001, with eye passing ~115 km NW from CNP). Looking at dry forests for the reference period, the historical variability was around 5% change from the mean EVI, with greater variability in the dry seasons of 2003–2004 (-13%) and 2004–2005 (-14%), which we hypothesize could be related to changes in hydrological and pollutant conditions, as discussed in Section 4.6.

The negative impacts on phenology (HDVA < 1) were much larger than for storm disturbances or fragmented ecosystems as documented elsewhere [22,24,44–46]. Our catastrophic damage class corresponds to 37 to 100% change in mean dry season EVI (Sentinel-2 and MODIS). Even the severe losses class has very high change in EVI (18 to 38% decline), which is well beyond observed changes in other impacted mangroves [45,46]. These are substantial changes in comparison to other land covers; for example, a tropical forest has ~10–20% annual variability in EVI, and a large soybean field may vary by roughly 75% in EVI from the peak growing season to the fallow season [39]. These findings illustrate the magnitude of impact on the ecosystem and persistence, with EVI changes in heavily affected areas lasting months to years.

Following the storm, mangroves showed remarkable declines in EVI in the initial months following Hurricane Irma (2017–2018) but also in the following dry seasons (Figure 10). EVI decline nearly doubled in the second dry season after the storm (2018–2019), before starting to show recovery in the third dry season (2019–2020) (Figure 10). Radabaugh et al. [16] stressed the importance of detecting delayed mortality to get a better idea of total losses by looking at damages out to 9 months post-hurricane. In this study, we focused on data from 3–9 months after Hurricane Irma; the MODIS EVI time series (Figure 10) suggests the true severity of the ecological losses may have been 14–17 months post-storm

(Figure 9). The EVI values (Figure 10) from 2020–2022 show a near a complete recovery within the mangrove and wetland land cover type. Based on field observations in CNP, we know these results are not indicative of the health of the entire mangrove population within the park. While the trends of these 10 points look promising, the catastrophically damaged mangroves may take over a decade to recover based on the local knowledge of this author team.

4.3. Sentinel-2 vs. MODIS

While single-scene before and after imagery can be incredibly useful qualitatively and quantitatively for hurricane analysis, programs such as Google Earth Engine allow for far greater accessibility to imagery and therefore greater accuracy of change detection. Instead of a snapshot in time, we compiled seasonal averages of pixel values over many years. The analysis in this paper shows the quantitative advantage of this increased accessibility; it is also very valuable visually. Many single-scene images were analyzed at the beginning of this project to get a sense of the extent of the damage using aerial imagery. Even when specifically looking at the dry season, cloud-free images are almost impossible to find in this part of the world.

MODIS has been a choice sensor for land cover change analysis on landscape or larger scales because of its temporal coverage, rapid repeat intervals, and moderate resolution. Our severity maps generated with MODIS and Sentinel-2 data show agreement in the severity of ecological damages, highlighting the most catastrophic damage as concentrated near the coast (Figure 8). MODIS data proved to be faster and simpler to process in GEE compared with Sentinel-2. MODIS is substantially faster to process (2.9 MB for the study area) compared with a ~150 times greater dataset for Sentinel-2 (445 MB).

After calculating the extent and severity of damage caused to three land cover types in this study, we found there to be an alarming amount of catastrophic and severe damage, specifically to mangrove and wetland populations. The confusion matrix in Table 4 highlights the fact that Sentinel-2 may provide a more refined estimate of damage at both high and low ends of the scale, which were comparatively under- and overestimated by MODIS.

Despite these discrepancies, both sensors showed comparable trends in damage—demonstrating that even with a short reference period, maps derived from Sentinel-2 provide a robust conclusion. It should be noted that the Sentinel-2 reference period was confirmed to have no extreme storm events within the two-year period. Before using a relatively short reference period, it is imperative to know that no significant events have occurred. Further, Sentinel-2's ability to detect smaller-scale damage is relevant to landowners or managers, such as in CNP.

In terms of estimates of damage, our MODIS analysis calculated more damage in the mangrove and wetland land covers compared with Sentinel-2 analysis. We believe a primary factor is an overestimation of the damaged area in the MODIS analysis due to mixed pixels (Figure 12) on the coasts and in the keys. Many MODIS pixels overlapped areas that are water upon visual inspection of higher-resolution imagery, including Sentinel-2 and Google Earth satellite view. This mixed pixel effect can also explain why we found a higher percentage of mangrove and wetlands to be catastrophically damaged with Sentinel-2. We found more area was categorized as moderate and mild damage in dry forest with Sentinel-2 than with MODIS. Here too, Sentinel-2's 10 m resolution picked up subtler and smaller changes that were lost as mixed pixel effects in MODIS (Figure 12). One disadvantage is that some Sentinel-2 detections in grasslands may pick up changes in crop rotations in small holder farms that are scattered on the landscape. This could give a false increase or decrease in EVI. Sentinel-2 provides a more accurate number when looking at damage to specific vegetation types.

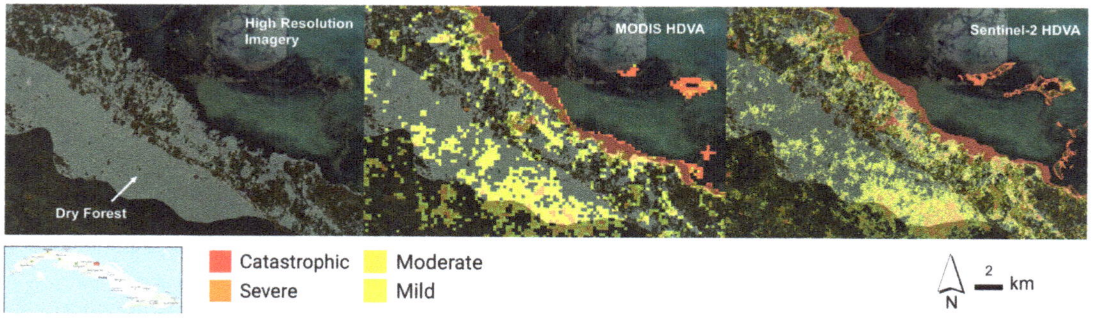

Figure 12. Visualization of MODIS and Sentinel-2 pixel size and abundance within the dry forest Class.

It is worth noting that areas calculated in this study vary by sensor, where MODIS tends to overestimate land area compared to Sentinel-2 because of the spatial resolutions affecting detections along the coastline. In terms of HDVA and damage classes, Sentinel-2 tends to detect more extreme damages (lower HDVA, greater severity class), which may be due to mixed pixel effects in MODIS.

4.4. HDVA Approach

Recent work by de Beurs et al. [25] mapped damages from a series of Caribbean hurricanes. They use a metric akin to a z-score to consider how an observation at or near the time of the hurricane varies from the pixel's long-term mean and standard deviation in its MODIS time series. Their approach systematically assesses hurricane damages in many places over several decades. This approach may yield different results than the HDVA because of the incorporation of the standard deviation and may be more challenging to interpret and communicate to stakeholders as a unitless metric. Both approaches utilized a historically observed time series to benchmark changes; however, de Beurs et al. [25] looked at anniversary dates (e.g., 1 January each year) and vegetation changes at/near the time of the hurricane. Our study used long-term observed seasonal (dry season) vegetation observations and extended the damage analysis out past the hurricane for the full time period available in the data record. We focused on a single hurricane and its sustained damages.

Focusing on Hurricane Irma in Cuba, our study and de Beurs et al. [25] detail the widespread extent of storm damage, extending several hundred kilometers along the coastal and inland zones. The differences in spatial patterns between these two studies need more investigation. In our MODIS analysis, there are some artificial detections along the coastline, although this does not appear in de Buers et al. [25], despite using the same dataset. de Buers does detect more inland (non-coastal) changes, while almost completely missing the damages to CNP.

The methods outlined in this paper can be used to show stark changes in EVI following an intense storm event, but they could also be applied to assess the extent of defoliation from milder storms. As we continue to see increased precipitation, sea level rise, and the spread of invasive species, this method could additionally be used to analyze slower degradation of mangrove populations as well as recovery rates.

Further refinement may be needed to determine the suitability of the quality labels (e.g., HDVA category breaks), as we found applying a quartile approach worked well for our data but might need refinement for other applications. For example, if we applied quartiles by land cover type instead of the entire study region, then we would have marginally damaged dry forests classed as "catastrophic", the same as mangroves with severe mortality. In the future, there may be a more robust break in categories that would suit many different ecosystems. Additional work to calibrate HDVA beyond vicarious validation with high-resolution imagery could enhance potential future applications of this metric but requires areas to be studied in detail before and after disturbances.

4.5. Caguanes National Park

Sentinel-2 provided a clear representation of the extent and severity of damage within CNP; this information is useful to park managers and local community members. Both MODIS and Sentinel-2 picked up widespread catastrophic damage to mangroves and wetlands within CNP. Using MODIS data, 98% of mangroves and wetlands were damaged, with 75% of that damage being catastrophic and 19% severe. Using Sentinel-2, 95% of mangroves and wetlands were found to be damaged, with 77% of that damage categorized as catastrophic and 11% severe. This 3% discrepancy is likely the result of MODIS overestimating the area along the coast due to mixed pixel effects (Figure 12).

Considering dry forest that was disturbed in CNP, MODIS picked up 76% damage, with 25% of the damage being catastrophic, 32% severe, 11% moderate, and 9% mild. Sentinel-2 picked up 80% damage to dry forest, with 40% of the damage being catastrophic, 20% severe, 13% moderate, and 8% mild. Both sensors highlighted more of the damage being concentrated in the catastrophic and severe categories, which is likely due to CNP's proximity to the eye of the storm. Even the dry forest, which we found to be less vulnerable when looking at the whole coast, was damaged more extensively at this coastal location. At this scale, data and maps derived from Sentinel-2 would provide local stakeholders a better idea of the location and severity of damage to vegetation. Planet data imagery of CNP from the following two years after the storm shows a healthier group of mangroves on the western side of the park, while the eastern side (where the storm primarily hit) is shown to be still heavily impacted but recovering (Figure 9).

Within CNP, we see fresh water inundation lasting several weeks and the slow demise of the mangroves based on field visits by our team. Our analysis demonstrates more persistent damages, which played out over weeks, months, and years following the hurricane, particularly for mangroves. It is possible that in other ecosystems, this approach would miss damages where refoliation is rapid (e.g., maples of North America). Future advances and availability of high-resolution imagery (~1 m) or incorporation of vegetation structure data will enhance our ability to remotely sense hurricane damages when ground-truth campaigns are not possible.

4.6. Land Cover and Land Use Legacies

Given our team's experiences living and working in the study area for decades, we think it is important to note some impacts of land cover and land use legacies that affect the historical period and perhaps the ecological impacts of Hurricane Irma.

In the time series of CNP EVI (Figure 10), we note a few interesting periods of small variations compared with the long-term mean, such as the dry season of 2001–2002, which likely was affected by Hurricane Michelle (5 November 2001, tracked across NW corner of study region).

We observed some historical variations in green leaf phenology within CNP (Figure 10). Upland (south) from CNP, the land use was agricultural historically, including industrial production of sugar cane and processing of sugar and torula yeast in several factories within 7 km. These factories did not have water treatment systems but used ferti-irrigation, as the manufacturing process used large volumes of water. The factories were phased out between 2000 and 2004, but contamination of water persisted over time, although it was concentrated in the early years of our study period. Additionally, modifications to the Pedraplen, a causeway connecting the island of Cuba with Cayo Santa María to the NW of CNP, were made due to its impediment of water exchange with the ocean, which caused increases in average salinity in the CNP mangroves, rising above 43 PSU. Hurricanes Ike and Paloma hit the region in 2008, and although they were not strong enough to do ecological damage to land cover, they did clear some of the prior industrial contamination from the coastal environments. After 2008, there was more natural rehabilitation of the ecosystems, although freshwater was scarce, due to the obstruction of drainage channels formerly used in sugar production in combination with dry or drought periods. Some coastal lagoons had unprecedented conditions, drying up completely, with salinity exceeding 67 PSU. Hurricane Irma brought the most massive deforestation of red and black mangroves in local memory, going back to 1933.

5. Conclusions

The methods outlined in this paper can be used to show stark changes in EVI following an intense storm event and can also be applied to assess the extent of defoliation from milder storms. As we continue to see increased precipitation, sea level rise, and the spread of invasive species, this method could additionally be used to analyze slower degradation of mangrove populations as well as recovery rates. The coordination of multitemporal and multisensor studies, with the communication of results in qualitative categories (e.g., catastrophic), is helpful for land managers and field scientists, particularly where computing infrastructure, field access, and expedition expenses limit field-based study. Future work should explore the use of the HDVA as well as category break values for damage classes.

Mangroves are essential to communities for coastal protection from tropical storm events due to their ability to act as barriers from wind, reduce storm surges by decreasing wave energy, and stabilizing the shore against erosion [49]. While there is consensus that healthy mangroves protect communities and reduce the number of lives lost to high-intensity storm events, there is concern about their ability to carry out these functions in a changing climate [49,50]. Today, Cuba's mangroves are still largely intact; however, even if they are not subject to direct human degradation, they are still vulnerable to the indirect impacts of climate change. With mangroves weakening from overexploitation, introduction of invasive species, and now the increase in intensity of hurricanes, we may witness a decrease in the ability of mangroves to stay intact and protect communities. Protection of mangroves is now more important than ever. Understanding the wide array of human- and climate-induced impacts on mangroves is an important contribution of remote sensing to future mangrove conservation.

Author Contributions: Conceptualization, H.C.T., G.L.G., N.H.L., A.F.M., D.Y.B.-E., I.H.R. and P.G.-D.; methodology, H.C.T. and G.L.G.; coding, H.C.T.; validation, H.C.T., G.L.G., N.H.L., A.F.M., D.Y.B.-E. and I.H.R.; formal analysis, H.C.T. and G.L.G.; resources, G.L.G.; data curation, H.C.T., G.L.G.; writing—original draft preparation, H.C.T. and G.L.G.; writing—review and editing, H.C.T., G.L.G., N.H.L., A.F.M., D.Y.B.-E., I.H.R. and P.G.-D.; visualization, H.C.T. and G.L.G. All authors have read and agreed to the published version of the manuscript.

Funding: This research was funded by the University of Vermont Environmental Studies Program Ian Worley Award granted to H.C.T. Phil Lasalle of the Norman Foundation and NASA LCLUC (80NSSC23K0537) supported G.L.G. The Gund Institute supported previous field courses in the region, including those by G.L.G. and P.G.-D., which led to this work.

Data Availability Statement: Publicly available datasets were analyzed in this study and can be found in the Google Earth Engine: code.earthengine.google.com. The new data presented in this study are available via https://doi.org/10.5281/zenodo.7826011.

Acknowledgments: We thank Planet for G.L.G.'s access to microsatellite data through their Education and Research Program for limited, non-commercial access to the Planetscope Imagery and Rapid Eye Archive. We acknowledge the infrastructure and resources of the Gund Institute, as well as their support in international collaborations and scientific exchange. We thank the Lasalle Foundation for support. We thank the reviewers for their helpful input.

Conflicts of Interest: The authors declare no conflict of interest. The funders had no role in the design of the study; in the collection, analyses, or interpretation of data; in the writing of the manuscript, or in the decision to publish the results.

References

1. Potter, C. Global Assessment of Damage to Coastal Ecosystem Vegetation from Tropical Storms. *Remote Sens. Lett.* **2014**, *5*, 315–322. [CrossRef]
2. Kuenzer, C.; Bluemel, A.; Gebhardt, S.; Quoc, T.V.; Dech, S. Remote Sensing of Mangrove Ecosystems: A Review. *Remote Sens.* **2011**, *3*, 878–928. [CrossRef]
3. Barr, J.G.; Engel, V.; Smith, T.J.; Fuentes, J.D. Hurricane Disturbance and Recovery of Energy Balance, CO_2 Fluxes and Canopy Structure in a Mangrove Forest of the Florida Everglades. *Agric. For. Meteorol.* **2012**, *153*, 54–66. [CrossRef]
4. Zhang, K.; Liu, H.; Li, Y.; Xu, H.; Shen, J.; Rhome, J.; Smith, T.J. The Role of Mangroves in Attenuating Storm Surges. *Estuar. Coast. Shelf Sci.* **2012**, *102–103*, 11–23. [CrossRef]
5. Das, S.; Vincent, J.R. Mangroves Protected Villages and Reduced Death Toll during Indian Super Cyclone. *Proc. Natl. Acad. Sci. USA* **2009**, *106*, 7357–7360. [CrossRef] [PubMed]
6. Liu, K.; Liu, L.; Liu, H.; Li, X.; Wang, S. Exploring the Effects of Biophysical Parameters on the Spatial Pattern of Rare Cold Damage to Mangrove Forests. *Remote Sens. Environ.* **2014**, *150*, 20–33. [CrossRef]
7. Giri, C.; Ochieng, E.; Tieszen, L.L.; Zhu, Z.; Singh, A.; Loveland, T.; Masek, J.; Duke, N. Status and Distribution of Mangrove Forests of the World Using Earth Observation Satellite Data. *Glob. Ecol. Biogeogr.* **2011**, *20*, 154–159. [CrossRef]
8. Donato, D.C.; Kauffman, J.B.; Murdiyarso, D.; Kurnianto, S.; Stidham, M.; Kanninen, M. Mangroves among the Most Carbon-Rich Forests in the Tropics. *Nat. Geosci.* **2011**, *4*, 293–297. [CrossRef]
9. Valiela, I.; Bowen, J.L.; York, J.K. Mangrove Forests: One of the World's Threatened Major Tropical EnvironmentsAt Least 35% of the Area of Mangrove Forests Has Been Lost in the Past Two Decades, Losses That Exceed Those for Tropical Rain Forests and Coral Reefs, Two Other Well-Known Threatened Environments. *BioScience* **2001**, *51*, 807–815. [CrossRef]
10. Duncan, C.; Owen, H.J.F.; Thompson, J.R.; Koldewey, H.J.; Primavera, J.H.; Pettorelli, N. Satellite Remote Sensing to Monitor Mangrove Forest Resilience and Resistance to Sea Level Rise. *Methods Ecol. Evol.* **2018**, *9*, 1837–1852. [CrossRef]
11. Sippo, J.Z.; Lovelock, C.E.; Santos, I.R.; Sanders, C.J.; Maher, D.T. Mangrove Mortality in a Changing Climate: An Overview. *Estuar. Coast. Shelf Sci.* **2018**, *215*, 241–249. [CrossRef]
12. Cahoon, D.; Hensel, P. *Hurricane Mitch: A Regional Perspective on Mangrove Damage, Recovery, and Sustainability*; US Department of the Interior, US Geological Survey: Reston, NA, USA, 2002.
13. Sherman, R.E.; Fahey, T.J.; Martinez, P. Hurricane Impacts on a Mangrove Forest in the Dominican Republic: Damage Patterns and Early Recovery1. *Biotropica* **2001**, *33*, 393–408. [CrossRef]
14. Reading, A.J. Caribbean Tropical Storm Activity over the Past Four Centuries. *Int. J. Climatol.* **1990**, *10*, 365–376. [CrossRef]
15. Zhang, K.; Thapa, B.; Ross, M.; Gann, D. Remote Sensing of Seasonal Changes and Disturbances in Mangrove Forest: A Case Study from South Florida. *Ecosphere Wash.* **2016**, *7*, e01366. [CrossRef]
16. Radabaugh, K.R.; Moyer, R.P.; Chappel, A.R.; Dontis, E.E.; Russo, C.E.; Joyse, K.M.; Bownik, M.W.; Goeckner, A.H.; Khan, N.S. Mangrove Damage, Delayed Mortality, and Early Recovery Following Hurricane Irma at Two Landfall Sites in Southwest Florida, USA. *Estuaries Coasts* **2020**, *43*, 1104–1118. [CrossRef]
17. Jimenez, J.A.; Lugo, A.E.; Cintron, G. Tree Mortality in Mangrove Forests. *Biotropica* **1985**, *17*, 177–185. [CrossRef]
18. Lugo, A.E. Effects and Outcomes of Caribbean Hurricanes in a Climate Change Scenario. *Sci. Total Environ.* **2000**, *262*, 243–251. [CrossRef]
19. Hayhoe, K.; Wuebbles, D.J.; Easterling, D.R.; Fahey, D.W.; Doherty, S.; Kossin, J.P.; Sweet, W.V.; Vose, R.S.; Wehner, M.F. *Chapter 2: Our Changing Climate. Impacts, Risks, and Adaptation in the United States: The Fourth National Climate Assessment, Volume II*; U.S. Global Change Research Program: Washington, DC, USA, 2018.
20. Alexandre, C.; Johary, R.; Catry, T.; Mouquet, P.; Révillion, C.; Rakotondraompiana, S.; Pennober, G. A Sentinel-1 Based Processing Chain for Detection of Cyclonic Flood Impacts. *Remote Sens.* **2020**, *12*, 252. [CrossRef]
21. Wang, L.; Jia, M.; Yin, D.; Tian, J. A Review of Remote Sensing for Mangrove Forests: 1956–2018. *Remote Sens. Environ.* **2019**, *231*, 111223. [CrossRef]
22. Zhang, C.; Durgan, S.D.; Lagomasino, D. Modeling Risk of Mangroves to Tropical Cyclones: A Case Study of Hurricane Irma. *Estuar. Coast. Shelf Sci.* **2019**, *224*, 108–116. [CrossRef]
23. Rossi, E.; Rogan, J.; Schneider, L. Mapping Forest Damage in Northern Nicaragua after Hurricane Felix (2007) Using MODIS Enhanced Vegetation Index Data. *GIScience Remote Sens.* **2013**, *50*, 385–399. [CrossRef]
24. Buitre, M.J.C.; Zhang, H.; Lin, H. The Mangrove Forests Change and Impacts from Tropical Cyclones in the Philippines Using Time Series Satellite Imagery. *Remote Sens.* **2019**, *11*, 688. [CrossRef]
25. De Beurs, K.M.; McThompson, N.S.; Owsley, B.C.; Henebry, G.M. Hurricane Damage Detection on Four Major Caribbean Islands. *Remote Sens. Environ.* **2019**, *229*, 1–13. [CrossRef]
26. Simard, M.; Fatoyinbo, L.; Smetanka, C.; Rivera-Monroy, V.H.; Castañeda-Moya, E.; Thomas, N.; Van der Stocken, T. Mangrove Canopy Height Globally Related to Precipitation, Temperature and Cyclone Frequency. *Nat. Geosci.* **2019**, *12*, 40–45. [CrossRef]
27. Taillie, P.J.; Roman-Cuesta, R.; Lagomasino, D.; Cifuentes-Jara, M.; Fatoyinbo, T.; Ott, L.E.; Poulter, B. Widespread Mangrove Damage Resulting from the 2017 Atlantic Mega Hurricane Season. *Environ. Res. Lett.* **2020**, *15*, 064010. [CrossRef]
28. Jamaluddin, I.; Thaipisutikul, T.; Chen, Y.-N.; Chuang, C.-H.; Hu, C.-L. MDPrePost-Net: A Spatial-Spectral-Temporal Fully Convolutional Network for Mapping of Mangrove Degradation Affected by Hurricane Irma 2017 Using Sentinel-2 Data. *Remote Sens.* **2021**, *13*, 5042. [CrossRef]

29. Ghorbanian, A.; Zaghian, S.; Asiyabi, R.M.; Amani, M.; Mohammadzadeh, A.; Jamali, S. Mangrove Ecosystem Mapping Using Sentinel-1 and Sentinel-2 Satellite Images and Random Forest Algorithm in Google Earth Engine. *Remote Sens.* **2021**, *13*, 2565. [CrossRef]
30. McCarthy, M.J.; Jessen, B.; Barry, M.J.; Figueroa, M.; McIntosh, J.; Murray, T.; Schmid, J.; Muller-Karger, F.E. Automated High-Resolution Time Series Mapping of Mangrove Forests Damaged by Hurricane Irma in Southwest Florida. *Remote Sens.* **2020**, *12*, 1740. [CrossRef]
31. McCarthy, M.J.; Jessen, B.; Barry, M.J.; Figueroa, M.; McIntosh, J.; Murray, T.; Schmid, J.; Muller-Karger, F.E. Mapping Hurricane Damage: A Comparative Analysis of Satellite Monitoring Methods. *Int. J. Appl. Earth Obs. Geoinf.* **2020**, *91*, 102134. [CrossRef]
32. Galford, G.L.; Fernandez, M.; Roman, J.; Monasterolo, I.; Ahamed, S.; Fiske, G.; González, P.; Kaufman, L. Cuban Land Use and Conservation, from Rainforests to Coral Reefs. *Bull. Mar. Sci.* **2018**, *94*, 171–191. [CrossRef]
33. Cangialosi, J.P.; Latto, A.S. Robbie Berg 2017 Atlantic Hurricane Season. Hurricane Irma. Available online: https://www.nhc.noaa.gov/data/tcr/index.php?season=2017&basin=atl (accessed on 12 February 2020).
34. IBTrACS—International Best Track Archive for Climate Stewardship. Available online: https://ibtracs.unca.edu/index.php?name=v04r00-1933234N13335 (accessed on 7 April 2023).
35. SNAP (Sistema Nacional de Áreas Protegidas). Sistema Nacional de Áreas Protegidas de Cuba 2021. Available online: https://www.gob.pe/institucion/sernanp/campa%C3%B1as/4340-sistema-nacional-de-areas-naturales-protegidas-por-el-estado (accessed on 17 July 2021).
36. Protected Areas (WDPA). Available online: https://www.protectedplanet.net/en/thematic-areas/wdpa (accessed on 25 September 2022).
37. Gorelick, N.; Hancher, M.; Dixon, M.; Ilyushchenko, S.; Thau, D.; Moore, R. Google Earth Engine: Planetary-Scale Geospatial Analysis for Everyone. *Remote Sens. Environ.* **2017**, *202*, 18–27. [CrossRef]
38. Didan, K. *MOD13Q1 MODIS/Terra Vegetation Indices 16-Day L3 Global 250m SIN Grid V006 [Data Set]*; NASA EOSDIS Land Processes DAAC: Sioux Falls, SD, USA, 2015. [CrossRef]
39. Galford, G.L.; Mustard, J.F.; Melillo, J.; Gendrin, A.; Cerri, C.C.; Cerri, C.E.P. Wavelet Analysis of MODIS Time Series to Detect Expansion and Intensification of Row-Crop Agriculture in Brazil. *Remote Sens. Environ.* **2008**, *112*, 576–587. [CrossRef]
40. García-Llamas, P.; Suárez-Seoane, S.; Fernández-Guisuraga, J.M.; Fernández-García, V.; Fernández-Manso, A.; Quintano, C.; Taboada, A.; Marcos, E.; Calvo, L. Evaluation and Comparison of Landsat 8, Sentinel-2 and Deimos-1 Remote Sensing Indices for Assessing Burn Severity in Mediterranean Fire-Prone Ecosystems. *Int. J. Appl. Earth Obs. Geoinf.* **2019**, *80*, 137–144. [CrossRef]
41. Huete, A.; Didan, K.; Miura, T.; Rodriguez, E.P.; Gao, X.; Ferreira, L.G. Overview of the Radiometric and Biophysical Performance of the MODIS Vegetation Indices. *Remote Sens. Environ.* **2002**, *83*, 195–213. [CrossRef]
42. Pezeshki, S.R.; DeLaune, R.D. Soil Oxidation-Reduction in Wetlands and Its Impact on Plant Functioning. *Biology* **2012**, *1*, 196–221. [CrossRef]
43. Lagomasino, D.; Fatoyinbo, T.; Castañeda-Moya, E.; Cook, B.D.; Montesano, P.M.; Neigh, C.S.R.; Corp, L.A.; Ott, L.E.; Chavez, S.; Morton, D.C. Storm Surge and Ponding Explain Mangrove Dieback in Southwest Florida Following Hurricane Irma. *Nat. Commun.* **2021**, *12*, 4003. [CrossRef]
44. Berlanga-Robles, A.C.; Ruiz-Luna, A.; Nepita Villanueva, M.R. Seasonal Trend Analysis (STA) of MODIS Vegetation Index Time Series for the Mangrove Canopy of the Teacapan-Agua Brava Lagoon System, Mexico. *GIScience Remote Sens.* **2019**, *56*, 338–361. [CrossRef]
45. Kai, K.H.; Shaghude, Y.W.; Kijazi, A.L.; Uiso, C.B. The Influence of Tropical Cyclones to the Plant Productivity Indices along the Coast of Tanzania. *Atmos. Clim. Sci.* **2021**, *11*, 589–613. [CrossRef]
46. Kanniah, K.D.; Kang, C.S.; Sharma, S.; Amir, A.A. Remote Sensing to Study Mangrove Fragmentation and Its Impacts on Leaf Area Index and Gross Primary Productivity in the South of Peninsular Malaysia. *Remote Sens.* **2021**, *13*, 1427. [CrossRef]
47. Elmore, A.J.; Mustard, J.F.; Manning, S.J.; Lobell, D.B. Quantifying Vegetation Change in Semiarid Environments: Precision and Accuracy of Spectral Mixture Analysis and the Normalized Difference Vegetation Index. *Remote Sens. Environ.* **2000**, *73*, 87–102. [CrossRef]
48. Fisher, J.I.; Mustard, J.F. Cross-Scalar Satellite Phenology from Ground, Landsat, and MODIS Data. *Remote Sens. Environ.* **2007**, *109*, 261–273. [CrossRef]
49. Sandilyan, S.; Kathiresan, K. Mangroves as Bioshield: An Undisputable Fact. *Ocean Coast. Manag.* **2015**, *103*, 94–96. [CrossRef]
50. Granek, E.F.; Ruttenberg, B.I. Protective Capacity of Mangroves during Tropical Storms: A Case Study from 'Wilma' and 'Gamma' in Belize. *Mar. Ecol. Prog. Ser.* **2007**, *343*, 101–105. [CrossRef]

remote sensing

MDPI

Article

Global Mangrove Watch: Monthly Alerts of Mangrove Loss for Africa

Pete Bunting [1,*], Lammert Hilarides [2], Ake Rosenqvist [3], Richard M. Lucas [1], Edmond Kuto [4], Yakhya Gueye [5] and Laye Ndiaye [5]

1 Department Geography and Earth Sciences, Aberystwyth University, Aberystwyth SY23 3DB, UK
2 Wetlands International, 6700 AL Wageningen, The Netherlands
3 solo Earth Observation (soloEO), Tokyo 104-0054, Japan
4 Wetlands International Eastern Africa, Nairobi PQ7X+5J7, Kenya
5 Wetlands International West Africa, Dakar MGXW+333, Senegal
* Correspondence: pfb@aber.ac.uk; Tel.: +44-1970-622615

Abstract: Current mangrove mapping efforts, such as the Global Mangrove Watch (GMW), have focused on providing one-off or annual maps of mangrove forests, while such maps may be most useful for reporting regional, national and sub-national extent of mangrove forests, they may be of more limited use for the day-to-day management of mangroves and for supporting the Global Mangrove Alliance (GMA) goal of halting global mangrove loss. To this end, a prototype change mangrove loss alert system has been developed to identify mangrove losses on a monthly basis. Implemented on the Microsoft Planetary Computer, the Global Mangrove Watch v3.0 mangrove baseline extent map for 2018 was refined and used to define the mangrove extent mask under which potential losses would be identified. The study period was from 2018 to 2022 due to the availability of Sentinel-2 imagery used for the study. The mangrove loss alert system is based on optimised normalised difference vegetation index (NDVI) thresholds used to identify mangrove losses and a temporal scoring system to filter false positives. The mangrove loss alert system was found to have an estimated overall accuracy of 92.1%, with the alert commission and omission estimated to be 10.4% and 20.6%, respectively. Africa was selected for the mangrove loss alert system prototype, where significant losses were identified in the study period, with 90% of the mangrove loss alerts identified in Nigeria, Guinea-Bissau, Madagascar, Mozambique and Guinea. The primary drivers of these losses ranged from economic activities that dominated West Africa and Northern East Africa (mainly agricultural conversion and infrastructure development) to climatic in Southern East Africa (primarily storm frequency and intensity). The production of the monthly mangrove loss alerts for Africa will be continued as part of the wider Global Mangrove Watch project, and the spatial coverage is expected to be expanded to other regions over the coming months and years. The mangrove loss alerts will be published on the Global Mangrove Watch online portal and updated monthly.

Keywords: mangroves; deforestation; mangrove loss; global mangrove watch; change detection; near real time; mangrove loss alerts; early warning; Sentinel-2; Africa

1. Introduction

Mangroves are a globally important [1] and dynamic [2] ecosystem providing a range of ecosystem services, including blue carbon [3,4], biodiversity [5], fisheries [6], coastal protection [7,8], and tourism [9]. Mangroves are an ecosystem under pressure [10,11] with significant losses occurring since the 1980s [12,13]. The Global Mangrove Alliance (GMA [14]) has set the goals of halting mangrove loss, increasing the amount of mangrove restoration and increasing mangrove protection through building awareness.

The Global Mangrove Watch (GMW) mangrove extent layers [15–18] cover eleven annual epochs between 1996 and 2020, and provide the most up to date historical extents of mangroves globally. They are available through the Global Mangrove Watch online

Portal [19]. However, given their annual update cycle, these layers may be of limited use for users looking to manage regions and protected areas containing mangrove forests. For efficient protection from human activities and informing on natural events disrupting mangroves, losses need to be identified as close to the loss event as possible and provided to users in a form they can easily use.

The GMW Portal [19] is subject to ongoing development, with additional functionality and improvements to hosted datasets being added at regular intervals. For example, an early version of the mangrove loss alerts was provided when the portal launched in July 2021 and has subsequently been updated with the results of this study in February 2023. This enables users without technical GIS or remote sensing skills to access and analyse mangrove loss alerts in near-real time. Users of the GMW datasets and the GMW Portal fall into several categories, including those who (a) seek a national or regional view of mangrove changes, with these tending to be academic or policy users, and (b) are tasked with managing and protecting small regions of mangroves (e.g., a national park, a single delta or section of coastline). Practitioners focused on small regions of mangroves would commonly have significant knowledge of the area in which they work. Therefore, global maps of annual mangrove changes may be of limited use. However, regularly visiting all parts of their area of interest is often not feasible. Therefore, being alerted to mangrove losses that have recently occurred can be useful, even if identifying illegal or unintended losses that could be stopped if caught in time.

For the reasons outlined above, the GMW Mangrove Loss Alerts were developed to provide a system for mapping mangrove losses in a timely manner, such that information can be provided to users allowing them to act accordingly [20]. To achieve this aim, the following objectives were identified for the system proposed within this article:

1. Provide updates at least monthly, thereby allowing end users to respond to ongoing activities or be informed of natural events that compromise the integrity or existence of mangroves.
2. Provide a low number of false positives; alerts might be ignored if false alerts are generated regularly.
3. Provide a basis for an operational system; implemented in a suitable computing platform with a maintainable code base.
4. To be computationally viable to scale to mangrove regions globally.

Africa was selected for prototype development and validation of the GMW Mangrove Loss Alerts system as it is a geographically discrete continent and sufficiently large in area to test scalability. It also comprises a range of mangrove types and conditions with areas, such as the Niger Delta known to have significant ongoing changes. There is significant variability between western and eastern Africa in terms of cloud cover and climate, with East Africa typically drier [21] and the South East witnessing a number of cyclones each year. Areas of West Africa are often characterised by significant cloud cover, allowing any limitations in the use of optical satellite data due to cloud cover to be identified. Finally, through GMA partners, such as Wetlands International, local knowledge of many regions could be collected and fed into optimising and validating the system. The monthly GMW Mangrove Loss Alerts dataset is available to users through the Global Mangrove Watch online Portal [19]. The version available on the GMW Portal was updated with the results of this study in February 2023 and will be updated on a monthly basis going forward.

2. Background

Many studies (e.g., [2,15,22–24]) have considered mapping land cover change, including forest loss, through remote sensing imagery using various sensors, modalities and methods. There are a number of approaches to mapping land cover change, but these fit within four broad categories, (1) map-to-map, (2) image-to-image, (3) map-to-image and (4) time series approaches. Map-to-map approaches (e.g., [25]) detect change by producing two or more classifications and comparing the resulting categorical maps. The advantages of such approaches are that they are independent of the datasets used to derive the maps, and therefore, different modalities can be used and compared. However, they can also find false positives due to errors within the classifications. Image-to-image approaches (e.g., [26]) require no supplementary data, such as a map and/or training data, but the imagery does need to be comparable. However, it is common for changes in image pixel values to not necessarily relate to a change in land cover but to changes in the season and/or condition of the land cover. Map-to-image approaches (e.g., [15]) require that each class has a defined response (e.g., reflectance or backscatter) within the remotely sensed imagery and that a change in land cover will cause that response to change such that it is identifiably different from the original class. These approaches often merge the advantages of the map-to-map and image-to-image approaches while mitigating many of the negatives. Time series approaches (e.g., [27–30]) model each pixel within a scene, either mapping the trend of each pixel or allowing the future pixel responses to be predicted and therefore compared to the observed response. A significant difference between the observed and predicted or the slope for the trend crossing a predetermined threshold allows for a change to be identified. These methodologies allow for seasonality and natural variations in response to be taken into account, thereby reducing the number of false positives for changes. Additionally, these methodologies can also assess long-term trends in response, providing the possibility for condition-based changes to be mapped (e.g., [31]). However, time series approaches are computationally expensive, requiring significant computational resources to fit the models and predict further observations. Computational costs can be reduced by reducing the frequency in which the models are re-fitted, such as re-fitting monthly or quarterly rather than after each observation. However, this can reduce the accuracy of the modelling approach.

There are two systems which provide publicly available forest loss alerts. The Global Forest Watch (GFW) alerts are provided from a combination of Global Land Analysis and Discovery lab (GLAD; [32]) and radar for detecting deforestation (RADD; [33]) systems. The Japan International Cooperation Agency (JICA) and Japan Aerospace Exploration Agency (JAXA) provide the JICA-JAXA Forest Early Warning System (JJ-FAST; [34]) alerts. While other systems have been proposed (e.g., [35,36]), many are based on lower spatial resolution datasets, such as MODIS, and/or have provided alerts for a specific country and therefore are not directly relevant to this study.

The GLAD alerts [32] are derived from USGS Landsat imagery with a pixel resolution of 30 m. A bagged tree machine learning classification algorithm is used and applied to each scene to identify possible areas of change. The system confirms a change if at least two from four consecutive scenes identify a pixel as a change. For a case study in Peru, the users accuracy was estimated at 86.5% $\pm$ 2.0 (13.5% commission) and producers as 67.0% $\pm$ 7.4 (33.0% omission).

The RADD [33] system uses Copernicus Sentinel-1 C-band SAR imagery with a spatial pixel resolution of 10 m and an updated period of 6–12 days. Although, with the failure of Sentinel-1B in December 2021, the RADD system is limited until the launch of Sentinel-1C, expected in 2023. The Sentinel-1 VH and VV backscatter are converted into per-pixel forest and non-forest probabilities using historical Gaussian mixture models. A non-forest threshold is then used to identify forest loss alerts, and Bayesian updating is applied to confirm alerts using subsequent observations. Using a case study from the Congo Basin, the users and producers accuracies were estimated as 97.6% (2.4% commission) and 95.0% (5% omission), respectively.

The JJ-FAST system [34] uses JAXA ALOS-2/PALSAR-2 ScanSAR L-band SAR data to provide deforestation alerts for 77 countries every 42 days with a spatial pixel resolution of 50 m. The approach has evolved with a number of versions of the algorithm released, all based on a system of thresholds. A number of global reference sites were used to define the thresholds. The users and producers accuracies for Version-3 of the alerts were estimated at 85.0% (15% commission) and 63.8% (36.2% omission), respectively.

The system proposed by this study used a map-to-image approach to identify mangrove loss alerts. A refined version of the GMW 2018 mangrove extent was used to define the mangrove extent, and a computationally simple normalised difference vegetation index (NDVI)-based threshold methodology was developed to identify the mangrove loss alerts on a monthly basis. Our system has similarities to the JJ-FAST approach in that a set of optimised thresholds are used to define change, rather than a machine learning-based classification or modelling approach. However, it is also similar to the GLAD alerts system as the proposed system uses multiple observations to confirm a change.

3. Methods

The analysis for the GMW Mangrove Loss Alerts was undertaken via the Microsoft Planetary Computer [37,38], which provides access to the many spatial datasets, including the full Sentinel-2 archive, hosted within an Azure storage bucket and searchable via a STAC API [39]. Using the freely provided jupyter-lab [40] Kubernetes [41] deployment, the Microsoft Planetary Computer allows access to a Dask [42] cluster to scale data analysis. For this study, the items identified through the STAC search API were loaded as a Xarray [43] into the Dask [42] cluster using the Open Data Cube (ODC) STAC API [44]. In the future, when computer requirements go beyond those provided in the free Kubernetes deployment, scaling can be achieved through the deployment of a custom Kubernetes deployment within the Microsoft Azure Cloud.

Analysis was undertaken on a 1×1 degree grid (Figure 1) with all data resampled into the WGS84 (EPSG:4326) coordinate system. A pixel size of 0.0002 degrees ($\sim$22 m) was used, resulting in output images for each 1×1 degree tile of 5000×5000 pixels. Therefore, each mangrove loss alert point represents a 0.0002×0.0002 degree pixel (i.e., 0.0484 ha) in the final results. In total, there were 199 tiles intersecting with mangrove regions in Africa (Figure 1).

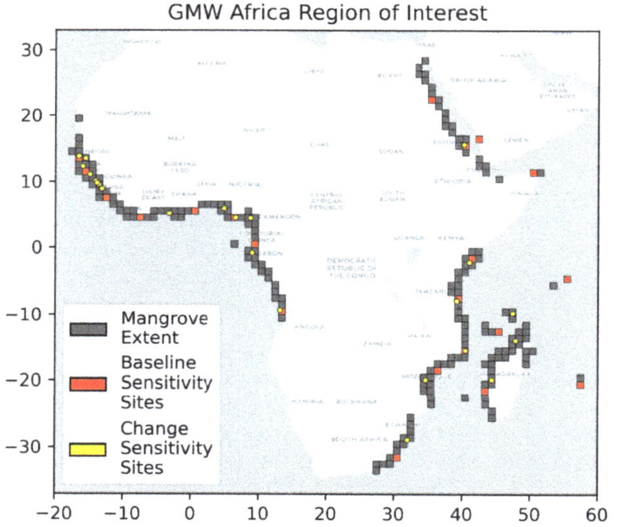

Figure 1. The regions of interest for Africa are defined by the 1×1 degree tiles intersecting with the mangrove regions across Africa, as defined by the GMW v3.0 2018 mangrove extent map. The figure also shows the locations of the sites used to optimise the parameters for the baseline refinement (Red) and change detection threshold (Yellow).

To generate the mangrove loss alerts, two main steps were required (Figure 2). Firstly, the refinement of the GMW v3.0 2018 mangrove extent mask to remove pixels affected by misregistration [15] within that dataset, when compared to the Sentinel-2 dataset. Secondly, a change analysis based on NDVI thresholding on a monthly basis producing monthly mangrove loss alerts. These two steps form the mangrove loss alerting system that is in use in support of the GMW, where the monthly change analysis executed each month allows the GMW portal to be updated with mangrove loss alerts for the previous month. For this study, a further analysis was undertaken to annotate the mangrove loss alerts with the drivers and pressures causing those losses and an accuracy assessment was carried out, with this also allowing comparison to the JICA-JAXA JJ-FAST and GFW forest loss alerts.

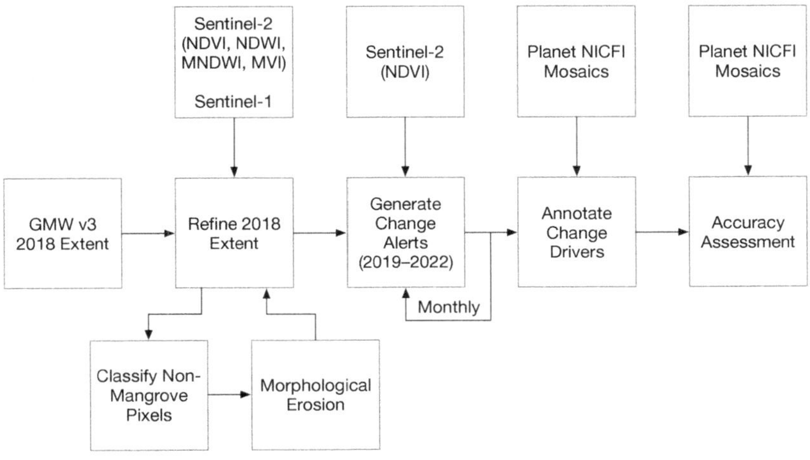

Figure 2. Flowchart illustrating the steps involved in the analysis. Acronym definitions: Global Mangrove Watch (GMW), Norway's International Climate and Forests Initiative Satellite Data Program (NICFI), normalised difference vegetation index (NDVI), normalised difference water index (NDWI), modified normalised difference water index (MNDWI), mangrove vegetation index (MVI).

3.1. Datasets

This analysis was undertaken using Copernicus Sentinel-2 optical and Sentinel-1 C-band radar data, where pre-processed versions (e.g., level-2 atmospheric corrected Sentinel-2 product) of these data were provided through the Microsoft Planetary Computer. The Sentinel-2A and -2B satellites were launched in 2015 and 2017, respectively, and 2018 was thus the first year for which full coverage using both Sentinel-2 satellites was available. The Sentinel-2 imagery spanning from January 2018 to October 2022 were provided corrected to surface reflectance and cloud masked using the Sen2Cor [45] tool. All scenes with a cloud cover of less than 50% were used. Additionally, all Sentinel-1 C-band SAR data acquired over the regions of interest in 2018 were used to refine the mangrove baseline, helping to remove non-mangrove pixels from the GMW 2018 mangrove mask. The Sentinel-1 data were provided radiometrically terrain corrected (RTC) with a correction algorithm based on Small [46], using the PlanetDEM digital elevation model (DEM).

3.2. Mangrove Baseline

To define the extent of mangroves across Africa, the GMW v3.0 mangrove layer for 2018 was used [15]. As outlined by Bunting et al. [15], the spatial baseline of these data were derived from L-band SAR global mosaics from the Japanese Earth Resources Satellite (JERS-1) synthetic aperture radar (SAR), the Advanced Land Observing Satellite (ALOS)-Phased Arrayed L-band SAR (PALSAR) and ALOS-2 PALSAR-2 missions from the Japan Aerospace Exploration Agency (JAXA) [47]. Due to inherent geolocation uncertainty in the version of the L-band SAR mosaics used for the generation of the GMW v3.0 layers, amounting up to 100 m in certain locations, some additional post-processing of the 2018 mangrove extent baseline was required to avoid false mangrove loss alerts in the mangrove/non-mangrove boundary regions occurring due to misregistration with the Sentinel-2 data. JAXA are subsequently (in 2022 and 2023) replacing all mosaic products with geolocation errors with re-processed products [48], which will be used for the next version (v4.0) of the GMW extent products. The GMW v4.0 products are scheduled for release in 2023.

The post-processing of the 2018 mangrove baseline was undertaken in two steps. Firstly, a classification of non-mangrove pixels was undertaken within the 2018 mangrove mask, and secondly, a morphological erosion was conducted to remove the remaining boundary non-mangrove pixels. The classification of non-mangrove pixels was undertaken using four optical indices: the normalised difference vegetation index (NDVI; Equation (1)), mangrove vegetation index (MVI; Equation (2) [49]), normalised difference water index (NDWI; Equation (3)), and the modified normalized difference water index (MNDWI; Equation (4)), as well as the minimum Sentinel-1 cross-polarisation (VH) backscatter for 2018.

$$NDVI = \frac{NIR - Red}{NIR + Red} \quad (1)$$

$$MVI = \frac{NIR - Green}{SWIR1 - Green} \quad (2)$$

$$NDWI = \frac{Green - NIR}{Green + NIR} \quad (3)$$

$$MNDWI = \frac{Green - SWIR1}{Green + SWIR1} \quad (4)$$

The four optical indices were calculated for each Sentinel-2 scene independently, and a count of the number of times a specified non-mangrove threshold (Table 1) was crossed (i.e., above or below a predefined threshold depending on the index) was made for each pixel and each index. These counts were subsequently thresholded (Table 1) to provide four masks of non-mangrove pixels, each of which was removed from the GMW v3 2018 mangrove extent mask. For the Sentinel-1 data, where low VH backscatter typically indicates open water, the minimum VH backscatter across all the scenes from 2018 was calculated for each pixel, and a threshold was applied to mask out non-mangrove pixels (e.g., water and mudflats) from the GMW v3 2018 mangrove baseline.

Table 1. Optimal thresholds identified for refining the mangrove extent for 2018. The index thresholds were applied to each scene to identify non-mangrove pixels. The count threshold was used to threshold the number of times a pixel has crossed the threshold combining the results from the individual scenes to create a single mask for each index. Acronym definitions: normalised difference vegetation index (NDVI), normalised difference water index (NDWI), mangrove vegetation index (MVI), Sentinel-1 vertical–horizontal (VH) polarisation.

Index	Index Threshold	Count Threshold
NDVI	<0.1	4
MVI	<0.1	4
NDWI	>-0.1	4
MNDWI	>0.15	4
Min. Sentinel-1 VH	<-19 [dB]	-

The optical index thresholds were derived through a sensitivity analysis using a set of reference masks (e.g., Figure 3c), where thresholds from -0.3 to 0.5, in steps of 0.05, were tested for each index. The count thresholds were also optimised with values tested from 1 to 10. For the Sentinel-1 data, VH backscatter values were extracted and intersected with the reference mangrove masks, and the optimal threshold differentiating mangroves and non-mangroves was selected. The reference mangrove masks (e.g., Figure 3c) were generated from 21 1×1 degree tiles randomly selected across the African coastline (Figure 1), within which 44 regions of approximately 5×5 km were identified. Within the 44 regions, pixels incorrectly identified as mangroves (i.e., false positives, primarily due to misregistration) within the GMW v3 2018 mask were digitised with a visual inspection of the Planet imagery for 2018, providing the reference dataset. The thresholds (Table 1) with the best agreement for removing the non-mangrove pixels within the 44 regions were then applied to the whole study area. Finally, a morphological erosion, using a 5 × 5 circular operator, was applied to the revised 2018 mangrove extent mask to remove any remaining edge effects due to misregistration. This operation removed approximately two pixels (∼40 m) from all the mangrove boundaries.

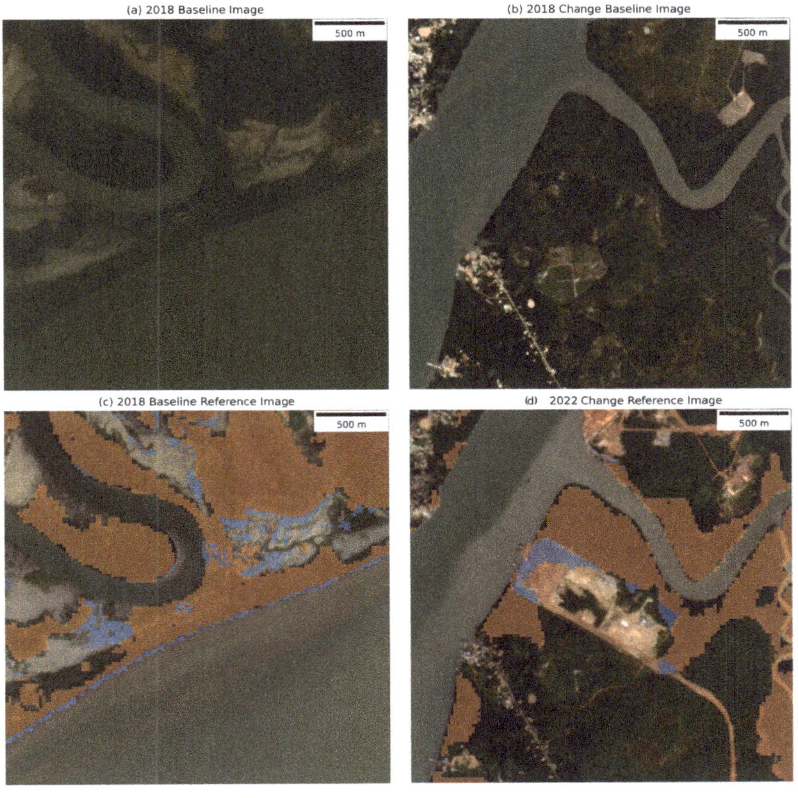

Figure 3. Examples of the reference images defined for the sensitivity analysis for identifying the baseline thresholds ((**c**); Table 1) and the change thresholds ((**d**); Table 2). (**a**,**b**) illustrate the 2018 Planet imagery for the area while (**c**) mangroves that should be within the 2018 baseline (orange) and pixels which should be removed from the mangrove mask by the thresholding (blue). (**d**) Mangroves that have not been lost (orange) or lost (blue) between 2018 and 2022. (Image courtesy: Planet Labs/NICFI).

These post-processing steps result in a refined mangrove baseline mask that minimises the errors associated with misregistration and avoids the associated false positives for change that might otherwise be found. However, it also means that true mangrove pixels are missing from the refined baseline mask and particularly those on the boundaries between mangrove and non-mangrove regions susceptible to erosion and other pressures leading to change.

Table 2. The upper part of the table provides the range of values tested during the sensitivity analysis to identify the optimal thresholds for the change analysis. The index threshold is the threshold to find a potential mangrove loss within an individual scene. The absolute difference threshold is the difference in the index between the current scene and the median pixel value for the month from the previous year. The score threshold is related to the number of times a change needs to have occurred for the mangrove loss alert to be confirmed. The bottom part of the table provides the resulting optimal thresholds identified through the sensitivity analysis. Acronym definitions: normalised difference vegetation index (NDVI), normalised difference water index (NDWI).

Index	Index Threshold	Abs. Diff Threshold	Score Threshold
	Sensitivity Analysis Test Values		
NDVI	0.05, 0.1, 0.15, 0.2, 0.25, 0.3	0.1, 0.15, 0.2, 0.25, 0.3	3, 5, 7, 9
NDWI	−0.2, −0.15, −0.1, −0.05, 0.0, 0.05	0.1, 0.15, 0.2, 0.25, 0.3	3, 5, 7, 9
	Optimal Thresholds		
NDVI	<0.25	0.15	5
NDWI	>0.05	0.25	7

3.3. Identification of Mangrove Loss

Only pixels within the refined 2018 mangrove mask were considered to identify mangrove loss alerts, and the analysis was executed for each month, providing monthly alerts of mangrove losses. Using fewer indices for the change analysis reduces the computational effort of the analysis. Therefore, the 44 reference regions used to identify non-mangrove thresholds were further analysed to identify the indices to be used for the change detection. Using each index as the base, Figure 4 illustrates the number of additional non-mangrove pixels identified by the other indices for the 44 reference regions. The NDVI and NDWI individually identified the most non-mangrove pixels within these reference regions (Figure 4), and these were therefore taken forward for the change analysis. Subsequently, a parameter optimisation for the change analysis (see below) found that only the NDVI was required for the identification of the mangrove loss alerts.

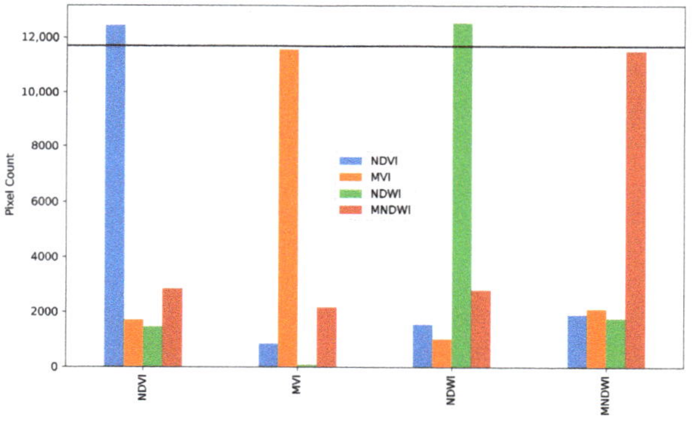

Figure 4. Importance of the indices used for the baseline refinement, represented as a count of the number of pixels identified as non-mangroves using the 44 reference sites using each index as a base. Using each index as the base (x-axis), the number of additional non-mangrove pixels identified using each of the other indices is shown. This illustrates that the NDVI and NDWI individually identified more non-mangrove pixels than the MNDWI and MVI pixels, but the MNDWI was more complementary to the other indices. Acronym definitions: normalised difference vegetation index (NDVI), normalised difference water index (NDWI), modified normalised difference water index (MNDWI), mangrove vegetation index (MVI).

The NDVI was thresholded to identify mangrove change from each Sentinel-2 scene. An absolute difference threshold was also applied to confirm changes, comparing the median NDVI of the same month from the previous year to each of the Sentinel-2 observations on a per-pixel basis. Additionally, a scoring system was used to help mitigate noise, such as from misclassified clouds or cloud shadows. The scoring system was applied (Figure 5) on a per-pixel basis and updated with each observation, where 1 was added to the score when a change was found, and 1 was subtracted if no change was found. The score cannot go below 0 or higher than the change score threshold (i.e., 5, identified through the threshold optimisation below). Therefore, to be confirmed as a change, a pixel needed multiple observations to reach the score threshold (i.e., 5) where the NDVI value was consistent with non-mangroves, and the pixel value had changed compared to 12 months earlier.

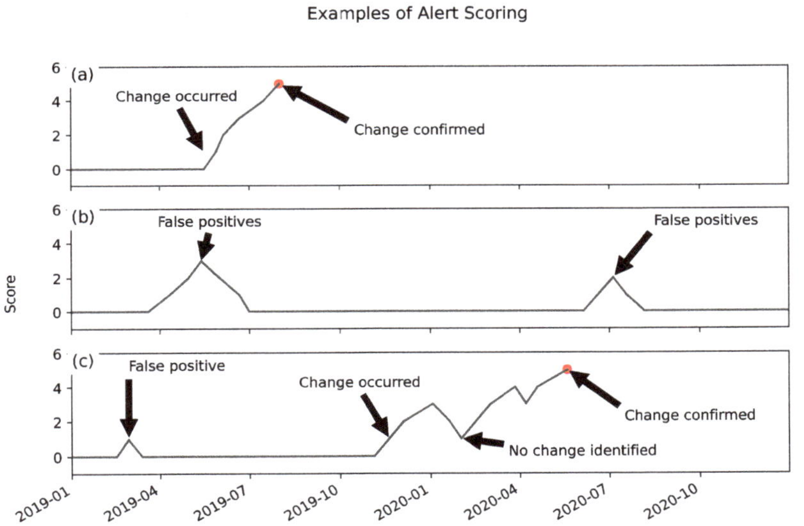

Figure 5. Examples for individual pixels illustrating the scoring system. Examples: (**a**) a change occurring, (**b**) false positives (e.g., misclassified clouds) ignored by the scoring system, and (**c**) a more complex example of a change occurring with a false positive and some observations which have missed identified the change.

The thresholds for identifying a change were defined through a sensitivity analysis, where 26 5 × 5 km regions were defined (Figure 1) and the pixels where changes had occurred from the 2018 baseline to October 2022 were digitised (e.g., Figure 3d). A set of values for each threshold (Table 2) were then applied, and those most accurately identifying the reference changes while minimising the identification of false positives were selected and applied for the analysis.

From the sensitivity analysis, it was found that the NDVI captured change more reliably than the NDWI. The NDVI also captured the majority of the changes identified by the NDWI. Therefore, to increase the computational efficiency of the processing, it was decided only to use the NDVI for the analysis.

It was observed that alerts for true losses of mangroves were generally clustered, while false positives tended to be isolated. Therefore, a process to remove isolated mangrove loss alerts was applied. The individual mangrove loss alert pixels were vectorised to points, creating the data product provided to end users from this analysis. Each mangrove loss alert point represents a single 0.0002 × 0.0002 degree (i.e., 22 × 22 m) pixel, for the month the alert was confirmed. The vectorised mangrove loss alert points were populated with the number of mangrove loss alerts within 222 m (10 pixels) and 444 m (20 pixels) radii, from either the month of confirmation or the months since the start of the analysis. For example, for the mangrove loss alerts confirmed in October 2022, the alerts from January 2019 to October 2022 will be used to find the isolated alerts in October 2022. Mangrove loss alert points were removed using the following criteria: (**a**) less than five alerts within the 444 m radius and less than four alerts within the 222 m radius, or (**b**) less than three alerts within the 222 m radius. The thresholds used to identify the isolated mangrove loss alerts were defined through a visual sensitivity analysis across various sites.

Finally, manual quality assurance (QA) was undertaken across the study area. However, it was found that edits were only required in Nigeria and Cameroon, where high cloud cover had resulted in a number of false positives from December 2020 to April 2021. This was attributed to a sequence of images where the cloud masking had failed to fully capture the cloud and shadows within the imagery. The QA process digitised polygons intersecting with alerts identified as false positives, which were removed from the dataset. The manual QA process took approximately 3 h and was therefore not considered a significant overhead. The QA process was simple to perform and is recommended to be routinely undertaken before every public release of new GMW Mangrove Loss Alerts on the Global Mangrove Watch Portal.

It should be emphasised that the mangrove loss alerts are designed to indicate the spatial density and temporal frequency of mangrove losses within a region and should not be used for area calculations. This is, in part, due to the filtering applied to the 2018 mangrove baseline, particularly the erosion required due to the inherent geolocation errors in the GMW v3.0 baseline due to the use of the old version of the JAXA L-band SAR mosaic dataset.

3.4. Accuracy Assessment

To assess the accuracy of the mangrove loss alerts, a 1 km hexagonal grid was constructed from which a random set of hexagons were selected. A hexagonal grid can have advantages over a square grid [50]. Specifically, the former helps to minimise sampling bias due to edge effects related to the grid shape due to the low perimeter-to-area ratio. Additionally, hexagonal grids can represent curved patterns within the data more naturally than square grids. The selected grid cells were visually annotated as to whether a mangrove loss had occurred, using the very-high-resolution Planet satellite mosaics provided through Norway's International Climate and Forests Initiative Satellite Data Program (NICFI) as a reference [51]. The annotation was a binary decision based on the presence or absence of a mangrove loss within the period of interest intersecting the 1 km hexagon. The 1 km grid was considered the scale at which users can be expected to use the mangrove loss alerts within their areas of interest. A 10% sample was taken (270) for the accuracy assessment of the hexagons intersecting with the mangrove loss alerts. An additional 1000 hexagons were randomly sampled from the whole mangrove area (107,146 hexagons) to assess false negatives.

The assessment was carried out without reference to the mangrove extent baseline with the mangrove extent interpreted from the Planet imagery. Using a mangrove extent baseline to constrain the accuracy assessment would bias the results, ignoring true mangrove loss alerts that may occur outside the GMW 2018 baseline. This is particularly important due to the refinement applied to the GMW 2018 baseline to mitigate the geolocation errors within that layer. This refinement is likely to result in some omission of mangrove regions and, therefore, errors of omission within the resulting mangrove loss alerts. To quantify the errors of omission directly resulting from the 2018 baseline, rather than the change detection approach, a further assessment of the reference hexagons was made only considering the losses within the refined 2018 baseline.

The methodology described in Bunting et al. [15] was used to validate that a sufficient number of reference samples were used to estimate the accuracy metrics accurately. The 1270 reference samples were split into five random subsets, each with 254 reference samples and cumulatively combined to assess whether the accuracy metrics significantly changed by adding further reference sets. Bootstrapping was used to estimate accuracy metrics and the 95% confidence interval [2], using a 30% sample with replacement and 1000 iterations.

To assess the temporal accuracy of the mangrove loss alerts, a sample of 100 alerts was randomly selected from the whole dataset. A reference date was derived for each mangrove loss alert using the NICFI Planet monthly mosaic imagery and compared to the date at which a change was first observed.

3.5. Causes of Change

During the analysis, it was noted that different pressures dominated the causes of mangrove loss within particular regions. To assess this, the NICFI Planet mosaics were used to visually assess and manually annotate the 2678 1 km hexagons to a dominant change pressure. The pressures considered were related to: agriculture (Figure 6a), infrastructure (Figure 6b), clearing (Figure 6c), erosion (Figure 6d) and 'other' (Figure 6e). While agriculture and infrastructure are sub-classes of clearing, they were labelled separately as they can be readily identified where field systems or roads have been built and are visible within the NICFI Planet mosaics. Erosion is a sub-class of natural processes, but identifiable through the context within the imagery. The 'other' class was used for situations where the cause of the change was unclear and where no clear anthropogenic signatures (e.g., straight lines) were present. It should also be noted that many processes occurring within the mangroves might naturally occur but have anthropogenic causes (e.g., changes in sediment supply due to deforestation or dam building or pollution events such as oil spills). Identifying these secondary pressures was not possible using satellite imagery alone, and therefore changes associated with these are likely to be within the 'other' class.

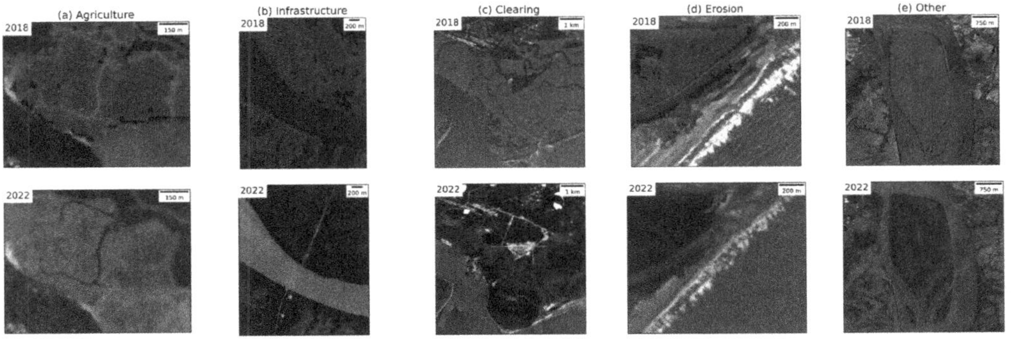

Figure 6. Examples from within the NICFI Planet mosaics of mangrove losses associated with (**a**) agriculture, (**b**) infrastructure, (**c**) clearing, (**d**) erosion and (**e**) 'other'. The green transparent areas within the first row illustrate the filtered GMW 2018 mangrove extent. (Image courtesy: Planet Labs/NICFI).

4. Results

From January 2019 to October 2022, 72,389 individual mangrove loss alerts were identified. As shown in Figure 7 and Table 3, Guinea-Bissau, Guinea and Nigeria represent the most significant hot spots of mangrove losses in West Africa, whilst, in East Africa, the most significant losses were found to occur in Madagascar and Mozambique.

However, the assessment indicates that the pressures causing the losses vary between the countries. For example, in Guinea and Guinea-Bissau, most alerts appear to be associated with agricultural expansion and development (Figure 8a). In contrast, in Nigeria, the development of infrastructure, particularly for oil and gas development [52], dominated (Table 3). Nigeria was also found to have significant losses in the 'other' class, with areas of mangroves appearing degraded and dying. These areas in Nigeria, allocated to the 'other' class, often coincide with areas of intense oil extraction. Therefore, these changes can likely be attributed to the significant pollution events reported to have occurred in the area [52] and be considered an effect of anthropogenic activities. Within northern Nigeria, there is also a substantive area of coastal erosion, as seen within Figure 8b, and this has had a significant impact on those who live within the area, resulting in a large number of homes being washed away over just the four year period of this study.

In Eastern Africa, particularly in Madagascar and Mozambique, the alerts identified are primarily associated with the 'other' class (Table 3), with some smaller areas of infrastructure development, such as in Kenya, where the alert system identified a new ferry port development (Figure 8c). These extensive areas of mangrove loss in Madagascar and Mozambique include areas of canopy dieback (e.g., Figure 8d), which are thought to be due to storm damage. However, there are many pressures and associated events and processes that can result in mangrove loss, including disturbed hydrology (e.g., through dam building and coastal engineering [53,54]), extreme weather events such as cyclones [55,56], pollution [52,57], sea-level changes [58] and seasonal (including lunar) cycles [59]. From the imagery alone, it is not possible to identify the cause.

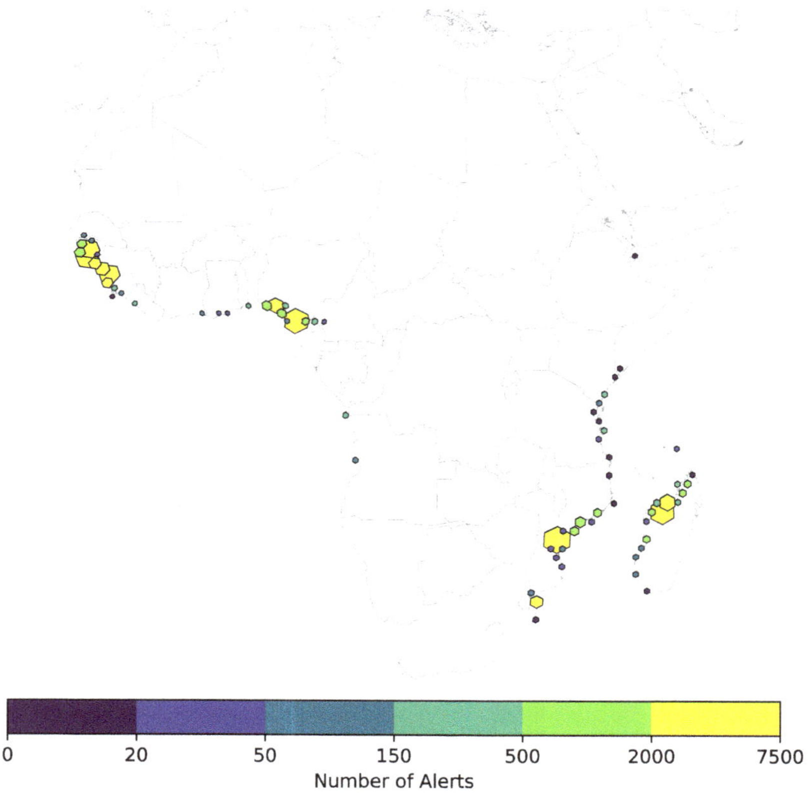

Figure 7. The mangrove loss alert hot spots throughout Africa. The hexagon sizes indicate the number of alerts, as does the colour.

Considering the number of alerts per mangrove area (Table 3), there appears to be no obvious relationship between the extent of mangroves within a country and the number of alerts. However, this does highlight some countries of interest, such as the Democratic Republic of the Congo, where just 12 alerts were identified with the 23,659 hectares of mangroves. In the Seychelles, 42 alerts were identified within only 382 ha of mangroves, with the majority thought to be due to storm damage.

Table 3. Summarising the mangrove loss alerts results on a per-country basis. Including the number of mangrove loss alerts identified within each country, the GMW 2018 mangrove extent for the country, the spatial density of mangrove loss alerts for the country, provided as the number of hectares per mangrove loss alert (Num. ha per Alert; mangrove extent/number of alerts) and the number of mangrove loss alerts assigned to each change pressure for each country.

Country	Number of Alerts	2018 Mangrove Extent (Hectares)	Num. ha per Alert	Agriculture	Infrastructure	Clearing	Erosion	Other
Nigeria	16,380	845,359	52	0	3266	3996	3611	5507
Guinea-Bissau	13,012	269,778	21	12,021	18	32	19	922
Madagascar	12,964	277,221	21	0	0	0	533	12,431
Mozambique	12,140	309,560	25	0	10	0	770	11,360
Guinea	10,722	222,774	21	6917	362	1759	616	1068
Sierra Leone	2659	157,629	59	59	99	2419	0	82
Senegal	1421	127,031	89	0	0	0	31	1390
Tanzania	845	111,542	132	818	11	0	5	11
Cameroon	500	196,877	394	0	14	0	433	53
Ghana	482	17,950	37	0	24	0	0	458
Kenya	327	54,328	166	0	76	0	0	251
Gambia	240	61,122	255	0	171	0	0	69
Angola	218	28,358	130	0	147	0	0	71
Benin	170	2941	17	0	0	6	0	164
Liberia	161	18,691	116	0	6	137	0	18
Côte d'Ivoire	65	5411	83	0	65	0	0	0
Seychelles	42	382	9	0	0	0	0	42
Democratic Republic of the Congo	12	23,659	1972	0	0	0	0	12
South Africa	11	2648	241	11	0	0	0	0
Djibouti	10	738	74	0	0	0	0	10
Somalia	8	3554	444	0	0	0	0	8
Total	72,389	2,737,553	38	19,826	4269	8349	6018	33,927

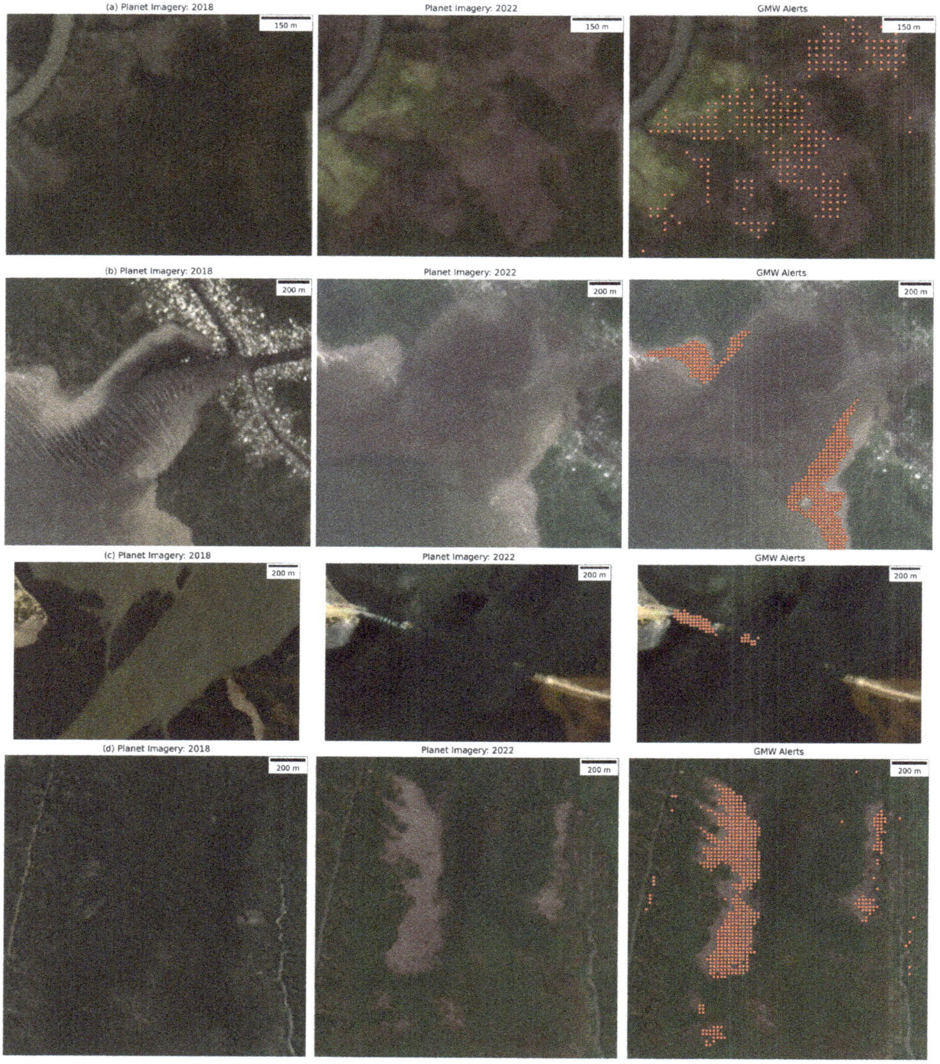

Figure 8. Examples of the mangrove loss alerts (red dots) for (**a**) agriculture conversion in Guinea-Bissau, (**b**) coastal erosion in Nigeria, (**c**) infrastructure development in Kenya, and (**d**) storm damage in Mozambique. Note, the occurrence of geolocation errors between the Planet imagery and the resulting mangrove loss alerts from the Sentinel-2 data, which are most evident in (**d**)).

Using the five countries with the most mangrove loss alerts, Figure 9a illustrates the frequency of alerts over the study period. For Nigeria, it can be seen that there is a pattern, with alerts mainly occurring from November to March. However, this was attributed to the low availability of cloud-free Sentinel-2 imagery rather than the timings of mangrove losses.

In some cases, individual events can result in a significant number of mangrove loss alerts. For example, in Mozambique, a significant number of alerts (8220; 68%) were identified in 2019, seeming to correspond to Cyclone Idai (Figure 10d [60]), which struck the region in March 2019. A similar but smaller peak was also noted in Madagascar, with 2703 (21%) alerts found in areas where Cyclone Idai might have impacted. The Mozambique

Channel witnessed 16 cyclones over the study period (Figure 10c), the most significant of which was Idai. It is also interesting to note that before the study period, four years of reduced cyclone activity occurred (Figure 10a,b), with fewer cyclones and low wind speeds. This might explain the high number of alerts identified within this region during the study period compared to the other countries in East and West Africa.

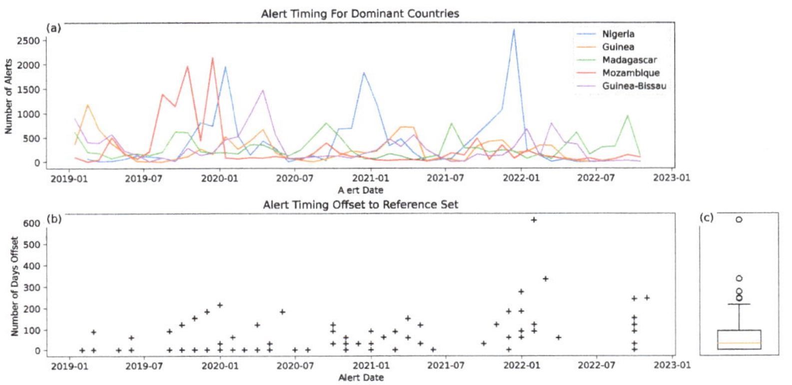

Figure 9. Analysis testing the timing of the alerts versus a reference dataset. (**a**) The number of alerts for the dominant change countries in Africa throughout the period of observation, (**b**) the observed date of change versus the number of days from the reference, and (**c**) a box plot for the number of days difference between the observed and reference date of the change (median = 92 days).

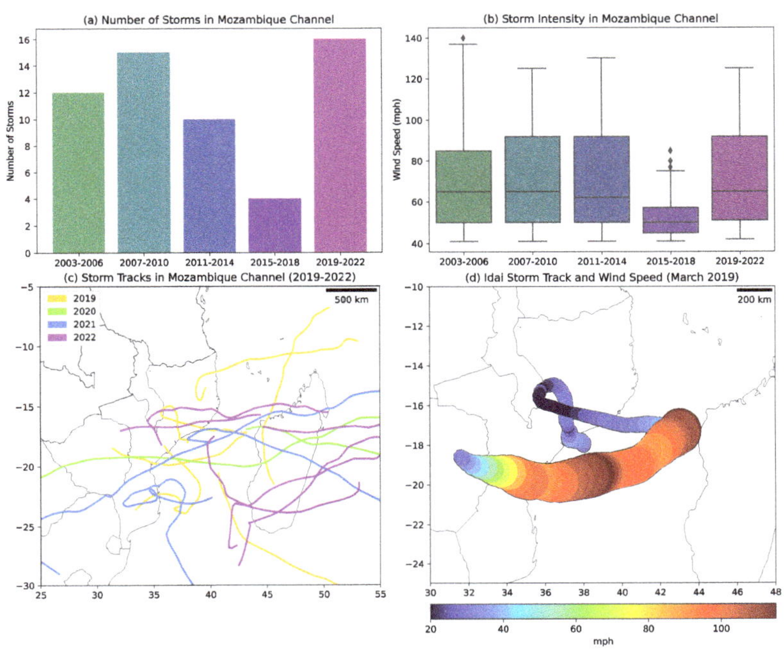

Figure 10. Cyclone activity in the Mozambique Channel: (**a**) the number of storms grouped by 4-year periods, (**b**) wind speed statistics for the 4-year clustered storms, (**c**) storm tracks for the storms that occurred during the 2019–2022 study period, and (**d**) storm track and wind speed of the Idai cyclone (March 2019), the most significant storm within the study period.

4.1. Accuracy Assessment

Based on the 1 km hexagon grid used for the assessment, the overall accuracy of the mangrove loss alerts was estimated to be 92.1% with a 95th confidence ranging from 89.0% to 95.7% (Table 4). The omission and commission of mangrove loss alerts were estimated to be 20.6% and 10.4%, respectively. This assessment was independent of the mangrove extent mask used for the analysis. However, if only pixels within the refined mangrove extent are considered for the accuracy assessment, the overall accuracy increased to 97.2% (94.8–98.8%), with the omission decreasing to 9.7%. This demonstrates that the refined mangrove extent baseline is the cause for approximately half the omission of mangrove losses, particularly those close to the mangrove edges (e.g., coastal erosion). The baseline can be expected to be improved in 2023 with the new GMW v4.0 mangrove extent dataset based on the reprocessed L-Band SAR global mosaic datasets from JAXA. For omissions primarily associated with the change detection, there are no clear patterns or systematic errors which have been observed.

Table 4. The accuracy assessment metrics using the 1 km hexagons to assess the accuracy of the mangrove loss alerts without reference to a mangrove baseline.

Metric	Median Estimate	95th Confidence
Overall Accuracy	92.1%	89.0–95.7%
Cohen Kappa	0.789	0.704–0.865
Balanced Overall Accuracy	88.0%	83.1–93.3%
Macro F1-Score	0.894	0.851–0.937
Weighted Avg. F1-Score	0.920	0.887–0.956
Matthews Correlation Coefficient	0.792	0.709–0.876
	Mangrove Loss Alerts	
F1-Score	0.842	0.769–0.907
Precision/Users	0.900	0.813–0.967
Recall/Producers	0.794	0.696–0.892
Omission	20.6%	10.8–30.4%
Commission	10.4%	3.3–18.6%
	No Mangrove Loss Class	
F1-Score	0.948	0.926–0.970
Precision/Producers	0.930	0.896–0.963
Recall/Users	0.968	0.940–0.989
Omission	3.2%	1.1–6.0%
Commission	7.0%	3.7–10.4%

Errors of commission were primarily found in narrow river channels included within the mangrove extent mask (e.g., Figure 11). While the baseline refinement removed many of these channels, some narrower channels with many mixed pixels remained. The change threshold to the previous year also reduced the number of these false positives (commissions) that were identified as mangrove loss alerts, but given the tidal nature of these systems, some change in reflectance can occur within these river channels without a loss of mangroves.

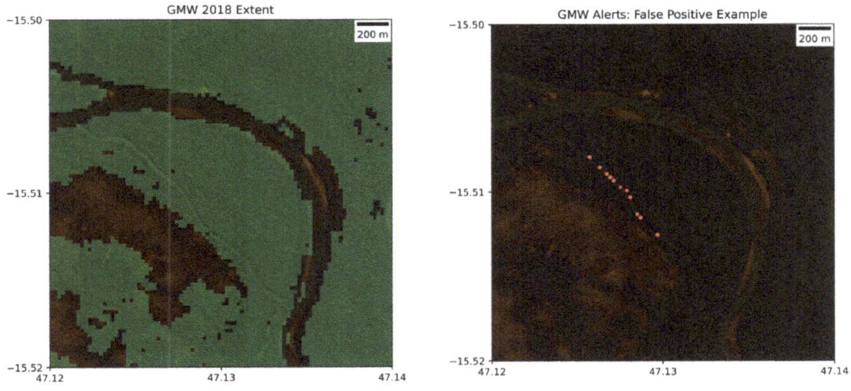

Figure 11. Illustrating a common cause of commission errors within the results from the mangrove loss alerts system, where the 2018 refined baseline (green regions) has not captured small river channels and, therefore, false positives (red points) for mangrove loss alerts are identified (Image courtesy: Planet Labs/NICFI).

To ensure that a sufficient number of samples were taken for the accuracy assessment, the accuracy metrics were calculated cumulatively using five subsets, each with 254 samples. Figure 12 illustrates that increasing the number of samples does not significantly alter the resulting accuracy metrics, with the mangrove loss alert class having the greatest variability. This is to be expected as change is rare. The variation in the mangrove loss alerts F1-score was from 0.952 (254 samples) to 0.943 with 1270 samples. The 95% confidence range decreased from 0.135 with 254 samples to 0.065 with 1270 samples. Therefore, it can be considered that adding further samples will not significantly alter the estimates of the individual class accuracies or overall accuracy but might reduce the range of the 95% confidence.

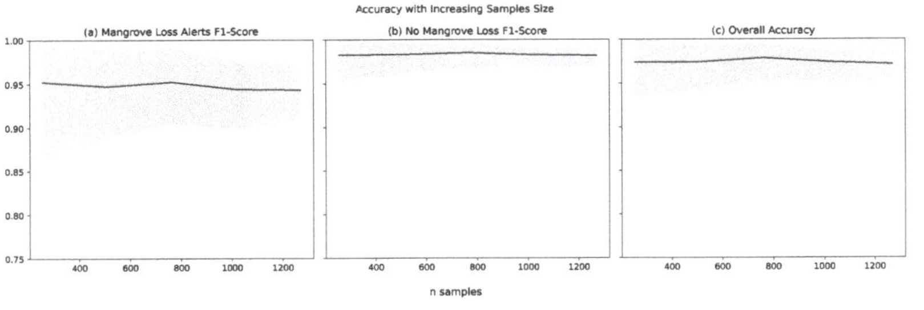

Figure 12. To illustrate that a sufficient number of accuracy samples has been used for the accuracy assessment, the reference data were subsampled and cumulatively combined to calculate the (**a**) the F1-score for the mangrove loss alerts, (**b**) the F1-score for pixels with no mangrove loss and (**c**) the overall accuracy. The grey areas illustrate the 95th confidence interval of the accuracy metric, where values closer to 1.0 indicate better accuracy. For all three metrics, including additional reference data does not significantly alter the accuracy estimated, and therefore, it can be concluded that sufficient reference data has been sampled.

4.2. Accuracy of Alert Timing

Using a sample of 100 reference points, an assessment was made of the alert timings, as identified by the system. Of the 100 reference points, five were found to be either incorrectly identified as changes or there was insufficient NICFI Planet imagery (e.g., due to cloud cover) to identify the timing of the loss. From the remaining 95 points, 75% of the alerts were identified within three months of the change identified within the NICFI Planet imagery (Figure 9c). As shown in Figure 9b, the offset from the alert date to the actual date the change occurred was largely consistent throughout the study period.

4.3. Comparison to JJ-FAST and GLAD Alerts

To compare the GMW Mangrove Loss Alerts (this study) with other early warning systems, such as the JICA-JAXA Forest Early Warning System in the Tropics (JJ-FAST) [34] and the Global Forest Watch (GFW) alerts (combined GLAD [32] and RADD [33] alerts), a further accuracy assessment was undertaken. The JJ-FAST and GFW alerts were intersected with the 1 km hexagonal grid used to assess the GMW Mangrove Loss Alert accuracy. The intersection identified 1665 hexagons for the GFW alerts and 1047 for the JJ-FAST alerts, compared to 2678 for the GMW Mangrove Loss Alerts. As shown in Table 5, the number of intersections between the different alert products is low, with 14% of the GFW alert hexagons intersecting with GMW Mangrove Loss Alerts and 5% for JJ-FAST.

Table 5. The 1 km hexagon intersections between the GMW Mangrove Loss Alerts (this study), the JJ-Fast alerts [34] and GFW alerts [32,33]. This revealed a low correspondence between the three datasets, with only 14% of the GMW Mangrove Loss Alerts intersecting with the GFW and only 5% of the JJ-FAST and GMW alerts intersecting.

	GMW	GFW	JJ-Fast
GMW	2678	397	121
GFW	14%	1665	90
JJ-Fast	5%	7%	1047

To assess the accuracy of the different alert systems for mangroves, the reference hexagons defined to assess the GMW Mangrove Loss Alerts were used. An additional 10% sample of hexagons was taken for the GFW and JJ-FAST alerts. These samples were assigned to mangrove change and no-change classes with reference to the NICFI Planet imagery. The outputs from the analysis are shown in Table 6, where the GMW Mangrove Loss Alerts are demonstrated to be more accurate for alerting to mangrove losses. It should be noted that the GMW Mangrove Loss Alerts are specifically focused on mangrove forests, while the GFW alerts and JJ-FAST alerts are generically applied to all forests. Given the larger area of terrestrial forests, these systems have naturally been developed and tested in these areas rather than mangroves. The low accuracy for JJ-FAST alerts over mangroves is primarily because there are few alerts within the mangrove extent. However, there are many terrestrial forest losses that the JJ-FAST system has correctly identified on the mangrove–terrestrial forest boundary. However, as a consequence of the lower 50 m pixel resolution and geolocation error in the older ALOS-2 PALSAR-2 ScanSAR products used to produce these alerts, there are cases of overlap with the mangrove mask. There are many false positives in the mangrove regions for the GFW alerts, which are often small in size and similar to the alerts that the distance-based filtering removed from the GMW Mangrove Loss Alerts dataset. The GFW alerts system also fails to detect a significant number of losses within the mangrove area, with an omission of 67.9%. However, it also identified some mangrove losses that the GMW Mangrove Loss Alerts system failed to identify.

The comparisons of accuracy indicate that the GMW Mangrove Loss Alerts system is better suited to detect mangrove losses; however, the JJ-FAST and GFW alert systems also work well within terrestrial forest areas. Therefore, while not appropriate for mapping mangrove loss per se, the JJ-FAST and GFW systems can provide important complementary information on the catchment area influencing the mangrove, with deforestation and degradation of terrestrial forests potentially impacting the mangroves.

Table 6. Comparison of the accuracy of mapping mangrove loss between the GMW Mangrove Loss Alerts (this study), the JJ-FAST alerts [34] and GFW alerts [32,33].

Metric	Median Estimate	95th Confidence
	GMW Mangrove Loss Accuracy	
F1-Score	0.842	0.769–0.907
Precision/Users	0.900	0.813–0.967
Recall/Producers	0.794	0.696–0.892
Omission	20.6%	10.8–30.4%
Commission	10.4%	3.3–18.6%
	GFW Mangrove Loss Accuracy	
F1-Score	0.40	0.298–0.504
Precision/Producers	0.534	0.400–0.673
Recall/Users	0.321	0.229–0.425
Omission	67.9%	57.5–77.1%
Commission	46.6%	32.7–60.0%
	JJ-Fast Mangrove Loss Accuracy	
F1-Score	0.148	0.063–0.244
Precision/Producers	0.269	0.117–0.448
Recall/Users	0.200	0.042–0.177
Omission	89.7%	82.3–95.8%
Commission	73.1%	55.2–88.3%

5. Discussion

This study has demonstrated that mangroves are under pressure in many areas of Africa. A significant number of losses have occurred between 2019 and 2022 in Nigeria, Guinea-Bissau and Guinea in West Africa and Madagascar and Mozambique in the East. However, the drivers of these mangrove losses vary between regions and therefore, management responses should also differ. For example, in West Africa, the drivers are largely economic, with mangroves being primarily removed for agricultural development and infrastructure. These economic activities can also produce significant secondary pressures, such as pollution events (e.g., [52]), which may cause the dieback of mangroves. To achieve the Global Mangrove Alliance (GMA) goals for mangrove conservation and restoration, these areas need increased protection and conservation to halt further losses, and where possible, regions of loss should be restored.

In East Africa, there are anthropogenic losses caused by agricultural development (e.g., Rufiji Delta in Tanzania), illegal logging (e.g., Mtwapa and Kilifi in Kenya) and infrastructure development (e.g., Mombasa in Kenya). However, the majority of mangrove loss alerts were focused around the Mozambique Channel, where it is likely that significant storm (e.g., Cyclone Idai) damage occurred within the mangrove regions. For these regions, further study is required to understand whether the changes witnessed between 2019 and 2022 are within the realms of normality, where recovery will naturally occur or whether these are part of a longer decline in mangroves.

Mangroves are naturally dynamic, and when in equilibrium, it would be expected that while some regions experience losses (e.g., erosion), others are expanding, or where regions experience dieback due to storm damage, others might be recovering. Future studies could use data, such as the historical GMW global extent layers, to assess whether regions are in equilibrium, at what spatial scale equilibrium occurs, and whether further protection or restoration efforts are required. Mangroves are complex systems, and future protections might also consider the wider ecosystem, such as the development of upstream infrastructures, including dams that can reduce sediment supply [53] or terrestrial deforestation or mining that can increase sediment supply [61].

5.1. Case Study from Guinea-Bissau

During the development of the mangrove loss alert system, an early version of the system found a dense patch of alerts in Guinea-Bissau (Figure 13) and a Wetlands International team from Guinea-Bissau were able to visit the site on 7 April 2021. The team found that a dam (Figure 14a) had been built to reclaim land for rice paddies, locally known as "bolanha's", and behind the dam, the mangroves were already being removed (Figure 14b). These bolanha's belonged to the villages of Ntus and Blom, within the Safim region and were previous rice paddies before 1993. However, they were forced to be abandoned in 1993 due to saltwater intrusion. An unsuccessful attempt was made to restore these fields in 1998, and the current intervention was started in 2017. In this case, the Guinea-Bissau government funded the project as part of a framework of food self-sufficiency for rural communities. Therefore, while the loss was identified and enabled a conversation, it could not prevent mangrove loss or allow restoration.

This case study highlights the complex nature of mangrove conservation, where many competing needs exist. Identification and notification of a mangrove loss, such as provided through this system and the GMW Portal [19], is an important first step but requires further 'on the ground' work to understand the causes and background.

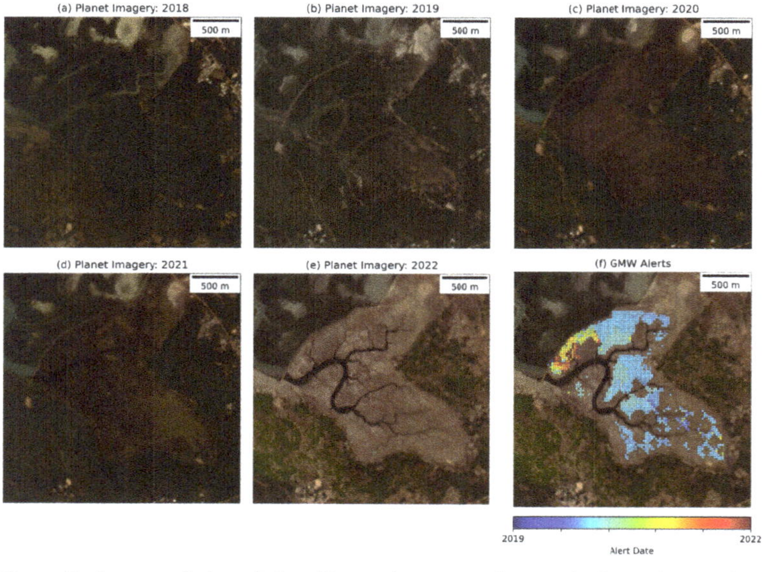

Figure 13. A case study from Guinea-Bissau where a new dam was built, resulting in the mangroves behind the dam being removed for rice cultivation. (**a**) illustrates the Planet imagery before the dam was built, (**b–e**) the dam was built and mangroves lost over the period and (**f**) the GMW Mangrove Loss Alerts identified over the period. (Image courtesy: Planet Labs/NICFI, 2018).

Figure 14. Example field photos taken during the field visits to the area shown in Figure 13. (**a**) The dam was built to hold back the tidal waters allowing for rice cultivation. (**b**) An area behind the dam where mangroves have started to be removed to be converted into rice paddies. Credit: Wetlands International; Taken: 7 April 2021.

5.2. Case Study from Kenya

Within Kenya, a number of mangrove loss alerts were identified within the county of Kilifi (Figure 15). Wetlands International staff were able to visit this area in November and December 2022 with support from the Kenya Forest Service. They verified the alerts as true losses of mangroves (Figure 16) and, through discussion with local authorities and community members, found that the cause of the mangrove losses was clearing for charcoal production. These were not areas that had witnessed extensive harvesting before, but with the COVID-19 pandemic, access to charcoal (the primary residential fuel source in the area) was reduced. During the pandemic lockdowns, local people were without work, and the cheap charcoal imported from other areas was unavailable. Additionally, others moved back home to the area from other regions where they had been working as jobs were no longer available. Therefore, local charcoal production was unofficially undertaken, and these areas of mangroves were identified as suitable harvestable wood sources. In addition, local authorities were unable to patrol due to the lockdowns, and consequently, these illegal harvesting activities went unchecked. At the time of the visit, local authorities had already started mangrove rehabilitation activities in the area (Figure 16).

This case study highlights the vulnerability of mangroves to changes in social and economic conditions. While the alerts provided through this study alone cannot stop such mangrove losses, they can help to identify that such losses are occurring. However, it also highlights the need for the time between a loss occurring and being identified to be as small as possible, with the potential for cumulative losses within as little as three months becoming significant.

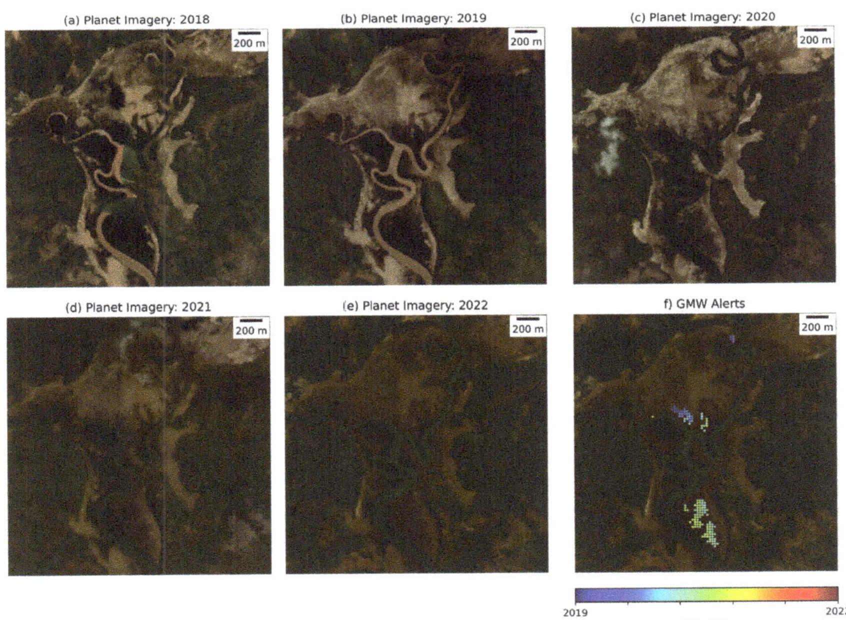

Figure 15. A case study from Kenya where the alerts identified mangroves being lost due to charcoal production. (**a**) Illustrates the planet imagery before the mangrove loss, (**b–e**) mangroves lost over the period and (**f**) the GMW Mangrove Loss Alerts identified over the period. (Image courtesy: Planet Labs/NICFI).

Figure 16. An area of mangroves which has been cut for charcoal production within the region shown in Figure 15. Rehabilitation activities have already started, with the mangrove seeding seen in this photo having been planted as part of that activity. Credit: Wetlands International; Taken: 1 December 2022.

5.3. Computational Scalability

Implemented using the Microsoft Planetary Computer [37,38], the system took approximately 18 h to run for all 199 tiles across the whole (2018–2022) time period. This was executed as five jobs, running simultaneously, each requesting a Dask cluster with up to 24 cores. The study area represents 13% of the 1521 1×1 degree tiles covering the mangrove regions globally. Therefore, using the same environment, the system could be executed globally in approximately one week. However, on a monthly basis, the execution will be faster, with only one month rather than 46 months to be processed. It would be expected that globally the monthly updates would take a few days of computing time.

5.4. Future Developments

Future developments might consider using Sentinel-1 imagery following the launch of Sentinel-1C, expected in 2023, particularly in areas such as the Niger Delta where persistent cloud cover limits the use of optical satellite data. Approaches might also consider whether Sentinel-1 and -2 could be used in combination. However, noting that L-band SAR from the ALOS PALSAR and ALOS-2 PALSAR-2 missions have been demonstrated to be well suited for mapping mangrove loss [62], looking to future missions, the forthcoming NASA-ISRO SAR (NISAR) mission is anticipated to provide global L-band SAR coverage every 12 days [63] and might be well suited for near real-time mangrove monitoring.

More complex change detection algorithms might be considered, such as the continuous monitoring of land disturbance (COLD) algorithm [30], which has already been demonstrated for mangrove applications [24,31]. These approaches make better use of the full time series with model fitting allowing regions with more variability in response (e.g., inter-tidal areas with open canopies) to be taken into account and reducing the number of false positives. Additionally, the long-term change trend in response might also be a useful metric, where changes can occur gradually over a longer period of time. The GMW Mangrove Loss Alerts system demonstrated here requires a significant mangrove change to occur within the 12-month period for a change to be identified. Considering the long-term trend after a change has been identified might also allow recovery (e.g., after a storm) to be identified and, therefore, cases where change occurs multiple times to be identified following a period(s) of recovery. Within the current system, once a change is identified, that pixel is removed from the mangrove mask, and a change cannot be flagged again.

6. Conclusions

This study is the first to have demonstrated a system for identifying mangrove losses on a monthly basis, with previous studies having focused on annual changes in mangrove extent. The study estimated an overall accuracy of 92.1%, the mangrove loss alert commission was estimated at 10.4% and an omission of 20.6%. The mangrove loss alerts have identified significant losses throughout Africa from 2019 to 2022, with Nigeria, Guinea-Bissau, Madagascar, Mozambique, and Guinea containing 90% of those alerts. The drivers of these changes ranged from those driven largely by economic (e.g., West Africa and northern East Africa) or climate (primarily changes in storm frequency and intensity in East Africa) factors, although both were evident across the continent. Future improvements in the quality of the mangrove loss alerts will be achieved through an updated GMW mangrove extent baseline, to be released in 2023. The results of this study are available through the Global Mangrove Watch (GMW) Portal [19] and will be expanding to further areas of interest over the coming months and years.

Author Contributions: Conceptualization, P.B., L.H., A.R. and R.M.L.; methodology, P.B.; software, P.B.; validation, P.B.; investigation, P.B. and L.H.; data curation, P.B. and L.H.; writing—original draft preparation, P.B.; writing—review and editing, P.B., L.H., A.R., R.M.L., E.K., Y.G. and L.N.; project administration, L.H. All authors have read and agreed to the published version of the manuscript.

Funding: The Global Mangrove Watch is funded by the Oak Foundation, the COmON Foundation, the National Philanthropic Trust, DOB Ecology, and the Dutch Postcode Lottery.

Data Availability Statement: All datasets are available from Zenodo https://doi.org/10.5281/zenodo.7459306 (accessed on 19 December 2022) while the Python code for the analysis is available on GitHub: https://github.com/globalmangrovewatch/gmw_planetary_computer_s2_alerts (accessed on 26 January 2023).

Acknowledgments: Microsoft and those who run the Planetary Computer are thanked for providing this system and the dataset curation. The three anonymous reviewers are also thanked for their comments and corrections, which significantly improved the manuscript before publication.

Conflicts of Interest: The authors declare no conflicts of interest.

References

1. Agaton, C.B.; Collera, A.A. Now or later? Optimal timing of mangrove rehabilitation under climate change uncertainty. *For. Ecol. Manag.* **2022**, *503*, 119739. [CrossRef]
2. Murray, N.J.; Worthington, T.A.; Bunting, P.; Duce, S.; Hagger, V.; Lovelock, C.E.; Lucas, R.; Saunders, M.I.; Sheaves, M.; Spalding, M.; et al. High-resolution mapping of losses and gains of Earth's tidal wetlands. *Science* **2022**, *376*, 744–749. [CrossRef] [PubMed]
3. Donato, D.C.; Kauffman, J.B.; Murdiyarso, D.; Kurnianto, S.; Stidham, M.; Kanninen, M. Mangroves among the most carbon-rich forests in the tropics. *Nat. Geosci.* **2011**, *4*, 293–297. [CrossRef]
4. Simard, M.; Fatoyinbo, L.; Smetanka, C.; Rivera-Monroy, V.H.; Castañeda-Moya, E.; Thomas, N.; Stocken, T.V.d. Mangrove canopy height globally related to precipitation, temperature and cyclone frequency. *Nat. Geosci.* **2019**, *12*, 40–45. [CrossRef]
5. Sievers, M.; Brown, C.J.; Tulloch, V.J.; Pearson, R.M.; Haig, J.A.; Turschwell, M.P.; Connolly, R.M. The Role of Vegetated Coastal Wetlands for Marine Megafauna Conservation. *Trends Ecol. Evol.* **2019**, *34*, 807–817. [CrossRef]
6. Ermgassen, P.S.z.; Mukherjee, N.; Worthington, T.A.; Acosta, A.; Araujo, A.R.d.R.; Beitl, C.M.; Castellanos-Galindo, G.A.; Cunha-Lignon, M.; Dahdouh-Guebas, F.; Diele, K.; et al. Fishers who rely on mangroves: Modelling and mapping the global intensity of mangrove-associated fisheries. *Estuar. Coast. Shelf Sci.* **2021**, *248*, 107159. [CrossRef]
7. Menéndez, P.; Losada, I.J.; Torres-Ortega, S.; Narayan, S.; Beck, M.W. The Global Flood Protection Benefits of Mangroves. *Sci. Rep.* **2020**, *10*, 4404. [CrossRef] [PubMed]
8. Hochard, J.P.; Barbier, E.B.; Hamilton, S.E. Mangroves and coastal topography create economic "safe havens" from tropical storms. *Sci. Rep.* **2021**, *11*, 15359. [CrossRef] [PubMed]
9. Spalding, M.; Parrett, C.L. Global patterns in mangrove recreation and tourism. *Mar. Policy* **2019**, *110*, 103540. [CrossRef]

10. Thomas, N.; Lucas, R.; Bunting, P.; Hardy, A.; Rosenqvist, A.; Simard, M. Distribution and drivers of global mangrove forest change, 1996–2010. *PLoS ONE* **2017**, *12*, e0179302. [CrossRef]
11. Friess, D.A.; Rogers, K.; Lovelock, C.E.; Krauss, K.W.; Hamilton, S.E.; Lee, S.Y.; Lucas, R.; Primavera, J.; Rajkaran, A.; Shi, S. The State of the World's Mangrove Forests: Past, Present, and Future. *Annu. Rev. Environ. Resour.* **2019**, *44*, 1–27. [CrossRef]
12. Wilkie, M.L.; Fortuna, S. *Status and Trends in Mangrove Area Extent Worldwide;* Technical Report; The Food and Agriculture Organization (FAO): Rome, Italy, 2003. Available online: http://www.fao.org/3/j1533e/J1533E02.htm (accessed on 12 December 2022).
13. Wilkie, M.L.; Fortuna, S. *Forest Resources Assessment (FRA) 2005 Thematic Study on Mangroves;* Technical Report; Forest Resources Division, Food and Agriculture Organization of the United Nations: Rome, Italy, 2007.
14. Global Mangrove Alliance. Available online: https://www.mangrovealliance.org (accessed on 12 December 2022).
15. Bunting, P.; Rosenqvist, A.; Hilarides, L.; Lucas, R.M.; Thomas, N.; Tadono, T.; Worthington, T.A.; Spalding, M.; Murray, N.J.; Rebelo, L.M. Global Mangrove Extent Change 1996–2020: Global Mangrove Watch Version 3.0. *Remote Sens.* **2022**, *14*, 3657. [CrossRef]
16. Bunting, P.; Rosenqvist, A.; Hilarides, L.; Lucas, R.M.; Thomas, N. Global Mangrove Watch: Updated 2010 Mangrove Forest Extent (v2.5). *Remote Sens.* **2022**, *14*, 1034. [CrossRef]
17. Bunting, P.; Rosenqvist, A.; Lucas, R.; Rebelo, L.M.; Hilarides, L.; Thomas, N.; Hardy, A.; Itoh, T.; Shimada, M.; Finlayson, M. *Global Mangrove Watch (1996–2016) Version 2.0;* Zenodo: Genève, Switzerland, 2019. [CrossRef]
18. Bunting, P.; Rosenqvist, A.; Lucas, R.; Rebelo, L.M.; Hilarides, L.; Thomas, N.; Hardy, A.; Itoh, T.; Shimada, M.; Finlayson, C. The Global Mangrove Watch—A New 2010 Global Baseline of Mangrove Extent. *Remote Sens.* **2018**, *10*, 1669. [CrossRef]
19. Global Mangrove Watch Portal. Available online: https://globalmangrovewatch.org (accessed on 12 December 2022).
20. Worthington, T.A.; Andradi-Brown, D.A.; Bhargava, R.; Buelow, C.; Bunting, P.; Duncan, C.; Fatoyinbo, L.; Friess, D.A.; Goldberg, L.; Hilarides, L.; et al. Harnessing Big Data to Support the Conservation and Rehabilitation of Mangrove Forests Globally. *One Earth* **2020**, *3*, 260. [CrossRef]
21. Camberlin, P. Climate of Eastern Africa. In *Oxford Research Encyclopedia of Climate Science;* Oxford University Press: Oxford, UK, 2018. [CrossRef]
22. Verburg, P.H.; Neumann, K.; Nol, L. Challenges in using land use and land cover data for global change studies. *Glob. Chang. Biol.* **2011**, *17*, 974–989. [CrossRef]
23. Hansen, M.C.; Potapov, P.V.; Moore, R.; Hancher, M.; Turubanova, S.A.; Tyukavina, A.; Thau, D.; Stehman, S.V.; Goetz, S.J.; Loveland, T.R.; et al. High-Resolution Global Maps of 21st-Century Forest Cover Change. *Science* **2013**, *342*, 850–853. [CrossRef]
24. Awty-Carroll, K.; Bunting, P.; Hardy, A.; Bell, G. Evaluation of the Continuous Monitoring of Land Disturbance Algorithm for Large-Scale Mangrove Classification. *Remote Sens.* **2021**, *13*, 3978. [CrossRef]
25. Serra, P.; Pons, X.; Saurí, D. Post-classification change detection with data from different sensors: Some accuracy considerations. *Int. J. Remote Sens.* **2003**, *24*, 3311–3340. [CrossRef]
26. Tewkesbury, A.P.; Comber, A.J.; Tate, N.J.; Lamb, A.; Fisher, P.F. A critical synthesis of remotely sensed optical image change detection techniques. *Remote Sens. Environ.* **2015**, *160*, 1–14. [CrossRef]
27. Verbesselt, J.; Hyndman, R.; Newnham, G.; Culvenor, D. Detecting trend and seasonal changes in satellite image time series. *Remote Sens. Environ.* **2010**, *114*, 106–115. [CrossRef]
28. Verbesselt, J.; Zeileis, A.; Herold, M. Near real-time disturbance detection using satellite image time series. *Remote Sens. Environ.* **2012**, *123*, 98–108. [CrossRef]
29. Zhu, Z.; Woodcock, C.E. Continuous change detection and classification of land cover using all available Landsat data. *Remote Sens. Environ.* **2014**, *144*, 152–171. [CrossRef]
30. Zhu, Z.; Zhang, J.; Yang, Z.; Aljaddani, A.H.; Cohen, W.B.; Qiu, S.; Zhou, C. Continuous monitoring of land disturbance based on Landsat time series. *Remote Sens. Environ.* **2020**, *238*, 111116. [CrossRef]
31. Awty-Carroll, K.; Bunting, P.; Hardy, A.; Bell, G. Using Continuous Change Detection and Classification of Landsat Data to Investigate Long-Term Mangrove Dynamics in the Sundarbans Region. *Remote Sens.* **2019**, *11*, 2833. [CrossRef]
32. Hansen, M.C.; Krylov, A.; Tyukavina, A.; Potapov, P.V.; Turubanova, S.; Zutta, B.; Ifo, S.; Margono, B.; Stolle, F.; Moore, R. Humid tropical forest disturbance alerts using Landsat data. *Environ. Res. Lett.* **2016**, *11*, 034008. [CrossRef]
33. Reiche, J.; Mullissa, A.; Slagter, B.; Gou, Y.; Tsendbazar, N.E.; Odongo-Braun, C.; Vollrath, A.; Weisse, M.J.; Stolle, F.; Pickens, A.; et al. Forest disturbance alerts for the Congo Basin using Sentinel-1. *Environ. Res. Lett.* **2021**, *16*, 024005. [CrossRef]
34. Watanabe, M.; Koyama, C.N.; Hayashi, M.; Nagatani, I.; Tadono, T.; Shimada, M. Refined algorithm for forest early warning system with ALOS-2/PALSAR-2 ScanSAR data in tropical forest regions. *Remote Sens. Environ.* **2021**, *265*, 112643. [CrossRef]
35. Hammer, D.; Kraft, R.; Wheeler, D. Alerts of forest disturbance from MODIS imagery. *Int. J. Appl. Earth Obs. Geoinf.* **2014**, *33*, 1–9. [CrossRef]
36. Shimabukuro, Y.E.; Santos, J.R.d.; Formaggio, A.R.; Duarte, V.; Rudorff, B.F.T. The Brazilian Amazon Monitoring Program: PRODES and DETER Projects. In *Global Forest Monitoring from Earth Observation;* Achard, F., Hansen, M.C., Eds.; Taylor & Francis: Abingdon, UK, 2016; Chapter 9, pp. 153–169.
37. Microsoft Planetary Computer. Available online: https://planetarycomputer.microsoft.com (accessed on 12 December 2022).
38. McFarland, M.; Emanuele, R.; Morris, D.; Augspurger, T. *Microsoft Open Source: Microsoft/PlanetaryComputer: October 2022 (2022.10.28);* Zenodo: Genève, Switzerland, 2022. [CrossRef]

39. STAC. STAC: SpatioTemporal Asset Catalogs. 2022. Available online: https://stacspec.org (accessed on 14 November 2022).
40. Project Jupyter. Available online: https://jupyter.org (accessed on 9 March 2023).
41. Kubernetes. Available online: https://kubernetes.io/ (accessed on 9 March 2023).
42. Rocklin, M. Dask: Parallel Computation with Blocked algorithms and Task Scheduling. In Proceedings of the 14th Python in Science Conference, Austin, TX, USA, 6–12 July 2015; pp. 130–136.
43. Hoyer, S.; Hamman, J. Xarray: N-D labeled arrays and datasets in Python. *J. Open Res. Softw.* **2017**, *5*. [CrossRef]
44. Open Data Cube STAC API. Available online: https://github.com/opendatacube/odc-stac (accessed on 18 December 2022).
45. Main-Knorn, M.; Pflug, B.; Louis, J.; Debaecker, V.; Müller-Wilm, U.; Gascon, F. Sen2Cor for Sentinel-2. *Image Signal Process. Remote Sens. XXIII* **2017**, *10427*, 1042704. [CrossRef]
46. Small, D. Flattening Gamma: Radiometric Terrain Correction for SAR Imagery. *IEEE Trans. Geosci. Remote Sens.* **2011**, *49*, 3081–3093. [CrossRef]
47. Shimada, M.; Itoh, T.; Motooka, T.; Watanabe, M.; Shiraishi, T.; Thapa, R.; Lucas, R. New global forest/non-forest maps from ALOS PALSAR data (2007–2010). *Remote Sens. Environ.* **2014**, *155*, 13–31. [CrossRef]
48. PALSAR-2 Mosaic Description v2.1.2. Available online: https://www.eorc.jaxa.jp/ALOS/en/dataset/pdf/DatasetDescription_PALSAR2_Mosaic_ver212.pdf (accessed on 18 December 2022).
49. Baloloy, A.B.; Blanco, A.C.; Ana, R.R.C.S.; Nadaoka, K. Development and application of a new mangrove vegetation index (MVI) for rapid and accurate mangrove mapping. *ISPRS J. Photogramm. Remote Sens.* **2020**, *166*, 95–117. [CrossRef]
50. Birch, C.P.; Oom, S.P.; Beecham, J.A. Rectangular and hexagonal grids used for observation, experiment and simulation in ecology. *Ecol. Model.* **2007**, *206*, 347–359. [CrossRef]
51. Norway's International Climate and Forests Initiative Satellite Data Program. Available online: https://www.planet.com/nicfi/ (accessed on 18 December 2022).
52. Osuji, L.C.; Erondu, E.S.; Ogali, R.E. Upstream Petroleum Degradation of Mangroves and Intertidal Shores: The Niger Delta Experience. *Chem. Biodivers.* **2010**, *7*, 116–128. [CrossRef] [PubMed]
53. Mentaschi, L.; Vousdoukas, M.I.; Pekel, J.F.; Voukouvalas, E.; Feyen, L. Global long-term observations of coastal erosion and accretion. *Sci. Rep.* **2018**, *8*, 12876. [CrossRef] [PubMed]
54. Winterwerp, J.C.; Albers, T.; Anthony, E.J.; Friess, D.A.; Mancheño, A.G.; Moseley, K.; Muhari, A.; Naipal, S.; Noordermeer, J.; Oost, A.; et al. Managing erosion of mangrove-mud coasts with permeable dams–lessons learned. *Ecol. Eng.* **2020**, *158*, 106078. [CrossRef]
55. Asbridge, E.; Lucas, R.; Rogers, K.; Accad, A. The extent of mangrove change and potential for recovery following severe Tropical Cyclone Yasi, Hinchinbrook Island, Queensland, Australia. *Ecol. Evol.* **2018**, *8*, 10416–10434. [CrossRef]
56. Krauss, K.W.; Osland, M.J. Tropical cyclones and the organization of mangrove forests: A review. *Ann. Bot.* **2019**, *125*, 213–234. [CrossRef]
57. Sandilyan, S.; Kathiresan, K. Decline of mangroves—A threat of heavy metal poisoning in Asia. *Ocean Coast. Manag.* **2014**, *102*, 161–168. [CrossRef]
58. Krauss, K.W.; McKee, K.L.; Lovelock, C.E.; Cahoon, D.R.; Saintilan, N.; Reef, R.; Chen, L. How mangrove forests adjust to rising sea level. *New Phytol.* **2014**, *202*, 19–34. [CrossRef] [PubMed]
59. Saintilan, N.; Lymburner, L.; Wen, L.; Haigh, I.D.; Ai, E.; Kelleway, J.J.; Rogers, K.; Pham, T.D.; Lucas, R. The lunar nodal cycle controls mangrove canopy cover on the Australian continent. *Sci. Adv.* **2022**, *8*, eabo6602. [CrossRef] [PubMed]
60. Charrua, A.B.; Padmanaban, R.; Cabral, P.; Bandeira, S.; Romeiras, M.M. Impacts of the Tropical Cyclone Idai in Mozambique: A Multi-Temporal Landsat Satellite Imagery Analysis. *Remote Sens.* **2021**, *13*, 201. [CrossRef]
61. Maina, J.; Moel, H.d.; Zinke, J.; Madin, J.; McClanahan, T.; Vermaat, J.E. Human deforestation outweighs future climate change impacts of sedimentation on coral reefs. *Nat. Commun.* **2013**, *4*, 1986. [CrossRef]
62. Thomas, N.; Lucas, R.; Itoh, T.; Simard, M.; Fatoyinbo, L.; Bunting, P.; Rosenqvist, A. An approach to monitoring mangrove extents through time-series comparison of JERS-1 SAR and ALOS PALSAR data. *Wetl. Ecol. Manag.* **2014**, *23*, 3–17. [CrossRef]
63. Rosen, P.A.; Kumar, R. NASA-ISRO SAR (NISAR) Mission Status. In Proceedings of the 2021 IEEE Radar Conference (RadarConf21), Atlanta, GA, USA, 7–14 May 2021; pp, 1–6. [CrossRef]

MDPI AG
Grosspeteranlage 5
4052 Basel
Switzerland
Tel.: +41 61 683 77 34

Remote Sensing Editorial Office
E-mail: remotesensing@mdpi.com
www.mdpi.com/journal/remotesensing

www.ingramcontent.com/pod-product-compliance
Lightning Source LLC
Chambersburg PA
CBHW061824210926
51850CB00006B/798
* 9 7 8 3 7 2 5 8 5 2 1 9 2 *